California SAXON MATH

Intermediate 6

Student Edition

Volume 2

Stephen Hake

A Harcourt Achieve Imprint

www.SaxonPublishers.com
1-800-284-7019

ACKNOWLEDGEMENTS

This book was made possible by the significant contributions of many individuals and the dedicated efforts of talented teams at Harcourt Achieve.

Special thanks to Chris Braun for conscientious work on Power Up exercises, Problem Solving scripts, and student assessments, and to Elizabeth Rivas and Bryon Hake for revision assistance on Intermediate 6. The long hours and technical assistance of John and James Hake were invaluable in meeting publishing deadlines. As always, the patience and support of Mary is most appreciated.

– Stephen Hake

Staff Credits

Editorial: Joel Riemer, Paula Zamarra, Hirva Raj, Smith Richardson, Pamela Cox, Michael Ota, Stephanie Rieper, Ann Sissac, Gayle Lowery, Robin Adams, David Baceski, Brooke Butner, Cecilia Colome, James Daniels, Leslie Bateman, Chad Barrett, Heather Jernt

Design: Alison Klassen, Joan Cunningham, Alan Klemp, Julie Hubbard, Lorelei Supapo, Andy Hendrix, Rhonda Holcomb

Production: Mychael Ferris-Pacheco, Jennifer Cohorn, Greg Gaspard, Donna Brawley, John-Paxton Gremillion

Manufacturing: Cathy Voltaggio, Kathleen Stewart

Marketing: Marilyn Trow, Kimberly Sadler

E-Learning: Layne Hedrick

ABOUT THE AUTHOR

Stephen Hake has authored six books in the **Saxon Math** series. He writes from 17 years of classroom experience as a teacher in grades 5 through 12 and as a math specialist in El Monte, California. As a math coach, his students won honors and recognition in local, regional, and statewide competitions.

Stephen has been writing math curriculum since 1975 and for Saxon since 1985. He has also authored several math contests including Los Angeles County's first Math Field Day contest. Stephen contributed to the 1999 National Academy of Science publication on the Nature and Teaching of Algebra in the Middle Grades.

Stephen is a member of the National Council of Teachers of Mathematics and the California Mathematics Council. He earned his BA from United States International University and his MA from Chapman College.

CONTENTS OVERVIEW

Student Edition, Volume 1

Table of Contents v

Letter from the Author xvii

How to Use Your Textbook xviii

Problem Solving Overview 1

Section 1 7
Lessons 1–10, Investigation 1

Section 2 71
Lessons 11–20, Investigation 2

Section 3 134
Lessons 21–30, Investigation 3

Section 4 201
Lessons 31–40, Investigation 4

Section 5 261
Lessons 41–50, Investigation 5

Section 6 325
Lessons 51–60, Investigation 6

English/Spanish Math Glossary R1

Index R49

Student Edition, Volume 2

Table of Contents v

Letter from the Author xvii

How to Use Your Textbook xviii

Section 7 385
Lessons 61–70, Investigation 7

Section 8 445
Lessons 71–80, Investigation 8

Section 9 504
Lessons 81–90, Investigation 9

Section 10 565
Lessons 91–100, Investigation 10

Section 11 629
Lessons 101–110, Investigation 11

Section 12 689
Lessons 111–114

English/Spanish Math Glossary R1

Index R49

Integrated and Distributed Units of Instruction

Section 1 • *Lessons 1–10, Investigation 1*

Lesson		Page	California Strands Focus
	Problem Solving Overview	1	
1	• Sequences	7	NS, MR
2	• Order of Operations, Part 1	11	NS, AF
3	• Properties of Operations	15	AF, MR
4	• Multistep Problems	23	NS, MR
5	• Unknown Numbers in Addition, Subtraction, Multiplication, and Division	29	AF, MR
6	• Translating and Evaluating Expressions and Equations, Part 1	35	NS, AF, MR
7	• Addition and Subtraction Word Problems	41	AF, MR
8	• Multiplication and Division Word Problems	48	AF, MR
9	• The Number Line: Negative Numbers	52	NS, MR
10	• Calculating the Mean	57	SDAP, MR
Investigation 1	• Investigating Fractions with Manipulatives **Activity** Using Fraction Manipulatives	65	NS, MR
Focus on Concepts A	• Why a Fraction Shows Division	69	NS, MR

California Strands Key:
NS = Number Sense
AF = Algebra and Functions
MG = Measurement and Geometry
SDAP = Statistics, Data Analysis, and Probability
MR = Mathematical Reasoning

TABLE OF CONTENTS

Section 2   *Lessons 11–20, Investigation 2*

Lesson		Page	California Strands Focus
11	• Factors and Prime Numbers Activity Prime Numbers	71	NS, MR
12	• Greatest Common Factor (GCF)	77	NS, MR
13	• Factors and Divisibility	82	NS, MR
14	• The Number Line: Fractions and Mixed Numbers Activity Inch Ruler to Sixteenths	87	NS, MR
15	• "Equal Groups" Problems with Fractions	95	NS, MR
16	• Ratio and Rate	100	NS, AF
17	• Adding and Subtracting Fractions That Have Common Denominators	105	NS, MR
18	• Writing Quotients and Interpreting Remainders	111	NS, MR
19	• Multiples	116	NS, MR
20	• Angles	120	MG, MR
Investigation 2	• Surveys and Samples	127	SDAP
Focus on Concepts B	• Sampling Errors in Data Collection and Display	132	SDAP

Section 3 — Lessons 21–30, Investigation 3

Lesson		Page	California Strands Focus
21	• Rounding and Estimation	134	NS, MR
22	• Adding and Subtracting Mixed Numbers with Common Denominators	141	NS, MR
23	• Perimeters and Areas of Rectangles	147	AF, MR
24	• Multiplying Fractions	155	NS, MR
Focus on Concepts C	• How to Use a Number Line to Multiply Fractions	161	NS, MR
25	• Reducing Fractions by Dividing by Common Factors	163	NS
26	• Least Common Multiple (LCM)	167	NS, MR
27	• Writing Percents as Fractions, Part 1	171	NS, MR
28	• Compare and Order Decimals, Part 1	175	NS, MR
29	• Relating Fractions and Decimals	181	NS, MR
30	• Circle Graphs	186	NS, MR
Investigation 3	• Data and Sampling	193	SDAP
Focus on Concepts D	• Do the Data Support the Claim?	199	SDAP, MR

California Strands Key:
NS = Number Sense
AF = Algebra and Functions
MG = Measurement and Geometry
SDAP = Statistics, Data Analysis, and Probability
MR = Mathematical Reasoning

TABLE OF CONTENTS

Section 4 *Lessons 31–40, Investigation 4*

Lesson		Page	California Strands Focus
31	• Subtracting Fractions and Mixed Numbers from Whole Numbers	201	NS, MR
32	• Squares and Square Roots	207	NS, AF
33	• Multiplying Decimal Numbers	212	AF, MR
34	• Using Zero as a Placeholder	217	AF, MR
35	• Finding a Percent of a Number	223	NS, MR
Focus on Concepts E	• Why Percent is Useful	229	NS, MR
36	• Reciprocals	231	NS, MR
37	• Renaming Fractions by Multiplying by 1	235	NS, MR
38	• Equivalent Division Problems	239	NS, MR
39	• Finding Unknowns in Fraction and Decimal Problems	245	AF, MR
40	• Compare and Order Decimals, Part 2	249	NS, MR
Investigation 4	• Statistics and Sampling	254	SDAP
Focus on Concepts F	• Which Measure Best Describes the Data?	260	SDAP

Section 5 *Lessons 41–50, Investigation 5*

Lesson		Page	California Strands Focus
41	• Dividing a Decimal Number by a Whole Number • Decimal Chart	261	AF, MR
Focus on Concepts G	• What Factors Affect the Results of a Survey?	267	SDAP
42	• Circumference **Activity** Circumference	269	AF, MG
43	• Subtracting Mixed Numbers with Regrouping, Part 1	275	NS, MR
44	• Dividing by a Decimal Number	279	AF, MR
45	• Dividing by a Fraction	285	NS
Focus on Concepts H	• Why Can We Invert and Multiply to Divide Fractions?	289	NS, MR
46	• Rounding Decimal Numbers	290	NS, MR
47	• Positive and Negative Decimals	295	NS, MR
Focus on Concepts I	• What is a Number?	301	NS, MR
48	• Reducing by Grouping Factors Equal to 1	303	NS, MR
49	• Dividing Fractions	307	NS, MR
Focus on Concepts J	• How Are Multiplying and Dividing Fractions Related?	313	NS, MR
50	• Multiply or Divide with Fractions	314	NS, MR
Investigation 5	• Graphing	319	AF, MR

California Strands Key:
NS = Number Sense
AF = Algebra and Functions
MG = Measurement and Geometry
SDAP = Statistics, Data Analysis, and Probability
MR = Mathematical Reasoning

TABLE OF CONTENTS

Section 6   *Lessons 51–60, Investigation 6*

Lesson		Page	California Strands Focus
51	• Simplifying Fractions	325	NS, MR
52	• Multiplying by Denominators to Add, Subtract, and Compare Fractions	329	NS
53	• Finding Common Denominators	335	NS
Focus on Concepts K	• How to Choose a Common Denominator	342	NS, MR
54	• Probability and Chance	343	SDAP
55	• Adding and Subtracting Fractions and Mixed Numbers	351	NS, MR
56	• Adding Three or More Fractions	357	NS, MR
57	• Writing Mixed Numbers as Improper Fractions	361	NS
58	• Subtracting Mixed Numbers with Regrouping, Part 2	367	NS
59	• Classifying Quadrilaterals	371	MG, MR
60	• Complementary and Supplementary Angles	376	AF, MG
Investigation 6	• Compound Experiments	380	SDAP

Section 7 *Lessons 61–70, Investigation 7*

Lesson		Page	California Strands Focus
61	• Prime Factorization	385	NS, MR
62	• Multiplying Mixed Numbers	391	NS, MR
63	• Using Prime Factorization to Reduce Fractions	397	NS, MR
64	• Dividing Mixed Numbers	401	NS
65	• Reducing Fractions Before Multiplying	406	NS, MR
Focus on Concepts L	• Why Does Reducing Before Multiplying Work?	411	NS, MR
66	• Parallelograms Activity Area of a Parallelogram	412	AF, MG
67	• Multiplying Three Fractions	419	NS, MR
68	• Writing Decimal Numbers as Fractions	425	NS, MR
69	• Writing Fractions and Ratios as Decimal Numbers	429	NS, SDAP
70	• Disjoint Events	435	SDAP
Investigation 7	• Experimental and Theoretical Probability Activity Probability Experiment	441	SDAP

California Strands Key:
NS = Number Sense
AF = Algebra and Functions
MG = Measurement and Geometry

SDAP = Statistics, Data Analysis, and Probability
MR = Mathematical Reasoning

TABLE OF CONTENTS

Section 8 *Lessons 71–80, Investigation 8*

Lesson		Page	California Strands Focus
71	• Exponents	445	NS, AF
72	• Order of Operations, Part 2	450	NS, AF
73	• Writing Fractions and Decimals as Percents, Part 1	455	NS, MR
74	• Comparing Fractions by Converting to Decimal Form	460	NS, MR
75	• Finding Unstated Information in Fraction Problems	464	SDAP, MR
76	• Metric System	469	AF, MR
77	• Area of a Triangle *Activity* Area of a Triangle	475	AF, MR
78	• Using a Constant Factor to Solve Ratio Problems	483	NS
79	• Arithmetic with Units of Measure	488	AF
80	• Independent and Dependent Events	494	SDAP
Focus on Concepts M	• Thinking About the Probability of More than One Outcome	499	SDAP, MR
Investigation 8	• Probability and Predicting	501	SDAP

Section 9   *Lessons 81–90, Investigation 9*

Lesson		Page	California Strands Focus
81	• U.S. Customary System	504	AF, MR
82	• Proportions	511	NS, MR
83	• Using Cross Products to Solve Proportions	516	NS, MR
Focus on Concepts N	• How to Use the Multiplicative Inverse to Solve Proportions	522	NS, MR
84	• Area of a Circle	523	AF, MG
Focus on Concepts O	• What is Pi?	528	MG, MR
85	• Finding Unknown Factors	529	AF, MR
86	• Using Proportions to Solve Ratio Word Problems	533	NS, MR
87	• Geometric Formulas	537	AF, MR
88	• Distributive Property Activity Perimeter Formulas	543	AF
89	• Reducing Rates Before Multiplying	549	NS, AF
90	• Classifying Triangles	553	MG, MR
Investigation 9	• Balanced Equations	559	AF, MR

California Strands Key:
NS = Number Sense
AF = Algebra and Functions
MG = Measurement and Geometry

SDAP = Statistics, Data Analysis, and Probability
MR = Mathematical Reasoning

TABLE OF CONTENTS

Section 10 *Lessons 91–100, Investigation 10*

Lesson		Page	California Strands Focus
91	• Using Properties of Equality to Solve Equations	565	AF, MR
92	• Writing Fractions and Decimals as Percents, Part 2	573	NS, MR
93	• Formulas and Substitution	578	AF
94	• Transversals	583	MG, MR
95	• Sum of the Angle Measures of Triangles and Quadrilaterals	589	MG
96	• Fraction-Decimal-Percent Equivalents	595	NS, MR
97	• Algebraic Addition Activity **Activity** Sign Game	599	NS, MR
98	• Addition of Integers	605	NS, MR
Focus on Concepts P	• What Adding Two Negative Integers Means	611	NS, MR
99	• Subtraction of Integers	612	NS, MR
100	• Ratio Problems Involving Totals	617	NS, MR
Investigation 10	• Similar Figures	623	NS, MG

Section 11   *Lessons 101–110, Investigation 11*

Lesson		Page	California Strands Focus
101	• Complex Shapes	**629**	MG, MR
102	• Using Proportions to Solve Percent Problems	**637**	NS
103	• Multiplying and Dividing Integers	**642**	NS, MR
Focus on Concepts Q	• Two Ways to Think About Multiplying Negative Integers	**647**	NS, MR
104	• Applications Using Division	**649**	SDAP, MR
105	• Unit Multipliers	**654**	NS, AF
106	• Writing Percents as Fractions, Part 2	**659**	NS, MR
107	• Order of Operations with Positive and Negative Numbers	**663**	NS, AF
108	• Evaluations with Positive and Negative Numbers	**669**	NS, AF
109	• Translating and Evaluating Expressions and Equations, Part 2	**673**	NS, AF
110	• Finding a Whole When a Fraction is Known	**679**	NS, MR
Investigation 11	• Scale Factor: Scale Drawings and Models	**684**	NS, AF

California Strands Key:
NS = Number Sense
AF = Algebra and Functions
MG = Measurement and Geometry

SDAP = Statistics, Data Analysis, and Probability
MR = Mathematical Reasoning

TABLE OF CONTENTS

Section 12 *Lessons 111–114*

Lesson		Page	California Strands Focus
111	• Finding a Whole When a Percent is Known	689	NS, MR
112	• Percent Word Problems	695	NS, MR
113	• Volume of a Cylinder	701	MG
114	• Volume of a Right Solid	705	MG

Dear Student,

We study mathematics because it plays a very important role in our lives. Our school schedule, our trip to the store, the preparation of our meals, and many of the games we play involve mathematics. The word problems in this book are often drawn from everyday experiences.

When you become an adult, mathematics will become even more important. In fact, your future may depend on the mathematics you are learning now. This book will help you to learn mathematics and to learn it well. As you complete each lesson, you will see that similar problems are presented again and again. *Solving each problem day after day is the secret to success.*

Your book includes daily lessons and investigations. Each lesson has three parts.

1. The first part is a Power Up that includes practice of basic facts and mental math. These exercises improve your speed, accuracy, and ability to do math *in your head.* The Power Up also includes a problem-solving exercise to help you learn the strategies for solving complicated problems.

2. The second part of the lesson is the New Concept. This section introduces a new mathematical concept and presents examples that use the concept. The Lesson Practice provides a chance for you to solve problems using the new concept. The problems are lettered a, b, c, and so on.

3. The final part of the lesson is the Written Practice. This section reviews previously taught concepts and prepares you for concepts that will be taught in later lessons. Solving these problems will help you practice your skills and remember concepts you have learned.

Investigations are variations of the daily lesson. The investigations in this book often involve activities that fill an entire class period. Investigations contain their own set of questions but do not include Lesson Practice or Written Practice.

Remember to solve every problem in each Lesson Practice, Written Practice, and Investigation. Do your best work, and you will experience success and true learning that will stay with you and serve you well in the future.

Temple City, California

HOW TO USE YOUR TEXTBOOK

Saxon Math Intermediate 6 is unlike any math book you have used! It doesn't have colorful photos to distract you from learning. The Saxon approach lets you see the beauty and structure within math itself. You will understand more mathematics, become more confident in doing math, and will be well prepared when you take high school math classes.

Power Yourself Up

Start off each lesson by practicing your basic skills and concepts, mental math, and problem solving. Make your math brain stronger by exercising it every day. Soon you'll know these facts by memory!

Learn Something New!

Each day brings you a new concept, but you'll only have to learn a small part of it now. You'll be building on this concept throughout the year so that you understand and remember it by test time.

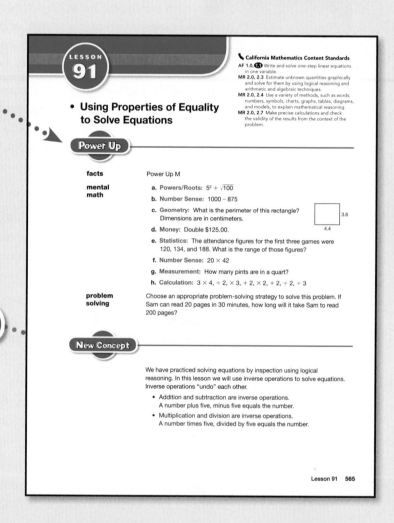

LESSON

91

California Mathematics Content Standards
AF 1.0, ⓫ Write and solve one-step linear equations in one variable.
MR 2.0, 2.3 Estimate unknown quantities graphically and solve for them by using logical reasoning and arithmetic and algebraic techniques.
MR 2.0, 2.4 Use a variety of methods, such as words, numbers, symbols, charts, graphs, tables, diagrams, and models, to explain mathematical reasoning.
MR 2.0, 2.7 Make precise calculations and check the validity of the results from the context of the problem.

• **Using Properties of Equality to Solve Equations**

Power Up

facts   Power Up M

mental math
a. Powers/Roots: $5^2 + \sqrt{100}$
b. Number Sense: $1000 - 875$
c. Geometry: What is the perimeter of this rectangle? Dimensions are in centimeters.   3.6   4.4
d. Money: Double $125.00.
e. Statistics: The attendance figures for the first three games were 120, 134, and 188. What is the range of those figures?
f. Number Sense: 20×42
g. Measurement: How many pints are in a quart?
h. Calculation: $3 \times 4, \div 2, \times 3, + 2, \times 2, + 2, \div 2, \div 3$

problem solving
Choose an appropriate problem-solving strategy to solve this problem. If Sam can read 20 pages in 30 minutes, how long will it take Sam to read 200 pages?

New Concept

We have practiced solving equations by inspection using logical reasoning. In this lesson we will use inverse operations to solve equations. Inverse operations "undo" each other.

• Addition and subtraction are inverse operations.
  A number plus five, minus five equals the number.

• Multiplication and division are inverse operations.
  A number times five, divided by five equals the number.

Lesson 91   565

Counting numbers that have exactly two factors are **prime numbers**. The first four prime numbers are 2, 3, 5, and 7. The only factors of a prime number are the number itself and 1. The number 1 is not a prime number because it has only one factor, itself.

Therefore, to determine whether a number is prime, we may ask ourselves the question, "Is this number divisible by any number other than the number itself and 1?" If the number is divisible by any other number, the number is not prime.

Example 4

The first four prime numbers are 2, 3, 5, and 7. What are the next four prime numbers?

We will consider the next several numbers and eliminate those that are not prime.

8, 9, 10, 11, 12, 13, 14, 15, 16, 17, 18, 19, 20

All even numbers have 2 as a factor. So no even numbers greater than two are prime numbers. We can eliminate the even numbers from the list.

8̶, 9, 1̶0̶, 11, 1̶2̶, 13, 1̶4̶, 15, 1̶6̶, 17, 1̶8̶, 19, 2̶0̶

Since 9 is divisible by 3, and 15 is divisible by 3 and by 5, we can eliminate 9 and 15 from the list.

8̶, 9̶, 1̶0̶, 11, 1̶2̶, 13, 1̶4̶, 1̶5̶, 1̶6̶, 17, 1̶8̶, 19, 2̶0̶

Each of the remaining four numbers on the list is divisible only by itself and by 1. Thus, the next four prime numbers after 7 are **11, 13, 17,** and **19.**

Prime Numbers

Thinking Skills

Conclude

Why don't the directions include "Circle the number 4" and "Circle the number 6"?

List the counting numbers from 1 to 100 (or use **Lesson Activity 8** Hundred Number Chart). Then follow these directions:

Step 1: Draw a line through the number 1. The number 1 is not a prime number.

Step 2: Circle the prime number 2. Draw a line through all the other multiples of 2 (4, 6, 8, and so on).

Step 3: Circle the prime number 3. Draw a line through all the other multiples of 3 (6, 9, 12, and so on).

Step 4: Circle the prime number 5. Draw a line through all the other multiples of 5 (10, 15, 20, and so on).

Step 5: Circle the prime number 7. Draw a line through all the other multiples of 7 (14, 21, 28, and so on).

Step 6: Circle all remaining numbers on your list (the numbers that do not have a line drawn through them).

Get Active!

Dig into math with a hands-on activity. Explore a math concept with your friends as you work together and use manipulatives to see new connections in mathematics.

Check It Out!

The Lesson Practice lets you check to see if you understand today's new concept.

When you have finished, all the prime numbers from 1 to 100 will be circled on your list.

Lesson Practice

List the factors of the following numbers:

a. 14   **b.** 15

c. 16   **d.** 17

Justify Which number in each group is a prime number? Explain how you found your answer.

e. 21, 23, 25

f. 31, 32, 33

g. 43, 44, 45

Classify Which number in each group is not a prime number?

h. 41, 42, 43

i. 31, 41, 51

j. 23, 33, 43

Prime numbers can be multiplied to make whole numbers that are not prime. For example, $2 \cdot 2 \cdot 3$ equals 12 and $3 \cdot 5$ equals 15. (Neither 12 nor 15 is prime.) Show which prime numbers we multiply to make these products:

k. 16   **l.** 18

m. Maria is ordering a table made of inlaid wood squares. She has twenty-four squares. How many different rectangular tables can she order if she wants to use all of her squares? Give the dimensions of each table.

Exercise Your Mind!

When you work the Written Practice exercises, you will review both today's new concept and also math you learned in earlier lessons. Each exercise will be on a different concept — you never know what you're going to get! It's like a mystery game — unpredictable and challenging.

As you review concepts from earlier in the book, you'll be asked to use higher-order thinking skills to show what you know and why the math works.

Written Practice   *Distributed and Integrated*

1. What is the sum of twelve thousand, five hundred and ten thousand, six hundred ten?
(RF12)

2. Susan B. Anthony worked to pass the 19th Amendment which gave women the right to vote. This Amendment passed in 1920 yet Susan B. Anthony died in 1906. How many years after she died did the Amendment pass?
(7)

* 3. Linda can run about 6 yards in one second. About how far can she run in 12 seconds? Write an equation and solve the problem.
(8)

* 4. A coin collector has a collection of two dozen rare coins. If the value of each coin is $1000, what is the value of the entire collection? Write an equation and solve the problem.
(8)

HOW TO USE YOUR TEXTBOOK

Become an Investigator!

Dive into math concepts and explore the depths of math connections in the Investigations.

Continue to develop your mathematical thinking through applications, activities, and extensions.

INVESTIGATION 3

✎ *California Mathematics Content Standards*

SDAP 1.0, 1.3 Understand how the inclusion or exclusion of outliers affects measures of central tendency.

SDAP 1.0, 1.4 Know why a specific measure of central tendency (mean, median, mode) provides the most useful information in a given context.

SDAP 2.0, 2.3 Analyze data displays and explain why the way in which the question was asked might have influenced the results obtained and why the way in which the results were displayed might have influenced the conclusions reached.

SDAP 2.0, 2.4 Identify data that represent sampling errors and explain why the sample (and the display) might be biased.

SDAP 2.0, 2.5 Identify claims based on statistical data and, in simple cases, evaluate the validity of the claims.

Focus on
Data and Sampling

In this investigation, we will continue our look at **statistics.** We will consider different types of data that we can gather, discuss methods of gathering the data, and look at ways to display the data.

Precious wondered which of three activities—team sports, dance, or walking/jogging—was most popular among her classmates. She gathered data by asking each classmate to select his or her favorite. Then she displayed the data with a **bar graph,** which displays numerical information in shaded rectangles to show comparisons.

Favorite Activities

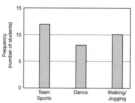

Analyze Which type of exercise is most popular among Precious's classmates?

Roger wondered how frequently the residents on his street visit the city park. He went to every third house on his street and asked, "How many times per week do you visit the city park?" Roger displayed the data he collected with a **line plot,** which shows individual data points. For each of the 16 responses, he placed an "x" above the corresponding number.

Number of Visits to the City Park per Week

```
        x
  x  x  x
  x  x  x  x           x
  x  x  x  x  x     x  x
  —————————————————————————
  0  1  2  3  4  5  6  7
```

California Mathematics Content Standards

NS **2.0**, **2.4** Determine the least common multiple and the greatest common divisor of whole numbers; use them to solve problems with fractions (e.g., to find a common denominator to add two fractions or to find the reduced form for a fraction).

MR 2.0, 2.4 Use a variety of methods, such as words, numbers, symbols, charts, graphs, tables, diagrams, and models, to explain mathematical reasoning.

• Prime Factorization

Power Up

facts   Power Up J

mental math

  a. Number Sense: 4×750

  b. Number Sense: $750 - 199$

  c. Money: $8.25 - $2.50

  d. Geometry: What is the perimeter of this rectangle?

  5 in.

  $12\frac{1}{2}$ in.

  e Measurement: The slide on the playground is 3 yards long. How many feet is that?

  f. Probability: What is the probability of rolling a 3 with one roll of a number cube?

  g. Statistics: Each day, Joel counts the glasses of water he drinks. On Sunday through Friday, he drank 6, 9, 8, 8, 10, 8, and 8 glasses, respectively. What is the mode of these numbers?

  h. Calculation: 6×10, $\div 3$, $\times 2$, $\div 4$, $\times 5$, $\div 2$, $\times 4$

problem solving

Choose an appropriate problem-solving strategy to solve this problem. A large piece of cardstock is 1 mm thick. If we fold it in half, and then fold it in half again, we have a stack of 4 layers of cardstock that is 4 mm high. If it were possible to continue folding the cardstock in half, how thick would the stack of layers be after 10 folds? Is that closest in height to a book, a table, a man, or a bus?

New Concept

Every whole number greater than 1 is either a prime number or a **composite number**. A prime number has *only two* factors (1 and itself), while a composite number has *more than two* factors. As we studied in Lesson 11, the numbers 2, 3, 5, and 7 are prime numbers. The numbers 4, 6, 8, and 9 are composite numbers. All composite numbers can be formed by multiplying prime numbers together.

$$4 = 2 \cdot 2$$
$$6 = 2 \cdot 3$$
$$8 = 2 \cdot 2 \cdot 2$$
$$9 = 3 \cdot 3$$

When we write a composite number as a product of its prime factors, we have written the **prime factorization** of the number. The prime factorizations of 4, 6, 8, and 9 are shown above. Notice that if we had written 8 as $2 \cdot 4$ instead of $2 \cdot 2 \cdot 2$, we would not have completed the prime factorization of 8. Since the number 4 is not prime, we would complete prime factorization by "breaking" 4 into its prime factors of 2 and 2.

In this lesson we will show two methods for factoring a composite number, **division by primes** and **factor trees.** We will use both methods to factor the number 60.

To factor a number using division by primes, we write the number in a division box and begin dividing by the smallest prime number that is a factor. The smallest prime number is 2. Since 60 is divisible by 2, we divide 60 by 2 to get 30.

$$\begin{array}{r} 30 \\ 2\overline{)60} \end{array}$$

Since 30 is also divisible by 2, we divide 30 by 2. The quotient is 15. Notice how we "stack" the divisions.

$$\begin{array}{r} 15 \\ 2\overline{)30} \\ 2\overline{)60} \end{array}$$

Although 15 is not divisible by 2, it is divisible by the next-smallest prime number, which is 3. Fifteen divided by 3 produces the quotient 5.

$$\begin{array}{r} 5 \\ 3\overline{)15} \\ 2\overline{)30} \\ 2\overline{)60} \end{array}$$

Five is a prime number. The only prime number that divides 5 is 5.

$$\begin{array}{r} 1 \\ 5\overline{)5} \\ 3\overline{)15} \\ 2\overline{)30} \\ 2\overline{)60} \end{array}$$

By dividing by prime numbers, we have found the prime factorization of 60.

$$60 = 2 \cdot 2 \cdot 3 \cdot 5$$

---

[1] Some people prefer to divide only until the quotient is a prime number. When using that procedure, the final quotient is included in the prime factorization of the number.

Example 1

Use division by primes to find the prime factorization of 36.

We begin by dividing 36 by its smallest prime-number factor, which is 2. We continue dividing by prime numbers until the quotient is 1.

$$
\begin{array}{r}
1 \\
3\overline{)3} \\
3\overline{)9} \\
2\overline{)18} \\
2\overline{)36}
\end{array}
$$

$$36 = 2 \cdot 2 \cdot 3 \cdot 3$$

To make a factor tree for 60, we simply think of any two whole numbers whose product is 60. Since 6×10 equals 60, we can use 6 and 10 as the first two "branches" of the factor tree.

The numbers 6 and 10 are not prime numbers, so we continue the process by factoring 6 into $2 \cdot 3$ and by factoring 10 into $2 \cdot 5$.

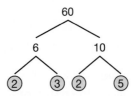

The circled numbers at the ends of the branches are all prime numbers. We have completed the factor tree. We will arrange the factors in order from least to greatest and write the prime factorization of 60.

$$60 = 2 \cdot 2 \cdot 3 \cdot 5$$

Example 2

Use a factor tree to find the prime factorization of 60. Use 4 and 15 as the first branches.

Some composite numbers can be divided into many different factor trees. However, when the factor tree is completed, the same prime numbers appear at the ends of the branches.

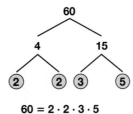

$$60 = 2 \cdot 2 \cdot 3 \cdot 5$$

a. **Classify** Which of these numbers are composite numbers?

19, 20, 21, 22, 23

b. Write the prime factorization of each composite number in problem **a.**

c. **Represent** Use a factor tree to find the prime factorization of 36.

d. Use division by primes to find the prime factorization of 48.

e. Write 125 as a product of prime factors.

f. **Generalize** Write the prime factorization of 10, 100, 1000, and 10,000. What patterns do you see in the prime factorizations of these numbers?

Written Practice   *Distributed and Integrated*

1. Fifty percent of the 60 questions on the test are multiple choice. Find the number
(35) of multiple-choice questions on the test.

2. **Analyze** Twelve of the 30 students in the class are boys.
(54)
a. What is the ratio of boys to girls in the class?

b. If each student's name is placed in a hat and one name is drawn, what is the probability that it will be the name of a girl?

3. **Analyze** Some railroad rails weigh 155 pounds per yard. How much would a
(8) 33-foot-long rail weigh?

4. **Multiple Choice** Which of the quadrilaterals below is *not* a rectangle?
(59)

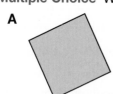

A   B   C   D

5. An angle measures 55°. What is:
(60)
a. the complement of the angle?

b. the supplement of the angle?

55°

6. The sum of five numbers is 200. What is the average of the numbers?
(10)

7. $\dfrac{100 + 75}{100 - 75}$
(2)

8. $1\dfrac{1}{5} + 3\dfrac{1}{2}$
(55)

*9. $\dfrac{1}{3} + \dfrac{1}{6} + \dfrac{1}{12}$
(56)

*10. $35\dfrac{1}{4} - 12\dfrac{1}{2}$
(58)

11. $\dfrac{4}{5} \times \dfrac{1}{2}$
(24)

12. $\dfrac{4}{5} \div \dfrac{1}{2}$
(49)

13. $0.25 \div 5$
(41)

14. $5 \div 0.25$
(44)

15. What is the product of the answers to problems 13 and 14?
(33)

16. **Multiple Choice** Which of the following is equal to $\frac{1}{2} + \frac{1}{2}$?
(49)

  A $\frac{1}{2} - \frac{1}{2}$ **B** $\left(\frac{1}{2}\right)^2$ **C** $\frac{1}{2} \div \frac{1}{2}$

* 17. (**Represent**) Use a factor tree to find the prime factorization of 30.
(61)

18. If three pencils cost a total of 75¢, how much would six pencils cost?
(4, 8)

19. Seven and one-half percent is equivalent to the decimal number 0.075. If the
(35) sales-tax rate is $7\frac{1}{2}$%, what is the sales tax on a \$10.00 purchase?

* 20. (**Analyze**) One side of a regular pentagon measures 0.8 meter. What is the
(RF24) perimeter of the regular pentagon?

21. Twenty minutes is what fraction of an hour?
(25)

22. The temperature dropped from 12°C to −8°C. This was a drop of how many
(9) degrees?

The bar graph below shows the weights of different types of cereals packaged in the same size boxes. Refer to the graph to answer problems **23–25.**

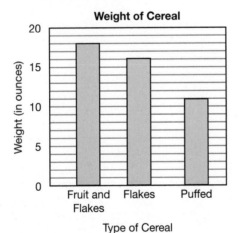

23. What is the range of the weights?
(Inv. 4)

24. What is the mean weight of the three types of cereal?
(Inv. 4)

* 25. (**Formulate**) Write a comparison word problem that relates to the graph,
(7) and then answer the problem.

* 26. (**Connect**) Use division by primes to find the prime factorization of 400.
(61)

27. (**Analyze**) Shandi covered the floor of a square room with 144 square floor
(32) tiles. How many floor tiles were along each wall of the room?

28. **Estimate** The weight of a 1-kilogram object is about 2.2 pounds. A large man
(RF22) may weigh 100 kilograms. About how many pounds is that?

29. Reduce: $\dfrac{5 \cdot 5 \cdot 5 \cdot 7}{2 \cdot 2 \cdot 2 \cdot 5 \cdot 5 \cdot 5}$
(48)

* **30.** **Multiple Choice** Which of these polygons is not a regular polygon?
(RF24)

A   △

B   ▭

C   ⬠

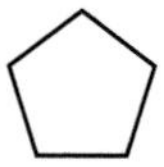

D   ⬡

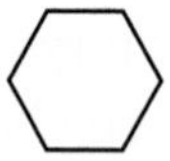

LESSON

62

• Multiplying Mixed Numbers

California Mathematics Content Standards

NS 2.0, 2.1 Convert one unit of measurement to another (e.g., from feet to miles, from centimeters to inches).

NS 2.0, 2.2 Demonstrate an understanding that *rate* is a measure of one quantity per unit value of another quantity.

MR 2.0, 2.1 Use estimation to verify the reasonableness of calculated results.

MR 3.0, 3.1 Evaluate the reasonableness of the solution in the context of the original situation

facts Power Up I

mental math

  a. Number Sense: 3×84

  b. Fractions: $\frac{1}{3}$ of $36.00

  c. Money: $\frac{\$25}{10}$

  d. Number Sense: A roll of quarters contains 40 quarters. How many quarters are in 5 rolls?

  e. Time: Convert 240 seconds into minutes.

  f. Statistics: Ali timed himself holding his breath underwater. In five tries, he held his breath for 26, 31, 31, 39, and 44 seconds, respectively. What was the median number of seconds?

  g. Algebra: If $z + 1 = 10$, what is z?

  h. Calculation: $6 \times 8, -4, \div 4, \times 2, +2, \div 6, \div 2$

problem solving Choose an appropriate problem-solving strategy to solve this problem. There are approximately 520 nine-inch long noodles in a 1-pound package of spaghetti. Placed end-to-end, how many feet of noodles are in a pound of uncooked spaghetti?

New Concept

Thinking Skills

Verify

How do we write a mixed number as a fraction?

Recall from Lesson 55 the three steps to solving an arithmetic problem with fractions.

Step 1: Put the problem into the correct shape (if it is not already).

Step 2: Perform the operation indicated.

Step 3: Simplify the answer if possible.

Remember that putting fractions into the correct shape for adding and subtracting means writing the fractions with common denominators. To multiply or divide fractions, we do not need to use common denominators. However, we must write the fractions in **fraction form.** This means we will write mixed numbers and whole numbers as improper fractions. We write a whole number as an improper fraction by making the whole number the numerator of a fraction with a denominator of 1.

Example 1

A length of fabric was cut into 4 equal sections. Each of 4 students received $2\frac{2}{3}$ yd of fabric. How much fabric was there before it was cut?

This is an equal groups problem. To find the original length of the fabric we multiply $2\frac{2}{3}$ yd by 4. First, we write $2\frac{2}{3}$ and 4 in fraction form.

$$\frac{8}{3} \times \frac{4}{1}$$

Second, we multiply the numerators to find the numerator of the product, and we multiply the denominators to find the denominator of the product.

$$\frac{8}{3} \times \frac{4}{1} = \frac{32}{3}$$

Third, we simplify the product by converting the improper fraction to a mixed number.

$$\frac{32}{3} = 10\frac{2}{3}$$

Before the fabric was cut it was **$10\frac{2}{3}$ yd** long.

(**Evaluate**) How can we use estimation to check whether our answer is reasonable?

Example 2

Reading Math

Recall that the **terms** of a fraction are the numerator and the denominator.

Multiply: $2\frac{1}{2} \times 1\frac{1}{3}$

First, we write the numbers in fraction form.

$$\frac{5}{2} \times \frac{4}{3}$$

Second, we multiply the terms of the fractions.

$$\frac{5}{2} \times \frac{4}{3} = \frac{20}{6}$$

Third, we simplify the product.

$$\frac{20}{6} = 3\frac{2}{6} = 3\frac{1}{3}$$

Sketching a rectangle on a grid is a way to check the reasonableness of a product of mixed numbers. To illustrate the product of $2\frac{1}{2}$ and $3\frac{1}{2}$, we use a grid that is at least 3 by 4 so that a $2\frac{1}{2}$ by $3\frac{1}{2}$ rectangle fits on the grid.

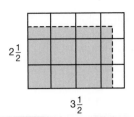

$2\frac{1}{2}$

$3\frac{1}{2}$

We sketch the rectangle and estimate the area. There are 6 full squares, 5 half squares, and a quarter square. Since 2 half squares equal a whole square, the area is about $8\frac{3}{4}$ square units.

Example 3

Oscar wants to tile his kitchen floor. He will pay $20 per square foot for someone to lay tile. Oscar's kitchen measures $20\frac{1}{4}$ feet by $9\frac{3}{4}$ feet. Estimate the amount he will pay for the job. Explain why an estimate is better than an exact calculation in this case.

The amount Oscar will pay is based on the area of the kitchen floor. To estimate the area we can round the length and width to 20 ft and 10 ft.

$$Area = lw$$

$$Area = (20 \text{ ft})(10 \text{ ft})$$

$$Area = 200 \text{ sq. ft}$$

Oscar will pay an $20 for each square foot. Since the floor covers 200 sq. ft, we multiply 20 and 200.

$$(\$20 \text{ per sq. ft})(200 \text{ sq. ft})$$

$$\$4000$$

Another way to estimate the cost is to calculate the floor area exactly, then round.

$$Area = lw$$

$$Area = (20\frac{1}{4} \text{ ft})(9\frac{3}{4} \text{ ft})$$

We write the mixed numbers as improper fractions.

$$Area = \frac{81}{4} \text{ ft} \times \frac{39}{4} \text{ ft}$$

$$Area = \frac{3159}{16} \text{ sq. ft}$$

Writing the area as a mixed number and rounding, we estimate the area to be 200 sq. ft.

$$Area = 197\frac{7}{16} \text{ sq. ft}$$

$$Area \approx 200 \text{ sq. ft}$$

Again, we estimate that Oscar will pay about $4000 for the tiling job.

An estimate is best in this case, since more tile is needed than that which would cover the floor. When tile is laid, excess tile is cut along the edges of the room. Since the area of the tile is usually greater than the floor area, knowing the exact area of the kitchen floor is not necessary.

Lesson Practice

Multiply:

a. $1\frac{1}{2} \times \frac{2}{3}$

b. $1\frac{2}{3} \times \frac{3}{4}$

c. $1\frac{1}{2} \times 1\frac{2}{3}$

d. $1\frac{2}{3} \times 3$

e. $2\frac{1}{2} \times 2\frac{2}{3}$

f. $3 \times 1\frac{3}{4}$

g. $3\frac{1}{3} \times 1\frac{2}{3}$

h. $2\frac{3}{4} \times 2$

i. $2 \times 3\frac{1}{2}$

j. Check the reasonableness of the products in e and h by sketching rectangles on a grid.

k. **Formulate** Write and solve a word problem about multiplying a whole number and a mixed number.

Written Practice *Distributed and Integrated*

* 1. Allison is making a large collage of a beach scene. She needs 2 yards of blue
(4, 62) ribbon for the ocean, $\frac{1}{2}$ yard of yellow ribbon for the sun, and $\frac{3}{4}$ yard of green ribbon for the grass. Ribbon costs $2 a yard. How much money will Allison need for ribbon?

2. **Estimate** A mile is 5280 feet. A nautical mile is about 6080 feet. A nautical
(7) mile is about how much longer than a mile?

3. **Verify** Instead of dividing $1.50 by $0.05, Malik formed an equivalent division
(38) problem by mentally multiplying both the dividend and the divisor by 100. Then he performed the equivalent division problem. What is the equivalent division problem Malik formed, and what is the quotient?

Find each unknown number:

4. $6 \text{ cm} + k = 11 \text{ cm}$
(5)

5. $8g = 9.6$
(39)

6. $\frac{7}{10} - w = \frac{1}{2}$
(39)

7. $\frac{3}{5} = \frac{n}{100}$
(37)

* 8. **Explain** The perimeter of a quadrilateral is 172 inches. What is the
(10, 59) average length of each side? Can we know for certain what type of quadrilateral this is? Why or why not?

9. $100.00 - ($46.75 + $9.68)$
(2)

10. $(2 \times 0.3) - (0.2 \times 0.3)$
(41)

* 11. **Analyze** $4\frac{1}{4} - 2\frac{7}{8}$
(58)

* 12. **Analyze** $2\frac{2}{3} \times \sqrt{9}$
(32, 62)

13. $3\frac{1}{3} + 2\frac{3}{4}$
(55)

* 14. $1\frac{1}{3} \times 2\frac{1}{4}$
(62)

15. $1.44 \div 60$
(41)

16. $\$6.00 \div \$0.15$
(44)

17. Five dollars was divided evenly among 4 people. How much money did each
(8) receive?

18. (Conclude) The area of a regular quadrilateral is 100 square inches. What is its
(59) perimeter? What is the name of the quadrilateral?

19. Write the prime factorization of
(61)
   a. 625

   b. 1000

* 20. What is the area of the rectangle shown below?
(23, 62)

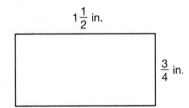

$1\frac{1}{2}$ in.

$\frac{3}{4}$ in.

21. Thirty-six of the 88 piano keys are black. What fraction of the piano keys
(25) are black?

* 22. (Represent) Draw a rectangular prism. Begin by drawing two congruent
(RF25) rectangles.

* 23. (Analyze) $1\frac{1}{2} \times \square = 1$
(36, 57)

24. There are 1000 meters in a kilometer. How many meters are in
(8) 2.5 kilometers?

25. (Connect) Which arrow could be pointing to 0.1 on the number line?
(28)

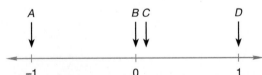

A   B C   D

−1   0   1

* 26. (Estimate) If the tip of the minute hand is 6 inches from the center of the clock,
(42) how far does the tip travel in one hour? Round the answer to the nearest inch.
(Use 3.14 for π.)

27. **Connect** A basketball is an example of what geometric solid?
(RF25)

28. Write 51% as a fraction. Then write the fraction as a decimal
(27, 29) number.

* 29. **Represent** What is the probability of rolling a prime number with
(11, 54) one toss of a number cube?

* 30. **Conclude** This quadrilateral has one pair of parallel sides. What
(59) kind of quadrilateral is it?

*Real-World
Connection*

Henrietta recently completed a driver's safety course, which means that
her insurance premiums will decrease. Before the course, Henrietta's
insurance cost $172 per month. It now costs $127 per month.

   a. How much was Henrietta paying for insurance per year?

   b. How much less is Henrietta now paying per year in premiums?

↘ *California Mathematics Content Standards*

NS 2.0, 2.4 Determine the least common multiple and the greatest common divisor of whole numbers; use them to solve problems with fractions (e.g., to find a common denominator to add two fractions or to find the reduced form for a fraction).

MR 1.0, 1.3 Determine when and how to break a problem into simpler parts.

• Using Prime Factorization to Reduce Fractions

facts   Power Up E

mental math

   a. Number Sense: 5×140 ($10 \times 140 \div 2$)

   b. Number Sense: $420 - 50$

   c. Money: $\frac{\$25}{100}$

   d. Functions: Masa can read 40 pages in one hour. At that rate, how many pages can he read in 4 hours?

   e. Decimals: 2×4.5

   f. Measurement: 500 centimeters are how many meters?

   g. Statistics: At the flower shop, the three bouquets contained 12, 12, and 6 roses, respectively. What was the mean number of roses per bouquet?

   h. Calculation: $5 \times 10, -20, +2, \div 4, +1, \div 3, -3$

problem solving

Choose an appropriate problem-solving strategy to solve this problem. Rhett chooses a marble at random from each of the four boxes below. From which box is he *most* likely to choose a blue marble?

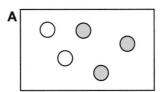

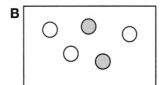

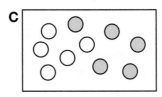

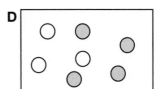

New Concept

Thinking Skills

Explain

What are two strategies for finding the prime factorization of a number?

One way to **reduce** fractions with large terms is to factor the terms and then reduce the common factors. To reduce $\frac{125}{1000}$, we could begin by writing the **prime factorizations** of 125 and 1000.

$$\frac{125}{1000} = \frac{5 \cdot 5 \cdot 5}{2 \cdot 2 \cdot 2 \cdot 5 \cdot 5 \cdot 5}$$

We see three pairs of 5's that can be reduced. Each $\frac{5}{5}$ reduces to $\frac{1}{1}$.

$$\frac{\overset{1}{\cancel{5}} \cdot \overset{1}{\cancel{5}} \cdot \overset{1}{\cancel{5}}}{2 \cdot 2 \cdot 2 \cdot \underset{1}{\cancel{5}} \cdot \underset{1}{\cancel{5}} \cdot \underset{1}{\cancel{5}}} = \frac{1}{8}$$

We multiply the remaining factors and find that $\frac{125}{1000}$ reduces to $\frac{1}{8}$.

Example 1

Thinking Skills

Discuss

When is it helpful to use prime factorization to reduce a fraction?

Reduce: $\frac{375}{1000}$

We write the prime factorization of both the numerator and the denominator.

$$\frac{375}{1000} = \frac{3 \cdot 5 \cdot 5 \cdot 5}{2 \cdot 2 \cdot 2 \cdot 5 \cdot 5 \cdot 5} = \frac{3 \cdot \overset{1}{\cancel{5}} \cdot \overset{1}{\cancel{5}} \cdot \overset{1}{\cancel{5}}}{2 \cdot 2 \cdot 2 \cdot \underset{1}{\cancel{5}} \cdot \underset{1}{\cancel{5}} \cdot \underset{1}{\cancel{5}}} = \frac{3}{8}$$

Then we reduce the common factors and multiply the remaining factors.

Notice that this method gives us the same result as dividing the numerator and denominator by the greatest common factor. In example 1, the greatest common factor is the product of the common prime factors, $5 \cdot 5 \cdot 5 = 125$.

$$\frac{375}{1000} = \frac{375 \div 125}{1000 \div 125} = \frac{3}{8}$$

By using the method of reducing common prime factors, we can avoid dividing by large numbers or reducing in multiple steps.

Example 2

Reduce: $\frac{180}{750}$

The fraction can by reduced by dividing the numerator and denominator by what number?

We write the prime factorization of the numerator and denominator; then we reduce.

$$\frac{180}{750} = \frac{2 \cdot 2 \cdot 3 \cdot 3 \cdot 5}{2 \cdot 3 \cdot 5 \cdot 5 \cdot 5} = \frac{\overset{1}{\cancel{2}} \cdot 2 \cdot \overset{1}{\cancel{3}} \cdot 3 \cdot \overset{1}{\cancel{5}}}{\underset{1}{\cancel{2}} \cdot \underset{1}{\cancel{3}} \cdot \underset{1}{\cancel{5}} \cdot 5 \cdot 5} = \frac{6}{25}$$

We also could have reduced by dividing the numerator and denominator by the greatest common factor, $2 \cdot 3 \cdot 5 = 30$.

$$\frac{180}{750} = \frac{180 \div 30}{750 \div 30} = \frac{6}{25}$$

Write the prime factorization of both the numerator and the denominator of each fraction. Then reduce each fraction.

a. $\dfrac{875}{1000}$

b. $\dfrac{48}{400}$

c. $\dfrac{125}{500}$

d. $\dfrac{36}{81}$

Written Practice

Distributed and Integrated

1. What is the difference between the sum of $\frac{1}{2}$ and $\frac{1}{4}$ and the product of $\frac{1}{2}$ and $\frac{1}{4}$?
(4, 53)

2. Bill ran a half mile in two minutes fifty-five seconds. How many seconds is that?
(8)

3. The gauge of a railroad—the distance between the two tracks—is usually 4 feet $8\frac{1}{2}$ inches. How many inches is that?
(RF11)

4. Three friends are painting a wall. If Felix paints $\frac{1}{3}$ of the wall, David paints $\frac{1}{4}$ of the wall, and Varujan paints $\frac{2}{5}$ of the wall, what fraction of the wall will be painted?
(56)

5. Convert $3\frac{5}{8}$ to an improper fraction.
(57)

6. In six games Yvonne scored a total of 108 points. How many points per game did she average?
(10)

* 7. Write the prime factorizations of 24 and 200. Then simplify $\frac{24}{200}$.
(63)

Find each unknown number:

* 8. $m - 5\frac{3}{8} = 1\frac{3}{16}$
(39, 55)

9. $3\frac{3}{5} + 2\frac{7}{10} = n$
(39, 55)

10. $25d = 0.375$
(39, 41)

11. $\dfrac{3}{4} = \dfrac{w}{100}$
(37)

* 12. $5\frac{1}{8} - 1\frac{1}{2}$
(58)

* 13. $3\frac{1}{3} \times 1\frac{1}{2}$
(62)

14. **Multiple Choice** Which of these quadrilaterals is *not* a parallelogram?
(59)

A   B  C   D

15. What is the area of a rectangle that is 4 inches long and $1\frac{3}{4}$ inches wide?
(23, 62)

16. $(3.2 + 1) - (0.6 \times 7)$
(2, 41)

17. $12.5 \div 0.4$
(44)

* 18. **Multiple Choice** The product 3.2×10 equals which of the following?
(RF23, 38)

   A $32 \div 10$ **B** $320 \div 10$ **C** $0.32 \div 10$

19. **Estimate** Find the sum of 6416, 5734, and 4912 to the nearest thousand.
(21)

* 20. **Verify** Instead of dividing 800 by 24, Arturo formed an equivalent division
(18, 38) problem by dividing both the dividend and the divisor by 8. Then he quickly
found the quotient of the equivalent problem. What is the equivalent problem
Arturo formed, and what is the quotient? Write the quotient as a mixed
number.

21. The perimeter of a square is 2.4 meters.
(32)
    a. How long is each side of the square?

    b. What is the area of the square?

22. What is the tax on an $18,000 car if the tax rate is 8%?
(35)

* 23. **Analyze** If the probability of an event occurring is 1 chance in a million, then
(54) what is the probability of the event not occurring?

24. **Explain** Why is a circle not a polygon?
(RF24)

25. If $\angle A$ and $\angle B$ are supplementary, find x.
(60)

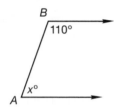

26. **Estimate** Use a ruler to find the length of this line segment to the nearest
(14) eighth of an inch.

_____

27. **Conclude** Which angle in this figure is an obtuse angle?
(20)

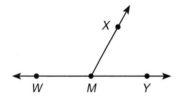

28. Write 3% as a fraction. Then write the fraction as a decimal number.
(27, 29)

* 29. **Connect** A shoe box is an example of what geometric solid?
(RF25)

30. Sunrise occurred at 6:20 a.m., and sunset occurred at 5:45 p.m. How many
(RF11, 7) hours and minutes were there from sunrise to sunset?

\ California Mathematics Content Standards

NS **2.0**, **2.1** Solve problems involving addition, subtraction, multiplication, and division of positive fractions and explain why a particular operation was used for a given situation.

NS **2.0**, **2.2** Explain the meaning of multiplication and division of positive fractions and perform the calculations (e.g., $\frac{5}{8} \div \frac{15}{16} = \frac{5}{8} \times \frac{16}{15} = \frac{2}{3}$).

• Dividing Mixed Numbers

facts   Power Up J

mental math

    a. Measurement: 4500 meters + 450 meters

    b. Money: $7.50 + $7.50

    c. Fractions: $\frac{1}{4}$ of $20.00

    d. Number Sense: $\frac{\$75}{10}$

    e. Statistics: The members of the Ross family have these ages: 41, 37, 16, 11, and 1. What is the median age of the family members?

    f. Algebra: If $8p = 32$, what is p?

    g. Estimation: Estimate the perimeter of this square in centimeters.

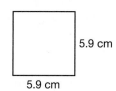

5.9 cm

5.9 cm

    h. Calculation: $6 \times 8, \div 2, + 1, \div 5, - 1, \times 4, \div 2$

problem solving

Choose an appropriate problem-solving strategy to solve this problem. Emily has a blue folder, a green folder, and a red folder. She uses one folder each for her math, science, and history classes. She does not use her blue folder for math. Her green folder is not used for science. She does not use her red folder for history. If her red folder is not used for math, what folder does Emily use for each subject? Make a table to show your work.

Recall the three steps to solving an arithmetic problem with fractions.

Step 1: Put the problem into the correct shape (if it is not already).

Step 2: Perform the operation indicated.

Step 3: Simplify the answer if possible.

In this lesson we will practice dividing mixed numbers. Recall from Lesson 62 that the correct shape for multiplying and dividing fractions is fraction form. So when dividing, we first write any mixed numbers or whole numbers as improper fractions.

Example 1

Shawna is pouring $2\frac{2}{3}$ cups of plant food into equal amounts to feed 4 plants. How much plant food is there for each plant?

Shawna is dividing $2\frac{2}{3}$ cups of plant food into four equal groups. We divide $2\frac{2}{3}$ by 4. We write the numbers as improper fractions.

$$\frac{8}{3} \div \frac{4}{1}$$

To divide, we find the number of 4's in 1. (That is, we find the reciprocal of 4.) Then we use the reciprocal of 4 to find the number of 4's in $\frac{8}{3}$.

$$1 \div \frac{4}{1} = \frac{1}{4}$$

$$\frac{8}{3} \times \frac{1}{4} = \frac{8}{12}$$

Math Language

Reciprocals are two numbers whose product is 1.

We simplify the answer.

$$\frac{8}{12} = \frac{2}{3}$$

There is $\frac{2}{3}$ **cup** of plant food for each plant. Notice that dividing a number by 4 is equivalent to finding $\frac{1}{4}$ of the number. Instead of dividing $2\frac{2}{3}$ by 4, we could have directly found $\frac{1}{4}$ of $2\frac{2}{3}$.

Example 2

Divide: $2\frac{2}{3} \div 1\frac{1}{2}$

We write the mixed numbers as improper fractions.

$$\frac{8}{3} \div \frac{3}{2}$$

Thinking Skills

Justify

Describe in your own words how to divide mixed numbers.

To divide, we find the number of $\frac{3}{2}$'s in 1. (That is, we find the reciprocal of $\frac{3}{2}$.) Then we use the reciprocal of $\frac{3}{2}$ to find the number of $\frac{3}{2}$'s in $\frac{8}{3}$.

$$1 \div \frac{3}{2} = \frac{2}{3}$$

$$\frac{8}{3} \times \frac{2}{3} = \frac{16}{9}$$

We simplify the improper fraction $\frac{16}{9}$ as shown below.

$$\frac{16}{9} = 1\frac{7}{9}$$

Example 3

Randy has a car detailing business. He can detail a car in $3\frac{1}{2}$ hours. How many cars can he detail in 8 hours?

This is an equal groups plot.

$$n \times 3\frac{1}{2} \text{ hr} = 8 \text{ hr}$$

To find the number of cars he can detail in 8 hours, we divide 8 by $3\frac{1}{2}$.

$$8 \div 3\frac{1}{2}$$

Rewrite 8 and $3\frac{1}{2}$ as improper fractions.

$$\frac{8}{1} \div \frac{7}{2}$$

To divide, we find the number of $\frac{7}{2}$'s in 1. Then we find the number of $\frac{7}{2}$'s in $\frac{8}{1}$.

$$1 \div \frac{7}{2} = \frac{2}{7}$$

$$\frac{8}{1} \times \frac{2}{7} = \frac{16}{7}$$

We simplify the improper fraction. $\frac{16}{7} = 2\frac{2}{7}$

In 8 hours, Randy can detail two cars and finish $\frac{2}{7}$ of the work on the third car.

Lesson Practice

Find each product or quotient:

a. $1\frac{3}{5} \div 4$ 　　　　　　　　**b.** $\frac{1}{4}$ of $1\frac{3}{5}$

c. $2\frac{2}{5} \div 3$ 　　　　　　　　**d.** $\frac{1}{3}$ of $2\frac{2}{5}$

e. **Generalize** Why is dividing by 4 the same as multiplying by $\frac{1}{4}$?

f. $1\frac{2}{3} \div 2\frac{1}{2}$　　　　　　**g.** $2\frac{1}{2} \div 1\frac{2}{3}$

h. $1\frac{1}{2} \div 1\frac{1}{2}$　　　　　　**i.** $7 \div 1\frac{3}{4}$

j. Gabriel has $2\frac{1}{4}$ hours to finish three projects. If he divides his time equally, what fraction of an hour can he spend on each project?

Written Practice　　　*Distributed and Integrated*

* 1. **(Model)** Draw a pair of parallel lines. Then draw a second pair of parallel lines
(20, 59)　that are perpendicular to the first pair. Trace the quadrilateral that is formed by the intersecting pairs of lines. What kind of quadrilateral did you trace?

* 2. **(Connect)** What is the quotient if the dividend is $\frac{1}{2}$ and the divisor is $\frac{1}{8}$?
(49)

3. The highest weather temperature recorded was 136°F in Africa. The lowest
(9)　was −129°F in Antarctica. How many degrees difference is there between these temperatures?

* 4. **(Estimate)** A dollar bill is about 6 inches long. Placed end to end, about how
(8)　many **feet** would 1000 dollar bills reach?

* 5. Write the prime factorization of both the numerator and the denominator of this
(63)　fraction. Then reduce the fraction.

$$\frac{45}{72}$$

* 6. **(Conclude)** In quadrilateral *QRST,* which segment appears to be parallel
(20)　to $\overline{RS}$?

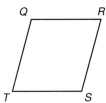

7. In 10 days Juana saved $27.50. On average, how much did she save per
(7)　day?

* 8. $\dfrac{1 \times 2 \times 3 \times 4 \times 5}{1 + 2 + 3 + 4 + 5}$　　　　**9.** $3\frac{1}{2} + 2\frac{3}{4} + 1\frac{5}{8}$
(2)　　　　　　　　　　　　　　　(56)

Find each unknown number:

10. $m + 1\frac{3}{4} = 5\frac{3}{8}$　　　　**11.** $\frac{3}{4} - f = \frac{1}{3}$
(39, 58)　　　　　　　　　　　(39, 53)

12. $\frac{2}{5}w = 1$　　　　　　**13.** $\frac{8}{25} = \frac{n}{100}$
(36)　　　　　　　　　　　(37)

* 14. $1\frac{2}{3} \div 2$　　　　　　*** 15.** $2\frac{2}{3} \times 1\frac{1}{5}$
(64)　　　　　　　　　　　(62)

16. $\dfrac{2.4}{0.08}$
(44)

17.
(32)
a. What is the perimeter of this square?

b. What is the area of this square?

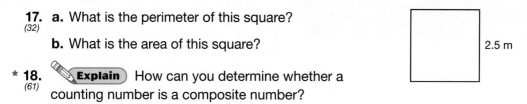

2.5 m

* 18.
(61)
✎ **Explain** How can you determine whether a counting number is a composite number?

* 19.
(61)
Represent Make a factor tree to find the prime factorization of 250.

20.
(RF24)
A stop sign has the shape of an eight-sided polygon. What is the name of an eight-sided polygon?

21.
(16, 25)
There were 15 boys and 12 girls in the class.

a. What fraction of the class was made up of girls?

b. What was the ratio of boys to girls in the class?

22.
(38, 62)
Verify Instead of dividing $4\frac{1}{2}$ by $1\frac{1}{2}$, Carla doubled both numbers before dividing mentally. What was Carla's mental division problem and its quotient?

23.
(36, 57)
What is the reciprocal of $2\frac{1}{2}$?

24.
(8)
There are 1000 grams in 1 kilogram. How many grams are in 2.25 kilograms?

25.
(RF6)
How many **millimeters** long is the line below?

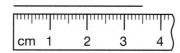

* 26.
(RF9)
Connect The length of $\overline{WX}$ is 53 mm. The length of $\overline{XY}$ is 35 mm. What is the length of $\overline{WY}$?

27.
(RF25)
Represent Draw a cylinder.

28.
(28)
Arrange these numbers in order from least to greatest:

0.1, 1, −1, 0

29.
(Inv. 3)
Why is this sample not representative? Why might the sample be biased? *The dog owners at a dog show were asked how many pets they owned.*

30.
(RF29)
How many smaller cubes are in the large cube shown below?

California Mathematics Content Standards

NS **2.0**, **2.4** Determine the least common multiple and the greatest common divisor of whole numbers; use them to solve problems with fractions (e.g., to find a common denominator to add two fractions or to find the reduced form for a fraction).

MR **1.0, 1.3** Determine when and how to break a problem into simpler parts.

• Reducing Fractions Before Multiplying

Power Up

facts

Power Up I

mental math

a. Number Sense: 3×65

b. Number Sense: $\frac{\$75}{100}$

c. Money: Double 75¢.

d. Time: How many minutes are in 5 hours?

e. Money: Alma handed the clerk a $20 bill for the magazine that cost $2.50. How much change should she receive?

f. Geometry: Angles A and B are supplementary angles. If m$\angle A = 120°$, what is m$\angle B$?

g. Statistics: In a common year, the months January to December have 31, 28, 31, 30, 31, 30, 31, 31, 30, 31, 30, and 31 days, respectively. What is the mode?

h. Calculation: $9 \times 9, -1, \div 2, +2, \div 6, +3, \div 10$

problem solving

Choose an appropriate problem-solving strategy to solve this problem. Use the digits 6, 7, and 8 to complete this multiplication problem:

$$\begin{array}{r} 23\_ \\ \times \quad \_ \\ \hline 166\_ \end{array}$$

New Concept

Before two or more fractions are multiplied, we might be able to reduce the fraction terms, even if the reducing involves different fractions. For example, in the multiplication below we see that the number 3 appears as a numerator and as a denominator in different fractions.

$$\frac{3}{5} \times \frac{2}{3} = \frac{6}{15} \qquad \frac{6}{15} \text{ reduces to } \frac{2}{5}$$

We may reduce the common terms (the 3s) before multiplying. We reduce $\frac{3}{3}$ to $\frac{1}{1}$ by dividing both 3's by 3. Then we multiply the remaining terms.

$$\frac{\overset{1}{\cancel{3}}}{5} \times \frac{2}{\underset{1}{\cancel{3}}} = \frac{2}{5}$$

By reducing before we multiply, we avoid the need to reduce after we multiply. Reducing before multiplying is also known as **canceling.**

Example 1

Simplify: $\frac{5}{6} \times \frac{1}{5}$

We reduce before we multiply. Since 5 appears as a numerator and as a denominator, we reduce $\frac{5}{5}$ to $\frac{1}{1}$ by dividing both 5's by 5. Then we multiply the remaining terms.

$$\frac{\overset{1}{\cancel{5}}}{6} \times \frac{1}{\underset{1}{\cancel{5}}} = \frac{1}{6}$$

Example 2

Simplify: $1\frac{1}{9} \times 1\frac{1}{5}$

First we write the numbers in fraction form.

$$\frac{10}{9} \times \frac{6}{5}$$

We mentally pair 10 with 5 and 6 with 9.

$$\frac{10}{9} \times \frac{6}{5}$$

We reduce $\frac{10}{5}$ to $\frac{2}{1}$ by dividing both 10 and 5 by 5. We reduce $\frac{6}{9}$ to $\frac{2}{3}$ by dividing both 6 and 9 by 3.

$$\frac{\overset{2}{\cancel{10}}}{\underset{3}{\cancel{9}}} \times \frac{\overset{2}{\cancel{6}}}{\underset{1}{\cancel{5}}} = \frac{4}{3}$$

We multiply the remaining terms. Then we simplify the product.

$$\frac{4}{3} = 1\frac{1}{3}$$

Thinking Skills

Discuss

Why might you want to reduce before you multiply?

Example 3

Thinking Skills

Justify

Explain how to divide any two fractions.

Simplify: $\frac{5}{6} \div \frac{5}{2}$

This is a division problem. We first find the number of $\frac{5}{2}$'s in 1. The answer is the reciprocal of $\frac{5}{2}$. We then use the reciprocal of $\frac{5}{2}$ to find the number of $\frac{5}{2}$'s in $\frac{5}{6}$.

$$1 \div \frac{5}{2} = \frac{2}{5}$$

$$\frac{5}{6} \times \frac{2}{5}$$

Now we have a multiplication problem. We cancel before we multiply.

$$\frac{\overset{1}{\cancel{5}}}{\underset{3}{\cancel{6}}} \times \frac{\overset{1}{\cancel{2}}}{\underset{1}{\cancel{5}}} = \frac{1}{3}$$

Note: We may cancel the terms of fractions only when multiplying. A division problem must be rewritten as a multiplication problem before we may cancel the terms of the fractions. We do not cancel the terms of fractions in addition or subtraction problems.

Lesson Practice

Reduce before multiplying:

a. $\frac{3}{4} \cdot \frac{4}{5}$ **b.** $\frac{2}{3} \cdot \frac{3}{4}$ **c.** $\frac{8}{9} \cdot \frac{9}{10}$

Write in fraction form. Then reduce before multiplying.

d. $2\frac{1}{4} \times 4$ **e.** $1\frac{1}{2} \times 2\frac{2}{3}$ **f.** $3\frac{1}{3} \times 2\frac{1}{4}$

Rewrite each division problem as a multiplication problem. Then reduce before multiplying.

g. $\frac{2}{5} \div \frac{2}{3}$ **h.** $\frac{8}{9} \div \frac{2}{3}$ **i.** $\frac{9}{10} \div 1\frac{1}{5}$

Written Practice

Distributed and Integrated

1. Alaska was purchased from Russia in 1867 for seven million, two hundred thousand dollars. Use digits to write that amount.
(RF12)

* 2. (Connect) How many eighth notes equal a half note?
(49)

* 3. (Verify) Instead of dividing $12\frac{1}{2}$ by $2\frac{1}{2}$, Shandra doubled both numbers and then divided. Write the division problem Shandra formed, as well as its quotient.
(38, 62)

Simplify before multiplying:

* 4. $\frac{5}{6} \cdot \frac{4}{5}$ *** 5.** $\frac{5}{6} \div \frac{5}{2}$ *** 6.** $\frac{9}{10} \cdot \frac{5}{6}$
(65)   (65)   (65)

7. What number is halfway between $\frac{1}{2}$ and 1 on the number line?
(14)

8. $\sqrt{100} + 10^2$
(32)

9. $3\frac{2}{3} + 4\frac{5}{6}$
(55)

10. $7\frac{1}{8} - 2\frac{1}{2}$
(58)

11. $4.37 + 12.8 + 6$
(RF21)

12. $0.46 \div 5$
(41)

13. $60 \div 0.8$
(44)

14. **Evaluate** What is the mean of the three numbers marked by the arrows on this
(10) decimal number line? (First estimate whether the mean will be more than 5 or less
than 5.)

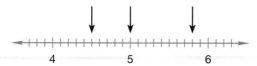

* 15. **Multiple Choice** The division problem 1.5 ÷ 0.06 is equivalent to which of the
(44) following?

    A 15 ÷ 6   **B** 150 ÷ 6   **C** 150 ÷ 60

16. There are 1000 milliliters in 1 liter. How many milliliters are in 3.8 liters?
(33)

Find each unknown number:

17. $\frac{2}{3} + n = 1$ **18.** $\frac{2}{3}m = 1$ **19.** $f - \frac{3}{4} = \frac{5}{6}$
(39)   (36)   (39, 53)

20. A pyramid with a triangular base has how many
(RF25)

   a. faces?

   b. edges?

   c. vertices?

Write the numbers in fraction form. Then simplify before multiplying.

* 21. $1\frac{2}{3} \times 1\frac{1}{5}$ *** 22.** $\frac{8}{9} \div 2\frac{2}{3}$
(65)   (65)

Refer to the line graph below to answer problems **23–25.**

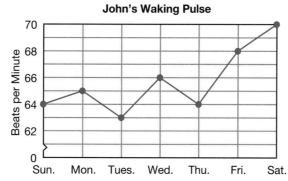

John's Waking Pulse

23. When John woke on Saturday, his pulse was how many beats per minute more
(Inv. 4) than it was on Tuesday?

24. On Monday John took his pulse for 3 minutes before marking the graph. How
(Inv. 4) many times did his heart beat in those 3 minutes?

* 25. ✏️ **Formulate** Write a question that relates to the graph and answer the
(Inv. 4) question.

* 26. **Analyze** Write the prime factorization of both the numerator and the
(63) denominator of this fraction. Then simplify the fraction.

$$\frac{72}{300}$$

In rectangle *ABCD* the length of $\overline{AB}$ is 2.5 cm, and the length of $\overline{BC}$ is 1.5 cm. Use this information and the figure below to answer problems **27–30.**

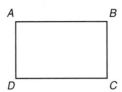

27. What is the perimeter of this rectangle?
(23)

28. What is the area of this rectangle?
(23)

* **29.** Name two segments perpendicular to $\overline{DC}$.
(59)

* **30.** If $\overline{BD}$ were drawn on the figure to divide the rectangle into two equal parts, what
(23) would be the area of each part?

Real-World Connection

Roland went to the local Super Store yesterday. He bought a new paint roller and roller pan for $8.97, a gallon of milk for $2.89, a magazine for $1.59, and two identical gallons of paint without marked prices. He paid a total of $47.83 before tax. Find the price for each gallon of paint.

California Mathematics Content Standards

NS **2.0**, **2.2** Explain the meaning of multiplication and division of positive fractions and perform the calculations (e.g., 5/8 ÷ 15/16 = 5/8 x 16/15 = 2/3).

MR 3.0, 3.3 Develop generalizations of the results obtained and the strategies used and apply them in new problem situations.

• Why Does Reducing Before Multiplying Work?

We can use mathematical ideas to save time and effort. Multiplying fractions is simple. Multiply the numerators. Multiply the denominators. Reduce the fraction answer, if possible. But sometimes the numbers are not easy to work with.

$$\frac{4}{9} \times \frac{15}{44} = \frac{4 \times 15}{9 \times 44} = \frac{60}{396}$$

Now we must reduce our answer. See how large the numbers are? Reducing this fraction is a difficult and long problem if the greatest common factor (GCF) is not used.

$$\frac{60}{396} = \frac{30}{198} = \frac{15}{99} = \frac{5}{33}$$

Besides finding the GCF, which was 12 for this problem, there is another way to make this problem easier. We can reduce the fractions before we multiply them. Why can we do this?

A fraction is division—one number divided by another number. Any factor in the numerator can be divided by any factor in the denominator. When we multiply fractions, the answer is a fraction, so we can either reduce before we multiply or after we multiply.

$$\frac{4}{9} \times \frac{15}{44} = \frac{\overset{1}{\cancel{4}} \times \overset{5}{\cancel{15}}}{\underset{3}{\cancel{9}} \times \underset{11}{\cancel{44}}} = \frac{1 \times 5}{3 \times 11} = \frac{5}{33}$$

- 4 and 44 have a common factor of 4
- 9 and 15 have a common factor of 3

Note how much easier it is to reduce the numbers when they are still factors. Both methods give the same answer, but by reducing the factors before multiplying, the numbers are smaller and easier to work with.

Discuss How are both methods alike and different?

Apply Try a few on your own. Show how you reduced each fraction.

a. $\frac{15}{52} \times \frac{13}{30}$ b. $\frac{6}{19} \times \frac{38}{50}$ c. $\frac{11}{12} \times \frac{13}{121}$

California Mathematics Content Standards

AF 3.0, 3.1 Use variables in expressions describing geometric quantities (e.g., *P* = 2w + 2l, *A* = 1/2bh, *C* = πd - the formulas for the perimeter of a rectangle, the area of a triangle, and the circumference of a circle, respectively).

MG 2.0, 2.1 Identify angles as vertical, adjacent, complementary, or supplementary and provide descriptions of these terms.

MG 2.0, 2.2 Use the properties of complementary and supplementary angles and the sum of the angles of a triangle to solve problems involving an unknown angle.

• Parallelograms

facts	Power Up H
mental math	**a. Number Sense:** 5×160
	b. Number Sense: 8×23
	c. Money: $1.75 + $1.75
	d. Number Sense: $\frac{$30}{10}$
	e. Fractions: The sale price was $\frac{1}{3}$ off the regular price of $60.00. What is $\frac{1}{3}$ of $60.00?
	f. Statistics: In his first four football games, Adrian scored 0, 1, 3, and 0 touchdowns. What was the mean number of touchdowns per game?
	g. Algebra: If $2 + p = 3.2$, what is p?
	h. Calculation: $8 \times 8, - 4, \div 2, + 3, \div 3, + 1, \div 6, \div 2$
problem solving	Choose an appropriate problem-solving strategy to solve this problem. In this figure a square and a regular pentagon share a common side. The area of the square is 25 square centimeters. What is the perimeter of the pentagon?

New Concept

In this lesson, we will learn about various properties of **parallelograms.** The following example describes some angle properties of parallelograms.

Example 1

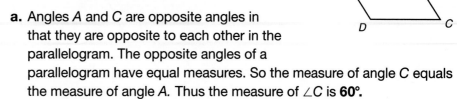

In parallelogram *ABCD*, the measure of angle *A* is 60°.

a. What is the measure of ∠C?

b. What is the measure of ∠B?

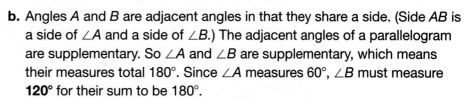

a. Angles *A* and *C* are opposite angles in that they are opposite to each other in the parallelogram. The opposite angles of a parallelogram have equal measures. So the measure of angle *C* equals the measure of angle *A*. Thus the measure of ∠C is **60°**.

b. Angles *A* and *B* are adjacent angles in that they share a side. (Side *AB* is a side of ∠A and a side of ∠B.) The adjacent angles of a parallelogram are supplementary. So ∠A and ∠B are supplementary, which means their measures total 180°. Since ∠A measures 60°, ∠B must measure **120°** for their sum to be 180°.

(**Model**) A flexible model of a parallelogram is useful for illustrating some properties of a parallelogram. A model can be constructed of brads and stiff tagboard or cardboard.

Lay two 8-in. strips of tagboard or cardboard over two parallel 10-in. strips as shown. Punch a hole at the center of the overlapping ends. Then fasten the corners with brads to hold the strips together.

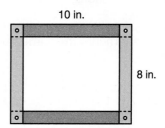

10 in.

8 in.

If we move the sides of the parallelogram back and forth, we see that opposite sides always remain parallel and equal in length. Though the angles change size, opposite angles remain equal and adjacent angles remain supplementary.

With this model we also can observe how the area of a parallelogram changes as the angles change. We hold the model with two hands and slide opposite sides in opposite directions. The maximum area occurs when the angles are 90°. The area reduces to zero as opposite sides come together.

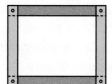
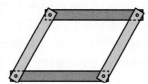
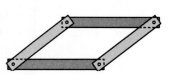

(**Discuss**) The area of a parallelogram changes as the angles change. Does the perimeter change?

The flexible model shows that parallelograms may have sides that are equal in length but areas that are different. To find the area of a parallelogram, we multiply two **perpendicular** measurements. We multiply the **base** by the **height** of the parallelogram.

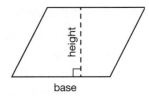

The base of a parallelogram is the length of one of the sides. The height of a parallelogram is the perpendicular distance from the base to the opposite side. The following activity will illustrate why the area of a parallelogram equals the base times the height.

Activity

Area of a Parallelogram

Materials needed:

- graph paper
- ruler
- pencil
- scissors

(**Represent**) Tracing over the lines on the graph paper, draw two parallel segments the same number of units long but shifted slightly as shown.

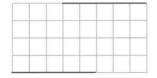

Then draw segments between the endpoints of the pair of parallel segments to complete the parallelogram.

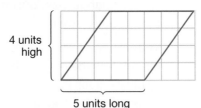

The base of the parallelogram we drew has a length of 5 units. The height of the parallelogram is 4 units. Your parallelogram might be different. How many units long and high is your parallelogram? Can you easily count the number of square units in the area of your parallelogram?

(**Model**) Use scissors to cut out your parallelogram.

Then select a line on the graph paper that is perpendicular to the first pair of parallel sides that you drew. Cut the parallelogram into two pieces along this line.

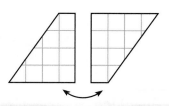

We will cut here.

Rearrange the two pieces of the parallelogram to form a rectangle. What is the length and width of the rectangle? How many square units is the area of the rectangle?

Our rectangle is 5 units long and 4 units wide. The area of the rectangle is 20 square units. So the area of the parallelogram is also 20 square units.

By making a perpendicular cut across the parallelogram and rearranging the pieces, we formed a rectangle having the same area as the parallelogram. The length and width of the rectangle equaled the base and height of the parallelogram. Therefore, by multiplying the base and height of a parallelogram, we can find its area.

Example 2

Find the area of this parallelogram:

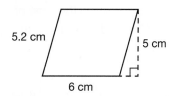

We multiply two perpendicular measurements, the base and the height. The height is often shown as a dashed line segment. The base is 6 cm. The height is 5 cm.

$$A = bh$$
$$A = 6 \text{ cm} \times 5 \text{ cm}$$
$$A = 30 \text{ sq. cm}$$

The area of the parallelogram is **30 sq. cm.**

Example 3

This parallelogram has unknown height x. Write an equation that expresses the area of the parallelogram in term of x.

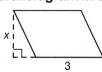

The area of a parallelogram is given by the formula.

$$A = bh$$

The base is 3, and the height is *x*. We replace *b* with *3* and *h* with *x*.

$$A = 3x$$

Lesson Practice

Conclude Refer to parallelogram *QRST* to answer problems **a–d**.

a. Which angle is opposite ∠Q?

b. Which angle is opposite ∠T?

c. Name two angles that are supplements of ∠T.

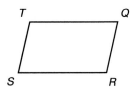

d. If the measure of ∠R is 100°, what is the measure of ∠Q?

Calculate the perimeter and area of each parallelogram:

e.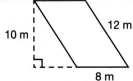

f.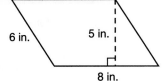

g. **Analyze** The formula for finding the area of a parallelogram is *A = bh*. This formula means

$$\text{Area} = \text{base} \times \text{height}$$

The base is the length of one side. The height is the perpendicular distance to the opposite side. Here we show the same parallelogram in two different positions, so the area of the parallelogram is the same in both drawings. What is the height in the figure on the right?

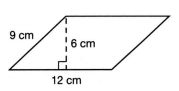 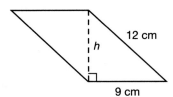

Written Practice

Distributed and Integrated

1. What is the least common multiple of 6 and 10?
(26)

2. **Analyze** The highest point on land is Mt. Everest, whose peak
(9) is 29,035 feet above sea level. The lowest point on land is the Dead Sea, which dips to 1371 feet below sea level. What is the difference in elevation between these two points?

3. The movie lasted 105 minutes. If the movie started at 1:15 p.m., at what time did
(RF11, 7) it end?

Analyze In problems **4–7,** simplify the fractions, if possible, before multiplying.

* 4. $\dfrac{2}{3} \cdot \dfrac{3}{8}$
(65)

* 5. $1\dfrac{1}{4} \cdot 2\dfrac{2}{3}$
(65)

* 6. $\dfrac{3}{4} \div \dfrac{3}{8}$
(65)

* 7. $4\dfrac{1}{2} \div 6$
(65)

8. $6 + 3\dfrac{3}{4} + 2\dfrac{1}{2}$
(55)

9. $5 - 3\dfrac{1}{8}$
(58)

10. $5\dfrac{1}{4} - 1\dfrac{7}{8}$
(58)

11. $(3.5)^2$
(33)

12. $15\overline{)\$75.00}$
(RF2)

13. $(1 + 0.6) \div (1 - 0.6)$
(41)

14. Quan ordered a $4.50 bowl of soup. The tax rate was $7\dfrac{1}{2}\%$ (which equals 0.075).
(35) He paid for the soup with a $20 bill.

    a. What was the tax on the bowl of soup?

    b. What was the total price including tax?

    c. How much money should Quan get back from his payment?

* 15. What is the name for the point on the coordinate plane that has the coordinates
(RF27) (0, 0)?

* 16. **Represent** Refer to the coordinate plane below to locate the points indicated.
(RF27)

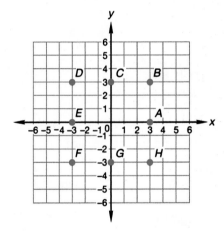

    Name the points that have the following coordinates:

    a. (−3, 3)   **b.** (0, −3)

    Identify the coordinates of the following points:

    c. *H*   **d.** *E*

Find each unknown number:

17. $1.2f = 120$
(39, 44)

18. $\dfrac{120}{f} = 1.2$
(39, 44)

* 19. Write the prime factorization of both the numerator and the denominator of this
(63) fraction. Then simplify the fraction.

$$\dfrac{64}{224}$$

20. The perimeter of a square is 6.4 meters. What is its area?
(32)

21. **Analyze** What fraction of this circle is not shaded?
(Inv. 1)

22. **Explain** If the radius of this circle is 1 cm, what is the circumference
(42) of the circle? (Use 3.14 for π.) How did you find your answer?

23. **Estimate** A centimeter is about as long as this segment:
(RF6)

_____

About how many centimeters long is your little finger?

24. **Connect** Water freezes at 32° Fahrenheit.
(10) The temperature shown on the thermometer is
how many degrees Fahrenheit above the freezing
point of water?

25. Ray watched TV for one hour. He determined
(25, 27) that commercials were shown 20% of that hour.
Write 20% as a reduced fraction. Then find the
number of minutes that commercials were shown
during the hour.

26. Name the geometric solid shown at right.
(RF25)

* 27. **Analyze** This square and regular triangle share a
(RF24) common side. The perimeter of the square is 24 cm.
What is the perimeter of the triangle?

* 28. **Multiple Choice** Choose the appropriate unit for the area of
(23) your state.

    A square inches   **B** square yards   **C** square miles

* 29. **a.** What is the perimeter of this parallelogram?
(66)
    b. What is the area of this parallelogram?

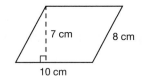

7 cm   8 cm

10 cm

* 30. **Conclude** In this figure $\angle BMD$ is a right angle.
(60) Name two angles that are

    a. supplementary.

    b. complementary.

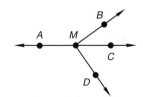

67

California Mathematics Content Standards

NS **2.0**, **2.1** Solve problems involving addition, subtraction, multiplication, and division of positive fractions and explain why a particular operation was used for a given situation.

NS **2.0**, **2.2** Explain the meaning of multiplication and division of positive fractions and perform the calculations (e.g., $\frac{5}{8} \div \frac{15}{16} = \frac{5}{8} \times \frac{16}{15} = \frac{2}{3}$).

MR **3.0**, **3.3** Develop generalizations of the results obtained and the strategies used and apply them in new problem situations.

• Multiplying Three Fractions

facts	Power Up J
mental math	**a. Number Sense:** 5×260
	b. Number Sense: $341 - 50$
	c. Money: $9.25 - 75¢
	d. Number Sense: $\frac{\$30}{100}$
	e. Estimation: Estimate the total price of 4 CDs that each cost \$14.97.
	f. Geometry: Angles C and D are complementary angles. If m$\angle C$ is 35°, what is m$\angle D$?
	g. Statistics: What is the range of 21, 44, 90, and 100?
	h. Calculation: $6 \times 6, -1, \div 5, \times 2, +1, \div 3, \div 2$

problem solving

Choose an appropriate problem-solving strategy to solve this problem. A famous conjecture states that any even number greater than two can be written as a sum of two prime numbers ($12 = 5 + 7$). Another states that any odd number greater than five can be written as the sum of three prime numbers ($11 = 7 + 2 + 2$). Write the numbers 10, 15, and 20 as the sums of primes. (The same prime number may be used more than once in a sum.)

We have learned three steps to take when performing pencil-and-paper arithmetic with fractions and mixed numbers:

Step 1: Write the problem in the correct **shape.**

Step 2: Perform the **operation.**

Step 3: Simplify the answer.

The letters S.O.S. can help us remember the steps as "shape," "operate," and "simplify." We summarize the S.O.S. rules we have learned in the following fractions chart.

Fractions Chart

	+ −	× ÷	
1. Shape	Write fractions with common denominators.	Write numbers in fraction form.	
2. Operate	Add or subtract the numerators.	×	÷
		Cancel.	Find reciprocal of divisor, then cancel.
		Multiply numerators. Multiply denominators.	
3. Simplify	Reduce fractions. Convert improper fractions.		

- Below the + and − symbols we list the steps for adding or subtracting fractions.
- Below the × and ÷ symbols, we list the steps for multiplying or dividing fractions.

The "shape" step for addition and subtraction is the same; we write the fractions with common denominators. Likewise, the "shape" step for multiplication and division is the same; we write both numbers in fraction form.

At the "operate" step, however, we separate multiplication and division. When multiplying fractions, we may reduce (cancel) before we multiply. Then we multiply the numerators to find the numerator of the product, and we multiply the denominators to find the denominator of the product. When dividing fractions, we first replace the divisor of the division problem with its reciprocal and change the division problem to a multiplication problem. We cancel terms, if possible, and then multiply.

The "simplify" step is the same for all four operations. We reduce answers when possible and convert answers that are improper fractions to mixed numbers.

To multiply three or more fractions, we follow the same steps we take when multiplying two fractions:

Step 1: We write the numbers in fraction form.

Step 2: We cancel terms by reducing numerator-denominator pairs that have common factors. Then we multiply the remaining terms.

Step 3: We simplify if possible.

> **Math Language**
>
> Recall that **canceling** means reducing before multiplying.

Example 1

Multiply: $\frac{2}{3} \times 1\frac{3}{5} \times \frac{3}{4}$

First we write $1\frac{3}{5}$ as the improper fraction $\frac{8}{5}$. Then we reduce where possible before multiplying. Multiplying the remaining terms, we find the product.

$$\frac{2}{3} \times \frac{\overset{2}{\cancel{8}}}{5} \times \frac{\overset{1}{\cancel{3}}}{\underset{1}{\cancel{4}}} = \frac{4}{5}$$

Example 2

Jim has drawn a parallelogram with base $1\frac{3}{4}$ inches and height $1\frac{1}{2}$ inches. He wants to draw a second parallelogram with an area that is $2\frac{2}{3}$ times the area of the first parallelogram. What will be the area of the new parallelogram?

The formula for area of parallelogram is base times height. We want to multiply the area by $2\frac{2}{3}$. First we write the mixed numbers as improper fractions. Then we reduce where possible before multiplying. Multiplying the remaining terms and then simplifying gives us the area.

$$A = 2\frac{2}{3} \cdot 1\frac{3}{4} \cdot 1\frac{1}{2}$$

$$A = \frac{\overset{1}{\cancel{8}}}{\underset{1}{\cancel{3}}} \cdot \frac{7}{\underset{1}{\cancel{4}}} \cdot \frac{\overset{1}{\cancel{3}}}{\underset{1}{\cancel{2}}}$$

$$A = 7$$

The area of the new parallelogram is **7 sq in.**

Example 3

What is area of a triangle whose base is $1\frac{3}{4}$ inches and whose height is $1\frac{1}{2}$ inches?

The formula for the area of a triangle is $\frac{1}{2}$ times base times height. First we write the mixed numbers as improper fractions. Then we reduce where possible before multiplying. Multiplying the remaining terms and then simplifying gives us the area.

$$A = \frac{1}{2} \cdot 1\frac{3}{4} \cdot 1\frac{1}{2}$$

$$A = \frac{1}{2} \cdot \frac{7}{4} \cdot \frac{3}{2}$$

$$A = \frac{21}{16}$$

$$A = 1\frac{5}{16}$$

The area of the triangle is **$1\frac{5}{16}$ sq in.**

Lesson Practice

a. Draw the fractions chart from this lesson.

b. Describe the three steps for adding fractions.

c. Describe the steps for dividing fractions.

Multiply:

d. $\dfrac{2}{3} \cdot \dfrac{4}{5} \cdot \dfrac{3}{8}$

e. $2\frac{1}{2} \times 1\frac{1}{10} \times 4$

Written Practice

* 1. What is the mean of 4.2, 2.61, and 3.6?
(10)

2. (Connect) Four tablespoons equals $\frac{1}{4}$ cup. How many tablespoons would equal
(49) one full cup?

3. The temperature on the moon ranges from a high of about 130°C to a low of
(9) about -110°C. This is a difference of how many degrees?

4. Four of the 12 marbles in the bag are blue. If one marble is taken from the bag,
(54) what is the probability that the marble is

    a. blue?   **b.** not blue?

    c. What word names the relationship between the events in **a** and **b?**

5. The diameter of a circle is 1 meter. The circumference is how many centimeters?
(RF6, (Use 3.14 for π.)
42)

6. (Connect) What fraction of a dollar is a nickel?
(25)

Find each unknown number:

7. $n - \frac{1}{2} = \frac{3}{5}$
(39, 53)

8. $1 - w = \frac{7}{12}$
(39)

9. $w + 2\frac{1}{2} = 3\frac{1}{3}$
(39, 55)

10. $1 - w = 0.23$
(39)

11. Write the standard decimal number for the following:
(RF22)

$$(6 \times 10) + \left(4 \times \frac{1}{10}\right) + \left(3 \times \frac{1}{100}\right)$$

12. Multiple Choice (Estimate) Which of these numbers is closest to 1?
(28)
    A -1 **B** 0.1   **C** 10

13. What is the largest prime number that is less than 100?
(11)

* 14. Multiple Choice Which of these figures is not a parallelogram?
(59)

    A   **C**

    B   **D**

15. (Connect) A loop of string two feet around is formed to make a square.
(32)
    a. How many inches long is each side of the square?

    b. What is the area of the square in square inches?

* 16. (Conclude) Figure *ABCD* is a rectangle.
(60)
    a. Name an angle complementary to $\angle DCM$.

    b. Name an angle supplementary to $\angle AMC$.

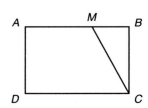

Refer to this menu and the information that follows to answer problems **17–19.**

Cafe Menu

Grilled Chicken Sandwich . . . $3.49	Juice:	Small $0.89
Green Salad $3.29		Medium . . . $1.09
Pasta Salad $2.89		Large $1.29

From this menu the Johnsons ordered two grilled chicken sandwiches, one green salad, one small juice, and two medium juices.

17. What was the total price of the Johnsons' order?
(RF1)

18. If 7% tax is added to the bill, and if the Johnsons pay for the food with a $20 bill,
(35) how much money should they get back?

19. ✎ **Formulate** Make up an order from the menu. Then calculate the bill, not including tax.
(RF1)

20. If $A = lw$, and if l equals 2.5 and w equals 0.4, what does A equal?
(6)

* **21.** Write the prime factorization of both the numerator and the denominator of this
(61) fraction. Then simplify the fraction.

$$\frac{72}{120}$$

Refer to the coordinate plane below to answer problems **22** and **23.**

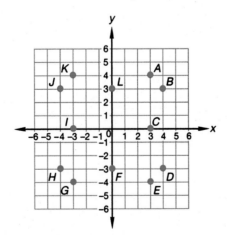

* **22.** Identify the coordinates of the following points:
(RF27)
  a. K   **b.** F

* **23.** Name the points that have the following coordinates:
(RF27)
  a. (3, −4)   **b.** (−3, 0)

* **24.** **Model** Draw a pair of parallel lines. Then draw a second pair of parallel
(59) lines perpendicular to the first pair of lines and about the same distance
apart. Trace the quadrilateral that is formed by the intersecting lines. Is the
quadrilateral a rectangle?

* **25.** $\dfrac{1}{2} \cdot \dfrac{5}{6} \cdot \dfrac{3}{5}$
(67)

* **26.** $3 \times 1\dfrac{1}{2} \times 2\dfrac{2}{3}$
(67)

27. $\dfrac{3}{4} \div 2$
(49)

* **28.** $1\dfrac{1}{2} \div 1\dfrac{2}{3}$
(64)

29. $(0.12)(0.24)$
(33)

30. $0.6 \div 0.25$
(44)

Real-World Connection

LaDonna had errands to run and decided to park her car in front of a parking meter rather than drive from store to store. She calculated that she would spend about 20 minutes in the post office and 10 minutes at the hardware store. Then she would spend 5 minutes picking up her clothes from the cleaner's and another 30 minutes eating lunch.

The sign on the meter read:

$0.25 = 15 minutes

$0.10 = 6 minutes

$0.05 = 3 minutes

a. How much time will LaDonna spend to finish doing her errands?

b. If the meter has ten minutes left, how much money will she need to put into the meter?

◥ *California Mathematics Content Standards*
NS **1.0**, **1.1** Compare and order positive and negative fractions, decimals, and mixed numbers and place them on a number line.
MR **2.0, 2.4** Use a variety of methods, such as words, numbers, symbols, charts, graphs, tables, diagrams, and models, to explain mathematical reasoning.

• Writing Decimal Numbers as Fractions

Power Up

facts   Power Up E

mental math

   a. Number Sense: 5×80

   b. Time: How many days are in 42 weeks? (Think "7×42.")

   c. Fractions: The dress was on sale for $\frac{1}{4}$ off the regular price of $48. What is $\frac{1}{4}$ of $48?

   d. Measurement: Esmerelda has walked 4300 feet. If she walks 980 more feet, she will have walked 1 mile. How many feet is 1 mile?

   e. Functions: What is the rule of this function?

x	−2	0	2	4
y	2	4	6	8

   f. Statistics: The five terrariums contained 2, 2, 5, 1, and 2 turtles, respectively. What was the mode?

   g. Algebra: If $100 - m = 75$, what is m?

   h. Calculation: $7 \times 8, -1, \div 5, \times 2, -1, \div 3, -8$

problem solving
Choose an appropriate problem-solving strategy to solve this problem. Megan has many gray socks, white socks, and black socks in a drawer. In the dark she pulled out two socks that did not match. How many more socks does Megan need to pull from the drawer to be certain to have a matching pair?

New Concept

We will review changing a decimal number to a fraction or mixed number. Recall from Lesson 29 that the number of places after the decimal point indicates the denominator of the decimal fraction (10 or 100 or 1000, etc.). The digits to the right of the decimal point make up the numerator of the fraction.

Example 1

Thinking Skills

Connect

What is the denominator of each of these decimal fractions: 0.2, 0.43, 0.658?

Write 0.5 as a common fraction.

We read 0.5 as "five tenths," which also names the fraction $\frac{5}{10}$. We reduce the fraction.

$$\frac{5}{10} = \frac{1}{2}$$

Example 2

Thinking Skills

Verify

How do you reduce $\frac{75}{100}$ to $\frac{3}{4}$?

Write 3.75 as a mixed number.

The whole-number part of 3.75 is 3, and the fraction part is 0.75. Since 0.75 has two decimal places, the denominator is 100.

$$3.75 = 3\frac{75}{100}$$

We reduce the fraction.

$$3\frac{75}{100} = 3\frac{3}{4}$$

Example 3

Order the following numbers:

$$-1.2 \qquad 2.4 \qquad -1\frac{1}{2} \qquad 2\frac{1}{5}$$

We rewrite the decimals as fractions then compare.

$$-1.2 = -1\frac{2}{10} = -1\frac{1}{5}$$
$$2.4 = 2\frac{4}{10} = 2\frac{2}{5}$$

Of the negative numbers $-1\frac{1}{2}$ and $-1\frac{1}{5}$, the number $-1\frac{1}{2}$ is the least.

We compare the positive numbers $2\frac{2}{5}$ and $2\frac{1}{5}$, and find that $2\frac{2}{5}$ (or 2.4) is the greatest.

We list the numbers in order from least to greatest.

$$-1\frac{1}{2}, -1.2, 2\frac{1}{5}, 2.4$$

Lesson Practice   Write each decimal number as a fraction or mixed number:

a. 12.5   **b.** 1.25   **c.** 0.125

d. 0.05   **e.** 0.24   **f.** 10.2

Written Practice   *Distributed and Integrated*

* 1. Minh measured the mass of a few coins and recorded them: 5.0 g, 3.9 g, 4.0 g.
(10)   Find the mean of this data.

2. The recipe for the health beverage called for $\frac{1}{3}$ cup of one juice, $\frac{3}{8}$ cup of another
(56) juice, and $\frac{1}{4}$ cup of water. What fraction of a cup is made by this recipe in all?

3. At the Independence Day celebration, apple pies were sliced into sixths. If there
(57) were $12\frac{4}{6}$ pies remaining, how many slices remained?

* 4. The spinner has 8 equal sections numbered 1 through 8. What is the probability
$(54, 61)$ that the spinner will stop on?

    a. a prime number?

    b. a composite number?

    c. the number 1?

5. Write the prime factorization of 76 and 84 and simplify $\frac{76}{84}$.
(63)

6. Simplify before multiplying: $\frac{75}{100} \times \frac{60}{90}$
(65)

Find each unknown number.

7. $x - \frac{1}{3} = \frac{2}{5}$
$(39, 53)$

8. $1 - y = \frac{5}{8}$
(39)

9. $1 - z = 0.11$
(39)

10. $w + 1\frac{1}{2} = 2\frac{1}{5}$
$(39, 55)$

11. Find the area of this parallelogram using the formula $A = b \cdot h$.
(66)

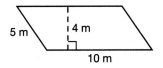

12. **Multiple Choice** Which of these numbers is closest to 1?
(28)

    A -0.2 **B** 2   **C** 0.22

13. Which prime number is even?
(11)

* 14. What is the name of a rectangle with sides of equal length?
(59)

15. **Multiple Choice** A farmer has 24 yards of fence that he can use to enclose a
(23) garden. Which of these options gives the greatest area?

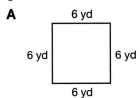
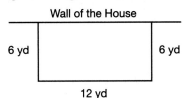
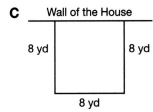

* 16. In parallelogram $ABCD$, $\angle A$ is supplementary to $\angle B$. $\angle B$ is
(60) supplementary to $\angle C$. Find the measure of $\angle C$.

17. Use the formula $A = b \cdot h$ to find the area of this
(66) parallelogram.

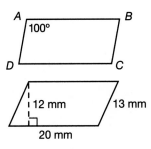

18. Write the prime factorizations of 52 and 65 and use them to
(63) simplify $\frac{52}{65}$.

19. Simplify these fractions before you multiply: $\frac{6}{8} \cdot \frac{6}{9} \cdot \frac{50}{100}$
(65)

20. Convert 0.85 to a reduced fraction.
(68)

21. Convert 1.6 to a mixed number.
(68)

Refer to the coordinate plane below to answer problems **22** and **23**.

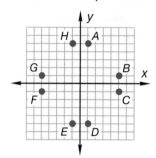

22. Identify the coordinates of the following points:
(RF27)

    a. *A*   **b.** *F*

23. Name the point for each of the following coordinates:
(RF27)

    a. $(-1, 5)$ **b.** $(5, -1)$

24. Name the sampling method used (Convenience, Random, Systematic, or
(Inv. 2) Volunteer).

        The third student in each row is asked a series of questions.

25. $\dfrac{1}{3} \cdot \dfrac{3}{8} \cdot \dfrac{4}{5}$ **26.** $3 \times 1\dfrac{1}{3} \times 2\dfrac{1}{4}$
(67)   (67)

27. $\dfrac{3}{5} \div 2$ *** 28.** $7\dfrac{11}{12} \div 7\dfrac{11}{12}$
(49)   (64)

29. $(0.15)(0.15)$ **30.** $0.3 \div 0.75$
(34)   (44)

California Mathematics Content Standards

NS **1.0**, **1.2** Interpret and use ratios in different contexts (e.g., batting averages, miles per hour) to show the relative sizes of two quantities, using appropriate notations (a/b, a to b, a:b).

SDAP 3.0, 3.2 Use data to estimate the probability of future events (e.g., batting averages or number of accidents per mile driven).

SDAP 3.0, **3.3** Represent probabilities as ratios, proportions, decimals between 0 and 1, and percentages between 0 and 100 and verify that the probabilities computed are reasonable; know that if *P* is the probability of an event, 1- *P* is the probability of an event not occurring.

LESSON 69

• Writing Fractions and Ratios as Decimal Numbers

Power Up

facts   Power Up I

mental math

    a. Number Sense: (5×180)

    b. Number Sense: 6×44

    c. Power Roots: $7^2 + 51$

    d. Money: Double $1.75.

    e. Money: How many $20 bills does it take to make $100?

    f. Money: How many $20 bills does it take to make $1000?

    g. Statistics: In three games Marta scored 2, 4 and 9 points. If Marta scores 2 points in the fourth game, will her mean score decrease or increase?

    h. Calculation: $6 \times 5, + 2, \div 4, \times 3, \div 4, - 2, \div 2, \div 2$

problem solving

Choose an appropriate problem-solving strategy to solve this problem. Nathan used a one-yard length of string to form a rectangle that was twice as long as it was wide. What was the area that was enclosed by the string?

New Concept

We learned earlier that a fraction bar indicates division. So the fraction $\frac{1}{2}$ also means "1 divided by 2," which we can write as $2\overline{)1}$. By attaching a decimal point and zero, we can perform the division and write the quotient as a decimal number.

$$\frac{1}{2} \longrightarrow \begin{array}{r} 0.5 \\ 2\overline{)1.0} \\ \underline{1\ 0} \\ 0 \end{array}$$

We find that $\frac{1}{2}$ equals the decimal number 0.5. To convert a fraction to a decimal number, we divide the numerator by the denominator.

Example 1

Convert $\frac{1}{4}$ to a decimal number.

The fraction $\frac{1}{4}$ means "1 divided by 4," which is $4\overline{)1}$. By attaching a decimal point and zeros, we can complete the division.

$$
\begin{array}{r}
0.25 \\
4\overline{)1.00} \\
\underline{8} \\
20 \\
\underline{20} \\
0
\end{array}
$$

Example 2

Convert $\frac{15}{16}$ to a decimal number.

The fraction $\frac{15}{16}$ means "15 divided by 16", which is $16\overline{)15}$. By attaching a decimal point and zeros, we can complete the division.

$$
\begin{array}{r}
0.9375 \\
16\overline{)15.0000} \\
\underline{14\,4} \\
60 \\
\underline{48} \\
120 \\
\underline{112} \\
80 \\
\underline{80} \\
0
\end{array}
$$

Example 3

Write $7\frac{2}{5}$ as a decimal number.

The whole number part of $7\frac{2}{5}$ is 7, which we write to the left of the decimal point. We convert $\frac{2}{5}$ to a decimal by dividing 2 by 5.

$$
\frac{2}{5} \longrightarrow
\begin{array}{r}
0.4 \\
5\overline{)2.0}
\end{array}
$$

Since $\frac{2}{5}$ equals 0.4, the mixed number $7\frac{2}{5}$ equals **7.4.**

Model Use a calculator to check the answer.

Example 4

Order the numbers from least to greatest.

$$\frac{3}{8} \qquad 0.3 \qquad 0.4$$

Write $\frac{3}{8}$ as a decimal in order to compare it to 0.3 and 0.4

$$
\frac{3}{8} \longrightarrow
\begin{array}{r}
0.375 \\
8\overline{)3.000} \\
\underline{2\,4} \\
60 \\
\underline{56} \\
40 \\
\underline{40} \\
0
\end{array}
$$

Since $\frac{3}{8}$ equals 0.375, we write 0.3 and 0.4 with trailing zeros to compare.

$$\frac{3}{8} = 0.375$$
$$0.3 = 0.300$$
$$0.4 = 0.400$$

We see that $\frac{3}{8}$ lies between 0.3 and 0.4. We list the numbers in order from least to greatest.are.

$$0.3 \qquad \frac{3}{8} \qquad 0.4$$

Converting ratios to decimal numbers is similar to converting fractions to decimal numbers.

Example 5

Dog breeders often calculate the length to height ratio of a dog. One dog has a length to height ratio of 11:10. Write this ratio as a decimal.

The ratio 11 to 10 can be written as a fraction.

$$\frac{11}{10}$$

We divide to write it as a decimal.

$$
\begin{array}{r}
1.1 \\
10\overline{)11.0} \\
\underline{10} \\
10 \\
\underline{10} \\
0
\end{array}
$$

The length to height ratio is **1.1**

Example 6

A number cube is rolled once. Express the probability of rolling an even number as a decimal number.

Probabilities are often expressed as decimal numbers between 0 and 1. Since three of the six numbers on a number cube are even, the probability of rolling an even number is $\frac{3}{6}$, which equals $\frac{1}{2}$.

We convert $\frac{1}{2}$ to a decimal by dividing 1 by 2.

$$
\begin{array}{r}
0.5 \\
2\overline{)1.0}
\end{array}
$$

Thus the probability of rolling an even number with one roll of a number cube is **0.5.**

Lesson Practice

Convert each fraction or mixed number to a decimal number:

a. $\frac{3}{4}$ b. $4\frac{1}{5}$ c. $\frac{1}{8}$

d. $\frac{7}{20}$ e. $3\frac{3}{10}$ f. $\frac{7}{25}$

g. $\frac{11}{16}$ **h.** $\frac{31}{32}$ **i.** $3\frac{24}{64}$

j. In a bag are three red marbles and two blue marbles. If Chad pulls one marble from the bag, what is the probability that the marble will be blue? Express the probability ratio as a fraction and as a decimal number.

Written Practice

Distributed and Integrated

* 1. **(Formulate)** Tyrone's temperature was 102°F. Normal body temperature is 98.6°F.
(7, 41) How many degrees above normal was Tyrone's temperature? Write an equation and solve the problem.

2. **(Formulate)** Jill has read 42 pages of a 180-page book. How many pages are left
(7) for her to read? Write an equation and solve the problem.

3. **(Formulate)** If Jill wants to finish the book in the next three days, then she
(8, 10) should read an average of how many pages per day? Write an equation and solve the problem.

* 4. Write 2.5 as a reduced mixed number.
(68)

* 5. Write 0.35 as a reduced fraction.
(68)

6. What is the total cost of a $12.60 item when $7\frac{1}{2}$% (0.075) sales tax is added?
(35)

* 7. $\frac{3}{4} \times 2 \times 1\frac{1}{3}$
(67)

* 8. **(Analyze)** $(100 - 10^2) \div 5^2$
(2, 32)

9. $3 + 2\frac{1}{3} + 1\frac{3}{4}$
(56)

10. $5\frac{1}{6} - 3\frac{1}{2}$
(58)

* 11. $\frac{3}{4} \div 1\frac{1}{2}$
(64)

12. $7 \div 0.4$
(44)

13. Explain why the way this question was worded might influence the result.
(Inv. 3) *"Most happy, successful people work hard. Do you plan to work hard?"*

14. The diameter of a quarter is about 2.4 cm.
(42)

     a. What is the circumference of a quarter? (Use 3.14 for π.)

     b. What is the ratio of the radius of the quarter to the diameter of the quarter?

Find each unknown number:

15. $25m = 0.175$
(39, 41)

16. $1.2 + y + 4.25 = 7$
(39)

17. Which digit is in the ten-thousands place in 123,456.78?
(28)

18. Arrange these numbers in order from least to greatest:
(53)

$$1, \frac{1}{2}, \frac{1}{10}, \frac{1}{4}, 0$$

19. Name the sampling method used:
(Inv. 2)
Contact the teammates whose e-mail addresses are on the roster to ask which gift should be given to the coach.

20. The store offered a 20% discount on all tools. The regular price of a hammer was $18.00.
(35)
    a. How much money is 20% of $18.00?

    b. What was the price of the hammer after the discount?

* 21. (Connect) The length of $\overline{AB}$ is 16 mm. The length of $\overline{AC}$ is 50 mm. What is the
(RF9)
length of $\overline{BC}$?

22. One half of the area of this square is shaded.
(32)
What is the area of the shaded region?

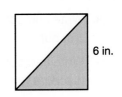

6 in.

23. (Verify) Is every square a rectangle?
(59)

24. Is it helpful to use a sample for the following situation? Explain.
(Inv. 2)
What color should the school be painted?

* 25. Convert $1\frac{3}{8}$ to a decimal.
(69)

Refer to this coordinate plane to answer problems **26** and **27.**

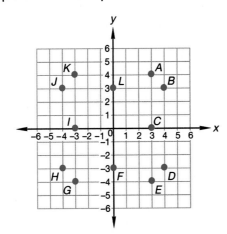

* 26. Identify the coordinates of the following points:
(RF27)
    a. *H*   **b.** *L*

* 27. Name the points that have the following coordinates:
(RF27)
    a. (−4, 3)   **b.** (3, 0)

* 28. If s equals 9, what does s^2 equal?
(6, 32)

29. Name an every day object that has the same shape as each of these geometric
(RF25) solids:

    a. cylinder   **c.** sphere

    b. rectangular prism   **d.** cube

* 30. (**Conclude**) The measure of $\angle W$ in parallelogram $WXYZ$ is 75°.
(66)
    a. What is the measure of $\angle X$?

    b. What is the measure of $\angle Y$?

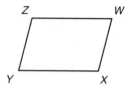

*Real-World
Connection*

Dexter works at a clothing store and receives a 15% employee discount.
He wants to buy a corduroy blazer that is on sale for 30% off the original
price of $79.

    a. Write and solve a proportion to find the price of the blazer without the employee
discount.

    b. Write and solve an equation to find the price Dexter would pay using his employee
discount.

• **Disjoint Events**

California Mathematics Content Standards

SDAP 3.0, **3.1** Represent all possible outcomes for compound events in an organized way (e.g., tables, grids, tree diagrams) and express the theoretical probability of each outcome.

SDAP 3.0, **3.3** Represent probabilities as ratios, proportions, decimals between 0 and 1, and percentages between 0 and 100 and verify that the probabilities computed are reasonable; know that if *P* is the probability of an event, 1- *P* is the probability of an event not occurring.

SDAP 3.0, **3.5** Understand the difference between independent and dependent events.

facts   Power Up J

mental math

    a. Number Sense: 5×280

    b. Number Sense: $476 + 99$

    c. Money: $4.50 + $1.75

    d. Time: Curtis read his book for 115 minutes. If he started reading at 9:35 a.m., when did he stop?

    e. Geometry: A square has a perimeter of 24 cm. What is the length of the sides of the square?

    f. Statistics: Consider this data set. Which is greater: the mean or the median? (Find your answer without calculating each measure.)

$$20, 20, 100$$

    g. Algebra: If $m = \frac{1}{2}$, what is $2m$?

    h. Calculation: $5 \times 10, \div 2, + 5, \div 2, - 5, \div 10, - 1$

problem solving

Choose an appropriate problem-solving strategy to solve this problem. The Crunch-O's cereal company makes two different cereal boxes. One is family size (12 in. high, 9 in. long, 2 in. wide) and the other is single-serving size (5 in. high, 3 in. long, 1 in. wide). Each of their boxes is made out of one piece of cardboard. To the right is a net of the family size box. Use this diagram to draw a net for the single-serving box.

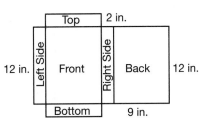

In a **probability** experiment, **events** are called disjoint if they cannot both happen at the same time. For example, if a bag of marbles contains 3 red, 4 blue, and 5 green marbles and if one marble will be drawn, the events "draw a red" and "draw a blue" are disjoint events since both colors cannot be selected in one draw.

$$\text{probability of "red" or "blue"} = \frac{3 + 4}{12}$$

$$= \frac{7}{12}$$

Or we can add the individual probabilities.

probability of "red" or "blue"	=	probability of "red"	+	probability of "blue"
	=	$\frac{3}{12}$	+	$\frac{4}{12}$
	=	$\frac{7}{12}$		

Example 1

Isaac holds a bag of 3 red, 4 blue, and 5 green marbles. He will draw one marble from the bag.

What is the probability that the marble he draws is blue or green? Write the probability as a decimal.

The events "blue" and "green" are disjoint.

$$\text{probability of "blue" or "green"} = \frac{4}{12} + \frac{5}{12}$$

$$= \frac{9}{12} = \frac{3}{4}$$

We divide to write the probability as a decimal.

$$\begin{array}{r} 0.75 \\ 4\overline{)3.00} \\ \underline{28} \\ 20 \\ \underline{20} \\ 0 \end{array}$$

The probability that Isaac draws a blue or green is **0.75**.

Model Draw a picture of this experiment. Explain why the probability you calculated is reasonable.

Example 2

Ilbea has 3 nickels, 5 dimes, and 2 quarters in her pocket. She takes one coin from her pocket.

Coin	Probability
Nickel	$\frac{3}{10}$ = 0.3
Dime	
Quarter	

a. **Fill in the table with probabilities expressed as decimals.**

b. **What is the probability that Ilbea picks a nickel or a dime? Express the probability as a decimal.**

a.

Coin	Probability
Nickel	$\frac{3}{10}$ = 0.3
Dime	$\frac{5}{10}$ = 0.5
Quarter	$\frac{2}{10}$ = 0.2

b. The probability that Ilbea picks a nickel or a dime is the sum of the two probabilities.

probability of "nickel" and "dime" = 0.3 + 0.5 = **0.8**

Analyze Suppose Ilbea has 3 nickels, 5 dimes, and 2 quarters in her pocket, and she draws *two* coins without replacement. What is the probability that she takes out a nickel and then a dime? Identify how this question is different than Example 2, and explain how you found the answer

Lesson Practice

a. This spinner is spun once. What is the probability that it stops on a blue region?

b. Risa rolls a number cube. She has to roll a 5 or a 6 to win the game. What is the probability that she will win the game on her turn?

c. Dean has four $1 bills, two $5 bills, and one $10 bill in his wallet. If he pulls out one bill without looking, what is the probability that it is a $5 or $10 bill?

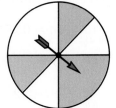

Written Practice   *Distributed and Integrated*

* 1. On a particular day in August, the high and low temperatures in a city in the San
(RF21, 7)   Gabriel Valley were 101.0° F and 68.5° F. What was the range in temperature that day (the difference between the high and low)?

2. Saul had $4\frac{1}{3}$ pounds of dough. Sonja had $3\frac{1}{4}$ pounds of dough. How many more
(55)   pounds did Saul have than Sonja?

3. Saul divided his $4\frac{1}{3}$ pounds of dough into piles of $1\frac{1}{12}$ pounds each. How many
(64)   piles can he make?

4. Use the formula $A = b \cdot h$ to find the area of this parallelogram:
(33, 66)

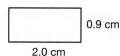

0.9 cm

2.0 cm

5. Write the prime factorizations of the numerator and denominator of this fraction,
(63) then simplify: $\frac{48}{72}$

* 6. What is the total cost of a $20,000 item when 8.25% (0.0825) sales tax is added?
(33, 35)

7. $\frac{2}{3} \times 3 \times 1\frac{1}{2}$
(67)

8. $(65 - 8^2) \times 12^2$
(2, 32)

9. $5 + 1\frac{1}{2} + 2\frac{1}{3}$
(56)

10. $10\frac{1}{4} - 3\frac{1}{2}$
(58)

11. $\frac{5}{6} \div 1\frac{4}{5}$
(64)

12. $10 \div 0.4$
(44)

13. Compare by first converting the fraction to a decimal.
(40, 69)

    a. $0.3 \bigcirc \frac{3}{10}$ **b.** $0.07 \bigcirc \frac{9}{100}$

* 14. The radius of the front wheel of a scooter is 3 in. The radius of the rear wheel of
(42) the scooter is 2 in. What is the difference between the *diameter* of the wheels?

Find each unknown number.

15. $13n = 0.26$
(39, 41)

16. $2.5 + x + 3.1 = 6$
(39)

17. Convert 2.4 to a mixed number and simplify.
(68)

18. Arrange these numbers in order from least to greatest.
(53)

$$\frac{3}{4}, \frac{1}{8}, \frac{1}{3}, 0$$

19. Write the prime factorization of 100.
(61)

20. A certain luxury car costs $90,000. In county A, the sales tax is 7.25% (0.0725). In
(4, 35) county B, the sales tax is 8.25% (0.0825). How much money is saved by buying the car in county A?

21. $\angle A$ and $\angle B$ are complementary. Find the measure of $\angle A$.
(60)

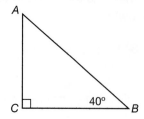

22. Half of the rectangle is shaded. What is the area of the shaded region?
(23)

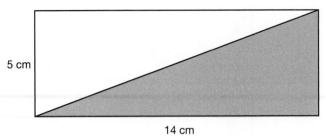

5 cm

14 cm

23. Ina insists that this shape is a square. Sergio suggests that it is a rectangle. Who
(59) is wrong?

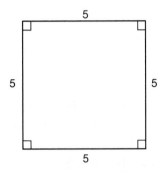

5

5

5

5

* **24.** The ratio of hits to at-bats is 9 to 24. Convert that ratio to a decimal.
(69)

25. Simplify these fractions before you multiply:
(67)
$$\frac{20}{25} \cdot \frac{20}{40} \cdot \frac{20}{100}$$

26. Convert $\frac{3}{20}$ to a decimal.
(67)

27. Four names are written on separate slips of paper and placed in a hat. If two are
(54, 70) drawn at random, how many possible outcomes are there?

28. The measure $\angle A$ in parallelogram *ABCD* is 63°. Find the measure of $\angle D$.
(67)

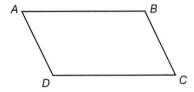

Refer to this coordinate plane to answer problems **29** and **30**.

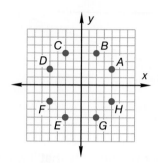

29. Find the coordinates of the following points:
(RF27)
    a. *G*   **b.** *E*

30. Name the points that have the following coordinates:
(RF27)
    a. (–4, 2)   **b.** (–4, –2)

Real-World Connection

On the first day of the month the balance in Rhonda's checking account is $500. Every day for the next fourteen days, she withdraws $30. On the fifteenth day Rhonda deposits two $712 checks. Over the next three days she makes three withdrawals that total $65.

Write an expression that represents the total amount in deposits. Write another expression that represents the total amount in withdrawals.

California Mathematics Content Standards

SDAP 3.0, 3.3 Represent probabilities as ratios, proportions, decimals between 0 and 1, and percentages between 0 and 100 and verify that the probabilities computed are reasonable; know that if P is the probability of an event, 1- P is the probability of an event not occurring.

SDAP 3.0, 3.4 Understand that the probability of either of two disjoint events occurring is the sum of the two individual probabilities and that the probability of one event following another, in independent trials, is the product of the two probabilities.

SDAP 3.0, 3.5 Understand the difference between independent and dependent events.

Focus on

Experimental and Theoretical Probability

In Lesson 54 we determined probabilities for the outcomes of experiments without actually performing the experiments. For example, in the case of rolling a number cube, the sample space of the experiment is {1, 2, 3, 4, 5, 6} and all six outcomes are equally likely. Each outcome therefore has a probability of $\frac{1}{6}$. In the spinner example we assume that the likelihood of the spinner landing in a particular sector is proportional to the area of the sector. Thus, if the area of sector A is twice the area of sector B, the probability of the spinner landing in sector A is twice the probability of the spinner landing in sector B.

Probability that is calculated by performing "mental experiments" (as we have been doing since Lesson 54) is called **theoretical probability.** Probabilities associated with many real-world situations, though, cannot be determined by theory. Instead, we must perform the experiment repeatedly or collect data from a sample experiment. Probability determined in this way is called **experimental probability.**

A survey is one type of probability experiment. Suppose a pizza company is going to sell individual pizzas at a football game. Three types of pizzas will be offered: cheese, tomato, and mushroom. The company wants to know how many of each type of pizza to prepare, so it surveys a representative sample of 500 customers.

The company finds that 175 of these customers would order cheese pizzas, 225 would order tomato pizzas, and 100 would order mushroom pizzas. To estimate the probability that a particular pizza will be ordered, the company uses **relative frequency.** This means they divide the frequency (the number in each category) by the total (in this case, 500).

	Frequency	Relative Frequency
Cheese	175	$\frac{175}{500} = 0.35$
Tomato	225	$\frac{225}{500} = 0.45$
Mushroom	100	$\frac{100}{500} = 0.20$

Notice that the sum of the three relative frequencies is 1. This means that the entire sample is represented. We can change the relative frequencies from decimals to percents.

$$0.35 \longrightarrow 35\% \qquad 0.45 \longrightarrow 45\% \qquad 0.20 \longrightarrow 20\%$$

Recall from Lesson 54 that we use the term *chance* to describe a probability expressed as a percent. So the company makes the following estimates about any given sale. The chance that a cheese pizza will be ordered is **35%.** The chance that a tomato pizza will be ordered is **45%.** The chance that a mushroom pizza will be ordered is **20%.** The company plans to make 3000 pizzas for the football game, so about 20% of the 3000 pizzas should be mushroom. How many pizzas will be mushroom?

Now we will apply these ideas to another survey. Suppose a small town has only four markets: Bob's Market, The Corner Grocery, Express Grocery, and Fine Foods. A representative sample of 80 adults was surveyed. Each person chose his or her favorite market: 30 chose Bob's Market, 12 chose Corner Grocery, 14 chose Express Grocery, and 24 chose Fine Foods.

1. (**Represent**) Present the data in a relative frequency table similar to the one for pizza.

2. Estimate the probability that in this town an adult's favorite market is Express Grocery. Write your answer as a decimal.

3. Estimate the probability that in this town an adult's favorite market is Bob's Market. Write your answer as a fraction in reduced form.

4. Estimate the chance that in this town an adult's favorite market is Fine Foods. Write your answer as a percent.

5. Suppose the town has 4000 adult residents. The Corner Grocery is the favorite market of about how many adults in the town?

6. (**Explain**) Estimate the probability that in this town an adult's favorite market is either Bob's Market or Corner Grocery. Explain your reasoning.

A survey is just one way of conducting a probability experiment. In the following activity we will perform an experiment that involves drawing two marbles out of a bag. By performing the experiment repeatedly and recording the results, we gather information that helps us determine the probability of various outcomes.

Activity

Probability Experiment

Materials needed:

- 6 marbles (4 green and 2 white)
- Small, opaque bag from which to draw the marbles
- Pencil and paper

The purpose of this experiment is to determine the probability that two marbles drawn from the bag at the same time will be green. We will create a relative frequency table to answer the question.

To estimate the probability, put 4 green marbles and 2 white marbles in a bag. Pair up with another student, and work through problems **7–9** together.

7. Choose one student to draw from the bag and the other to record results. Shake the bag; then remove two objects at the same time. Record the result by marking a tally in a table like the one below. Replace the marbles and repeat this process until you have performed the experiment exactly 25 times.

Outcome	Tally
Both green	
Both white	
One of each	

8. Use your tally table to make a relative frequency table. (Divide each row's tally by 25 and express the quotient as a decimal.)

9. Estimate the probability that both marbles drawn will be green. Write your answer as a reduced fraction and as a decimal.

If, for example, you drew two green marbles 11 times out of 25 draws, your best estimate of the probability of drawing two green marbles would be

$$\frac{11}{25} = \frac{44}{100} = 0.44$$

But this is only an estimate. The more times you draw, the more likely it is that the estimate will be close to the theoretical probability. It is better to repeat the experiment 500 times than 25 times. Thus, combining your results with other students' results is likely to produce a better estimate. To combine results, add everyone's tallies together; then calculate the new frequency.

10. Calculate the theoretical probability that two green marbles are drawn in this experiment. Compare the theoretical probability with the experimental probability that you found as a class.

Investigate Further

a. **Represent** Ask 10 other students the following question: "What is your favorite sport: baseball, football, soccer, or basketball?" Record each response. Create a relative frequency table of your results. Share the results of the survey with your class.

b. **Analyze** In groups, conduct an experiment by drawing two counters out of a bag containing 3 green counters and 3 white counters. Each group should perform the experiment 30 times. Record each group's tallies in a frequency table like the one shown on the next page.

	Both Green		Both White		One of Each	
	Tally	Rel. Freq.	Tally	Rel. Freq.	Tally	Rel. Freq.
Group 1						
Group 2						
Group 3						
Group 4						
Group 5						
Group 6						
Whole Class						

Calculate the relative frequency for each group by dividing the tallies by 30 (the number of times each group performed the experiment). Then combine the results from all the groups. To combine the results, add the tallies in each column and write the totals in the last row of the table. Then divide each of these totals by the *total* number of times the experiment was performed (equal to the number of groups times 30). The resulting quotients are the whole-class relative frequencies for each event. Discuss your findings.

On the basis of their own data, which groups would guess that the probabilities were less than the "Whole class" data indicate? Which groups would guess that the probabilities were greater than the whole class's data indicate?

c. Choose a partner and roll two number cubes 100 times. Each time, observe the sum of the upturned faces, and fill out a relative frequency table like the one below. The sample space of this experiment has 11 outcomes.

(Predict) Are the outcomes equally likely? If not, which outcomes are more likely and which are less likely?

Sum	2	3	4	5	6	7	8	9	10	11	12
Frequency											
Relative Frequency											

After the experiment and calculation, estimate the probability that the sum of a roll will be 8. Estimate the probability that the sum will be at least 10. Estimate the probability that the sum will be odd.

California Mathematics Content Standards

NS 2.0, 2.3 Solve addition, subtraction, multiplication, and division problems, including those arising in concrete situations, that use positive and negative integers and combinations of these operations.

AF 1.0, 1.4 Solve problems manually by using the correct order of operations or by using a scientific calculator.

• Exphonents

Power Up

facts   Power Up I

mental math

    a. Number Sense: 5×480

    b. Number Sense: $367 - 99$

    c. Number Sense: 8×43

    d. Money: Tanner gave the clerk a $10 bill for the pen. He received $8.75 in change. How much did the pen cost?

    e. Measurement: How many inches is $10\frac{1}{2}$ feet?

    f. Geometry: A square has an area of 25 in^2. What is the length of each side?

    g. Statistics: Consider this data set. Which is greater: the mean or the median? (Find your answer without calculating each measure.)

$$30, 80, 80$$

    h. Calculation: $8 \times 9, + 3, \div 3, \times 2, - 10, \div 5, + 3, \div 11$

problem solving

Choose an appropriate problem-solving strategy to solve this problem. Tyjon used 14 blocks to build this three-layer pyramid. How many blocks would he need to build a six-layer pyramid? How many blocks would he need for the bottom layer of a nine-layer pyramid?

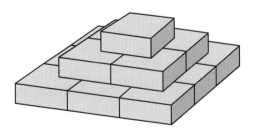

New Concept

Since Lesson 32 we have used the exponent 2 to indicate that a number is multiplied by itself.

$$5^2 \text{ means } 5 \times 5$$

Exponential expressions are expressions indicating that the **base** is to be used as a factor the number of times shown by the exponent. Exponents indicate repeated multiplication, so

$$5^3 \text{ means } 5 \times 5 \times 5$$

$$5^4 \text{ means } 5 \times 5 \times 5 \times 5$$

The exponent indicates how many times the base is used as a factor.

We read numbers with exponents as **powers.** Note that when the exponent is 2, we usually say "squared," and when the exponent is 3, we usually say "cubed." The following examples show how we read expressions with exponents:

5^2 "five to the second power" or "five squared"

10^3 "ten to the third power" or "ten cubed"

3^4 "three to the fourth power"

2^5 "two to the fifth power"

Example 1

Compare: $3^4 \bigcirc 4^3$

We find the value of each expression.

3^4 means $3 \times 3 \times 3 \times 3$, which equals 81.

4^3 means $4 \times 4 \times 4$, which equals 64.

Since 81 is greater than 64, we find that 3^4 is greater than 4^3.

$$3^4 > 4^3$$

Example 2

Write the prime factorization of 1000, using exponents to group factors.

Using a factor tree or division by primes, we find the prime factorization of 1000.

$$1000 = 2 \times 2 \times 2 \times 5 \times 5 \times 5$$

We group the three 2s and the three 5s with exponents.

$$1000 = 2^3 \times 5^3$$

Example 3

Simplify: $100 - 10^2$

We perform operations with exponents before we add, subtract, multiply, or divide. Ten squared is 100. So when we subtract 10^2 from 100, the difference is zero.

$$100 - 10^2$$

$$100 - 100 = 0$$

Lesson Practice   Find the value of each expression:

  a. 10^4 **b.** $2^3 + 2^4$ **c.** $2^2 \times 5^2$

  d. Write the prime factorization of 72 using exponents.

  e. $5 - 2^2$ **f.** $4 + 4^2$ **g.** $3^2 - 1$ **h.** $7^2 + 1$

Written Practice   *Distributed and Integrated*

* 1. *(4, 71)* **(Analyze)** What is the difference when five squared is subtracted from four cubed?

* 2. *(8)* On LeAnne's map, 1 inch represents a distance of 10 miles. If Dallas, TX and Fort Worth, TX are 3 inches apart on the map, approximately how many miles apart are they?

* 3. *(69)* Convert $2\frac{3}{4}$ to a decimal number.

* 4. *(54, 69)* Tito spins the spinner once.

  a. What is the sample space of the experiment?

  b. What is the probability that he spins a number greater than 1? Express the probability ratio as a fraction and as a decimal.

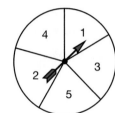

* 5. *(68)* Write 0.24 as a reduced fraction.

6. *(7)* **(Formulate)** Steve hit the baseball 400 feet. Lesley hit the golf ball 300 yards. How many feet farther did the golf ball travel than the baseball? After converting yards to feet, write an equation and solve the problem.

7. *(6)* If $A = bh$, and if b equals 12 and h equals 8, then what does A equal?

* 8. *(71)* Compare: $3^2 \bigcirc 3 + 3$

9. *(56)* $\frac{1}{2} + \frac{2}{3} + \frac{1}{6}$

10. *(58)* $3\frac{1}{4} - 1\frac{7}{8}$

11. *(67)* $\frac{5}{8} \cdot \frac{3}{5} \cdot \frac{4}{5}$

12. *(62)* $3\frac{1}{3} \times 3$

13. *(64)* $\frac{3}{4} \div 1\frac{1}{2}$

14. *(2, 41)* $(4 + 3.2) - 0.01$

15. *(70)* Abby wants to roll a 1 or 2 on a number cube. Find the probability.

16. *(8)* **(Formulate)** LaFonda bought a dozen golf balls for $10.44. What was the cost of each golf ball? Write an equation and solve the problem.

17. *(21)* Estimate the product of 81 and 38.

18. In four days Jamar read 42 pages, 46 pages, 35 pages, and 57 pages. What was
(10) the mean number of pages he read per day?

19. What is the least common multiple of 6, 8, and 12?
(26)

20. $24 + c + 96 = 150$
(5)

21. Write the prime factorization of both the numerator and the denominator of this
(63) fraction. Then reduce the fraction.

$$\frac{40}{96}$$

* 22. (**Analyze**) If the perimeter of this square is 40 centimeters, then
(42)

    a. what is the diameter of the circle?

    b. what is the circumference of the circle? (Use 3.14 for π.)

23. Twenty-four of the three dozen cyclists rode mountain bikes. What fraction of the
(25) cyclists rode mountain bikes?

* 24. (**Explain**) Why are some rectangles not squares?
(59)

25. (**Connect**) Which arrow could be pointing to $\frac{3}{4}$?
(14)

26. (**Conclude**) In quadrilateral *PQRS*, which segment appears to be
(20)   **a.** parallel to $\overline{PQ}$?

    b. perpendicular to $\overline{PQ}$?

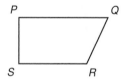

27. (**Analyze**) The figure at right shows a cube with edges 3 feet long.
(RF25)   **a.** What is the area of each face of the cube?

    b. What is the total surface area of the cube?

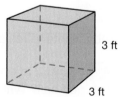

3 ft

3 ft

Refer to this coordinate plane to answer problems **28** and **29**.

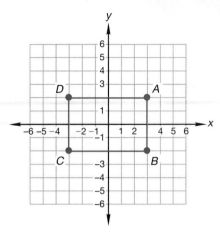

* 28. Identify the coordinates of the following points:
(RF27)
     a. C   **b.** origin

29. (**Connect**) One pair of parallel segments in rectangle *ABCD* is $\overline{AB}$ and $\overline{DC}$. Name
(59)
a second pair of parallel segments.

30. Farmer Ruiz planted corn on 60% of his 300 acres. Find the number of acres
(35)
planted with corn.

California Mathematics Content Standards
NS **2.0**, **2.3** Solve addition, subtraction, multiplication, and division problems, including those arising in concrete situations, that use positive and negative integers and combinations of these operations.
AF **1.0, 1.3** Apply algebraic order of operations and the commutative, associative, and distributive properties to evaluate expressions; and justify each step in the process.
AF **1.0, 1.4** Solve problems manually by using the correct order of operations or by using a scientific calculator.

• Order of Operations, Part 2

facts	Power Up H
mental math	**a. Money:** $3 \times \$0.99$
	b. Money: $\$20.00 - \$9.99$
	c. Money: One package of batteries costs $2.50. What is the cost for 10 packages of batteries?
	d. Fractions: $\frac{1}{3}$ of $6.60
	e. Statistics: What is the mean of $25, $25, and $50, and $60?
	f. Algebra: If $c = 12.5$, what is $2c$?
	g. Geometry: What is the perimeter of this square?
	h. Calculation: $2 \times 2, \times 2, \times 2, \times 2, - 2, \div 2$

2.5 cm

problem solving	Choose an appropriate problem-solving strategy to solve this problem. Kioko was thinking of two numbers whose mean was 24. If one of the numbers was half of 24, what was the other number?

New Concept

Recall that the four operations of arithmetic are addition, subtraction, multiplication, and division. We can also raise numbers to powers. When more than one operation occurs in the same expression, we perform the operations in the order listed below.

Order Of Operations

1. Perform operations within parentheses (or other symbols of inclusion), before simplifying outside of the parentheses.

2. Simplify powers.

3. Multiply and divide in order from left to right.

4. Add and subtract in order from left to right.

Math Language

Note: **Symbols of inclusion** set apart portions of an expression so they may be evaluated first. These include parentheses, and the division bar in a fraction.

The initial letter of each word in the sentence "Please excuse my dear Aunt Sally" reminds us of the order of operations:

Please	Parentheses (or other symbols of inclusion)
Excuse	Exponents
My Dear	Multiplication and division (left to right)
Aunt Sally	Addition and subtraction (left to right)

Example 1

Simplify: $2 + 4 \times 3 - 4 \div 2$

Justify each step.

We multiply and divide in order from left to right before we add or subtract.

$2 + 4 \times 3 - 4 \div 2$	Problem
$2 + 12 - 2$	Multiplied and divided
12	Added and subtracted

Example 2

Simplify: $(5 - 2)^2 + 9 \times 10$

Justify each step.

Thinking Skills

Explain

Why are the parentheses needed in the expression $4 \times (3 + 7)$?

We simplify within the parentheses before applying the exponent.

$(5 - 2)^2 + 9 \times 10$	Problem
$3^2 + 9 \times 10$	Simplified within the parentheses
$9 + 9 \times 10$	Applied the exponent
$9 + 90$	Multiplied
99	Added

Example 3

Simplify: $\dfrac{3^2 + 3 \times 5}{2}$

Justify each step.

A division bar may serve as a symbol of inclusion, like parentheses. We simplify above and below the bar before dividing.

$\dfrac{3^2 + 3 \times 5}{2}$	Problem
$\dfrac{9 + 3 \times 5}{2}$	Applied exponent
$\dfrac{9 + 15}{2}$	Multiplied above
$\dfrac{24}{2}$	Added above
12	Divided

Example 4

Evaluate: $a + ab$ if $a = 3$ and $b = 4$

We will begin by writing parentheses in place of each variable. This step may seem unnecessary, but many errors can be avoided if this is always our first step.

$$a + ab$$
$$(\ ) + (\ )(\ ) \qquad \text{parentheses}$$

Then we replace a with 3 and b with 4.

$$a + ab$$
$$(3) + (3)(4) \qquad \text{substituted}$$

We follow the order of operations, multiplying before adding.

$$(3) + (3)(4) \qquad \text{problem}$$
$$3 + 12 \qquad \text{multiplied}$$
$$\mathbf{15} \qquad \text{added}$$

Calculators with *algebraic-logic* circuitry are designed to perform calculations according to the order of operations. Calculators without algebraic-logic circuitry perform calculations in sequence. You can test a calculator's design by selecting a problem such as that in Example 1 and entering the numbers and operations from left to right, concluding with an equal sign. If the problem in Example 1 is used, a displayed answer of 12 indicates an algebraic-logic design.

Lesson Practice

Generalize Simplify.

    a. $5 + 5 \times 5 - 5 \div 5$

    b. $50 - 8 \times 5 + 6 \div 3$

    c. $24 - 8 - 6 \times 2 \div 4$

    d. $\dfrac{2^3 + 3^2 + 2 \times 5}{3}$

    e. $(5^2 - 5) + 5 \times 6$

Conclude Evaluate:

    f. $ab - bc$ if $a = 5$, $b = 3$, and $c = 4$

    g. $x - xy$ if $x = 2$ and $y = \dfrac{1}{2}$

Written Practice   *Distributed and Integrated*

* **1.** What is the difference when three squared is subtracted from three cubed?
(71)

* 2. California borders three other states: Oregon, Nevada, and Arizona. According to
(8) the rounded data from the 2000 census, approximately how many more people
does California have than the other three states combined? (California – 34
million; Arizona – 5 million; Oregon – 3 million; Nevada – 2 million)

3. Convert $3\frac{2}{5}$ to a decimal number.
(69)

4. If a number cube is rolled, find the probability of rolling a number divisible by 5.
(54)

5. Write 0.12 as a reduced fraction.
(68)

6. A sign on the highway said a turnoff was 200 feet ahead. About how many yards
(8, 21) ahead is that? (Round to the nearest yard.)

* 7. If $A = bh$ and if b equals 7 and h equals 6, then what does A equal?
(6)

* 8. Compare $5^2 \bigcirc 5 + 4^2$
(71)

9. $\frac{1}{5} + \frac{1}{4} + \frac{3}{20}$
(56)

10. $2\frac{1}{4} - 1\frac{3}{8}$
(58)

11. $\frac{3}{4} \times \frac{4}{5} \times \frac{5}{8}$
(67)

12. $13\frac{1}{4} \times 4$
(62)

13. $\frac{5}{6} \div 1\frac{1}{6}$
(64)

14. $(3 + 1.1) - 0.13$
(41)

15. Find the area:
(66)

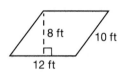

12 ft

16. Before sales tax was added, the price for the four-pack was $10.80. What was
(8) the cost per item?

17. Estimate the product of 68 and 91.
(21)

18. The values of the items were 10, 20, 30, 40, 70, 30, and 80. What is the mean
(10) of these values?

19. $66 + x + 55 = 133$
(5)

20. Write the prime factorization of both the numerator and the denominator of this
(63) fraction. Then reduce the fraction.

$$\frac{30}{84}$$

21. What is the least common multiple of 8, 12, and 16?
(26)

22. If the diameter of the circle is 10 cm, then:
(32, 42)
    a. Find the area of square *ABCD*.

    b. What fraction of the square is shaded?

    c. Find the area of the shaded triangle.

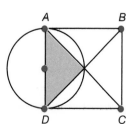

* 23. **a.** If a score is twenty, how many are four score and seven?
(7, 8)

b. What year is "four score and seven years" after 1776?

24. **Multiple Choice** A square has the characteristics of each of the following but
(59) one. Which is not another name for a square?

   A Quadrilateral   **B** Parallelogram   **C** Rectangle   **D** Trapezoid

25. Plot $1\frac{2}{3}$ on a number line.
(14)

26. **Multiple Choice** Line *l* and line *m* are:
(20)

   A Parallel

   B Perpendicular

   C Intersecting

   D None of the above

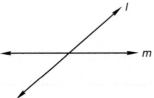

27. Find the perimeter of this
(RF7) figure. Dimensions are in feet.

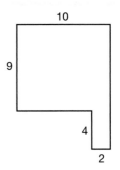

28. Reduce these fractions before multiplying: $\frac{400}{600} \times \frac{9}{12} \times \frac{10}{15}$
(65)

29. The complement of $\angle A$ is 31°.
(60)

   a. Find the measure of $\angle A$.

   b. Find the supplement of $\angle A$.

* 30. Pallando has read 37% of the 1000 page book. How many pages
(35) has he read?

California Mathematics Content Standards
NS **1.0**, **1.4** Calculate given percentages of quantities and solve problems involving discounts at sales, interest earned, and tips.
MR **2.0, 2.2** Apply strategies and results from simpler problems to more complex problems.

• Writing Fractions and Decimals as Percents, Part 1

Power Up

facts	Power Up J
mental math	**a. Number Sense:** 4×112
	b. Money: $\$2.99 + \$1.99$
	c. Money: A bag of chips costs $0.99. How much would four bags cost?
	d. Money: Double $3.50.
	e. Money: The total price for the set of ten stamps is $2. What is the cost per stamp?
	f. Measurement: The skyscraper was 900 feet tall. How many yards is that?
	g. Statistics: Miranda has 4 pet snakes. They have lengths of 9 cm, 12 cm, 18 cm, and 40 cm. What is the range of lengths?
	h. Calculation: $3 \times 3, \times 3, + 3, \div 3, - 3, \times 3$

problem solving

Choose an appropriate problem-solving strategy to solve this problem. How many different ways can Dwayne spin a total of 6 if he spins each spinner once?

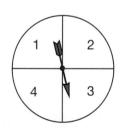

New Concept

A percent is actually a fraction with a denominator of 100. Instead of writing the denominator 100, we can use a percent sign (%). So $\frac{25}{100}$ equals 25%.

Example 1

Write $\frac{3}{100}$ as a percent.

A percent is a fraction with a denominator of 100. Instead of writing the denominator, we write a percent sign. We write $\frac{3}{100}$ as **3%**.

Example 2

Thinking Skills

Explain

How do we write an equivalent fraction?

Write $\frac{3}{10}$ as a percent.

First we will write an equivalent fraction that has a denominator of 100.

$$\frac{3}{10} = \frac{?}{100}$$

We multiply $\frac{3}{10}$ by $\frac{10}{10}$.

$$\frac{3}{10} \cdot \frac{10}{10} = \frac{30}{100}$$

We write the fraction $\frac{30}{100}$ as **30%**.

Example 3

Of the 30 students who took the test, 15 earned an A. What percent of the students earned an A?

Fifteen of the 30 students earned an A. We write this as a fraction and reduce.

$$\frac{15}{30} = \frac{1}{2}$$

To write $\frac{1}{2}$ as a fraction with a denominator of 100, we multiply $\frac{1}{2}$ by $\frac{50}{50}$.

$$\frac{1}{2} \cdot \frac{50}{50} = \frac{50}{100}$$

The fraction $\frac{50}{100}$ equals **50%**.

Example 4

Write 0.12 as a percent.

The decimal number 0.12 is twelve hundredths.

$$0.12 = \frac{12}{100}$$

Twelve hundredths is equivalent to **12%**.

Example 5

Write 0.08 as a percent.

The decimal 0.08 is eight hundredths.

$$0.08 = \frac{8}{100}$$

Eight hundredths is equivalent to **8%**.

Example 6

Write 0.8 as a percent.

The decimal number 0.8 is eight tenths. If we place a zero in the hundredths place, the decimal is eighty hundredths.

$$0.8 = 0.80 = \frac{80}{100}$$

Eighty hundredths equals **80%**.

Notice that when a decimal number is converted to a percent, the decimal point is shifted two places to the right. In fact, shifting the decimal point two places to the right is a quick and useful way to write decimal numbers as percents.

Lesson Practice

Write each fraction as a percent:

a. $\frac{31}{100}$ b. $\frac{1}{100}$ c. $\frac{1}{10}$

d. $\frac{3}{50}$ e. $\frac{7}{25}$ f. $\frac{2}{5}$

g. Twelve of the 30 students earned a B on the test. What percent of the students earned a B?

h. Jorge correctly answered 18 of the 20 questions on the test. What percent of the questions did he answer correctly?

Write each decimal number as a percent:

i. 0.25   j. 0.3   k. 0.05

l. 1.0   m. 0.7   n. 0.15

Written Practice   *Distributed and Integrated*

1. **Connect** What is the reciprocal of two and three fifths?
(36, 57)

* 2. What time is one hour thirty-five minutes after 2:30 p.m.?
(RF11, 7)

3. A 1-pound box of candy cost $4.00. What was the cost per ounce
(8) (1 pound = 16 ounces)?

4. **Estimate** Freda bought a sandwich for $4.00 and a drink for 94¢.
(35) Her grandson ordered a meal for $6.35. What was the total price of all three items when 8% sales tax was added? Explain how to use estimation to check whether your answer is reasonable.

5. If the chance of rain is 50%, then what is the chance it will not rain?
(54)

6. Devonte randomly selects a marble from a bag of 6 white and 9 green marbles.
(70) Find the probability he selects green.

* 7. **a.** Write $\frac{3}{4}$ as a decimal number.
(69, 73)

    b. Write the answer to part **a** as a percent.

* 8. **a.** Write $\frac{3}{20}$ as a fraction with a denominator of 100.
(73)

    b. Write $\frac{3}{20}$ as a percent.

* 9. Write 12% as a reduced fraction. Then write the fraction as a
(27, 69) decimal number.

Find each unknown number:

10. $\frac{7}{10} = \frac{n}{100}$
(37)

11. $5 - m = 3\frac{1}{8}$
(39)

12. $1 - w = 0.95$
(39)

13. $m + 1\frac{2}{3} = 3\frac{1}{6}$
(39, 58)

14. $\left(\frac{1}{2} + \frac{1}{3}\right) - \frac{1}{6}$
(55)

* 15. $3\frac{1}{2} \times 1\frac{1}{3} \times 1\frac{1}{2}$
(67)

16. $(0.43)(2.6)$
(33)

17. $0.26 \div 5$
(41)

* 18. Natrick correctly answered 17 of the 20 questions on the test. What percent of
(73) the questions did Natrick answer correctly?

19. (Estimate) The diameter of the big tractor tire was about 5 feet. As the tire
(42) rolled one full turn, the tire rolled about how many feet? Round the answer to the
nearest foot. (Use 3.14 for π.)

20. $\frac{4 - 2^2}{2}$
(72)

21. Write the prime factorization of both the numerator and denominator of $\frac{18}{30}$. Then
(63) simplify the fraction.

22. What is the greatest common factor of 18 and 30?
(12)

23. **Multiple Choice** If the product of two numbers is 1, then the two numbers are
(36) which of the following?

    A equal   **B** reciprocals   **C** opposites   **D** prime

24. (Verify) Why is every rectangle a quadrilateral?
(59)

25. If b equals 8 and h equals 6, what does $\frac{bh}{2}$ equal?
(6)

* 26. (Represent) Find the prime factorization of 400 using a factor tree.
(61, 71) Then write the prime factorization of 400 using exponents.

* 27. (Represent) Draw a coordinate plane on graph paper. Then
(RF27) draw a rectangle with vertices located at (3, 1), (3, −1), (−1, 1),
and (−1, −1).

28. Refer to the rectangle drawn in problem **27** to answer parts **a**
(23) and **b** below.

    a. What is the perimeter of the rectangle?

    b. What is the area of the rectangle?

* **29.** **a.** What is the perimeter of this parallelogram?
(66)

     b. What is the area of this parallelogram?

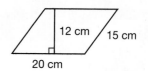

30. (**Model**) Draw two parallel segments of different lengths.
(59) Then form a quadrilateral by drawing two segments that connect
the endpoints of the parallel segments. Is the quadrilateral a rectangle?

Real-World
Connection

The Moon is Earth's only natural satellite. The average distance from
Earth to the Moon is approximately 620^2 kilometers. This distance is
about 30 times the diameter of Earth.

    a. Simplify 620^2 kilometers.

    b. Find the diameter of Earth. Round your answer to the nearest kilometer.

California Mathematics Content Standards
**NS ①.⓪, ①.① Compare and order positive and negative fractions, decimals, and mixed numbers and place them on a number line.
MR 2.0, 2.4 Use a variety of methods, such as words, numbers, symbols, charts, graphs, tables, diagrams, and models, to explain mathematical reasoning.

- # Comparing Fractions by Converting to Decimal Form

facts	Power Up I
mental math	**a.** **Number Sense:** $430 + 270$
	b. **Money:** One notebook costs $0.99. How much would five notebooks cost?
	c. **Money:** The burrito cost $1.98. Gabriella paid with a $5 bill. How much change should she receive?
	d. **Fractions:** $\frac{1}{4}$ of $2.40
	e. **Geometry:** The two angles shown are supplementary angles. What is m $\angle A$?

- **f.** **Statistics:** The daily rainfall amounts for Monday through Friday were 0, 0, 3, 2, and 0 centimeters. What is the mean of the amounts?

- **g.** **Algebra:** If $v = 25$, what is $\frac{75}{v}$?

- **h.** **Calculation:** $5 \times 5, -5, \times 5, \div 2, +5, \div 5$

problem solving

Choose an appropriate problem-solving strategy to solve this problem. A 60 in.-by-104 in. rectangular tablecloth was draped over a rectangular table. Eight inches of the 104-inch length of cloth hung over the left edge of the table, 3 inches over the back, 4 inches over the right edge, and 7 inches over the front.

In which directions (left, back, right, and/or forward) and by how many inches should the tablecloth be shifted so that equal amounts of cloth hang over opposite edges of the table? What are the dimensions of the table?

We have compared fractions by drawing pictures of fractions and by writing fractions with common denominators. Another way to compare fractions is to convert the fractions to decimal form.

Example 1

Thinking Skills

Explain

How do we compare decimal numbers?

Compare these fractions. First convert each fraction to decimal form.

$$\frac{3}{5} \bigcirc \frac{5}{8}$$

We convert each fraction to a decimal number by dividing the numerator by the denominator.

$$\frac{3}{5} \longrightarrow 5\overline{)3.0}^{\,0.6} \qquad \frac{5}{8} \longrightarrow 8\overline{)5.000}^{\,0.625}$$

We write both numbers with the same number of decimal places. Then we compare the two numbers.

$$0.600 < 0.625$$

Since 0.6 is less than 0.625, we know that $\frac{3}{5}$ is less than $\frac{5}{8}$.

$$\frac{3}{5} < \frac{5}{8}$$

Example 2

Compare: $\frac{3}{4} \bigcirc 0.7$

First we write the fraction as a decimal.

$$\frac{3}{4} \longrightarrow 4\overline{)3.00}^{\,0.75}$$

Then we compare the decimal numbers.

$$0.75 > 0.70$$

Since 0.75 is greater than 0.7, we know that $\frac{3}{4}$ is greater than 0.7.

$$\frac{3}{4} > 0.7$$

Lesson Practice

Change the fractions to decimals to compare these numbers:

a. $\frac{3}{20} \bigcirc \frac{1}{8}$ **b.** $\frac{3}{8} \bigcirc \frac{2}{5}$ **c.** $\frac{15}{25} \bigcirc \frac{3}{5}$

d. $0.7 \bigcirc \frac{4}{5}$ **e.** $\frac{2}{5} \bigcirc 0.5$ **f.** $\frac{3}{8} \bigcirc 0.325$

* 1. **(Connect)** What is the product of ten squared and two cubed?
(71)

2. **(Connect)** What number is halfway between 4.5 and 6.7?
(28)

3. **(Formulate)** It is said that one year of a dog's life is the same as 7 years
(8) of a human's life. Using that thinking, a dog that is 13 years old is how
many "human" years old? Write an equation and solve the problem.

* 4. Compare. First convert each fraction to decimal form.
(74)
$$\frac{2}{5} \bigcirc \frac{1}{4}$$

* 5. **a.** What fraction of this circle is shaded?
(69, 73)

b. Convert the answer from part **a** to a decimal number.

c. What percent of this circle is shaded?

6. Leonardo spins a spinner with 5 equal sections numbered 1–5. Find the
(70) probability he spins a number greater than 3.

* 7. **a.** Convert $2\frac{1}{2}$ to a decimal number.
(68, 69)

b. Write 3.75 as a simplified mixed number.

* 8. **a.** Write 0.04 as a simplified fraction.
(68, 73)

b. Write 0.04 as a percent.

9. **(Verify)** Instead of dividing 200 by 18, Sam found half of each number and
(38) then divided. Show Sam's division problem and write the quotient as a mixed
number.

10. $6\frac{1}{3} + 3\frac{1}{4} + 2\frac{1}{2}$ **11.** $\frac{4}{5} = \frac{?}{100}$
(56)   (37)

* 12. **(Analyze)** $\left(2\frac{1}{2}\right)\left(3\frac{1}{3}\right)\left(1\frac{1}{5}\right)$ *** 13.** $5 \div 2\frac{1}{2}$
(67)   (64)

Find each unknown number:

14. $6.7 + 0.48 + n = 8$ **15.** $12 - d = 4.75$
(39)   (39)

16. 0.35×0.45 **17.** $4.3 \div 100$
(33)   (RF23, 32)

18. Find the median of these numbers:
(Inv. 4)
$$0.3, 0.25, 0.313, 0.2, 0.27$$

19. **(Estimate)** Find the sum of 3926 and 5184 to the nearest thousand.
(21)

20. **(List)** Name all the prime numbers between 40 and 50.
(11)

* 21. Twelve of the 25 students in the class earned As on the test. What percent of the
(73) students earned A's?

Refer to the triangle to answer problems **22** and **23**.

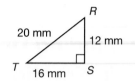

22. What is the perimeter of this triangle?
(RF7)

* 23. (**Analyze**) Angles *T* and *R* are complementary. If the
(60) measure of ∠*R* is 53°, then what is the measure of ∠*T*?

24. (**Estimate**) About how many **millimeters** long is this line segment?
(RF6)

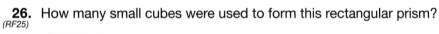

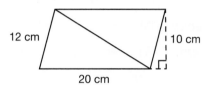

* 25. This parallelogram is divided into two congruent triangles.
(66)

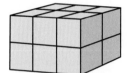

    a. What is the area of the parallelogram?

    b. What is the area of one of the triangles?

26. How many small cubes were used to form this rectangular prism?
(RF25)

* 27. (**Represent**) Sketch a coordinate plane on graph paper. Graph point
(RF27) *A* (1, 2), point *B* (−3, −2), and point *C* (1, −2). Then draw segments to
connect the three points. What type of polygon is figure *ABC*?

28. (**Conclude**) In the figure drawn in problem 27,
(20)   **a.** which segment is perpendicular to $\overline{AC}$?

    b. which angle is a right angle?

29. If *b* equals 12 and *h* equals 9, what does $\frac{bh}{2}$ equal?
(6)

* 30. Draw a pair of parallel lines. Draw a third line perpendicular to the parallel lines.
(59) Complete a quadrilateral by drawing a fourth line that intersects but is not
perpendicular to the pair of parallel lines. Trace the quadrilateral that is formed.
Is the quadrilateral a rectangle?

California Mathematics Content Standards

SDAP 3.0, **3.1** Represent all possible outcomes for compound events in an organized way (e.g., tables, grids, tree diagrams) and express the theoretical probability of each outcome.

SDAP 3.0, **3.3** Represent probabilities as ratios, proportions, decimals between 0 and 1, and percentages between 0 and 100 and verify that the probabilities computed are reasonable; know that if *P* is the probability of an event, 1- *P* is the probability of an event not occurring.

MR 1.0, 1.1 Analyze problems by identifying relationships, distinguishing relevant from irrelevant information, identifying missing information, sequencing and prioritizing information, and observing patterns.

• Finding Unstated Information in Fraction Problems

facts	Power Up J
mental math	**a. Number Sense:** 504×6
	b. Measurement: 625 meters – 250 meters
	c. Money: $2.50 + $1.99
	d. Money: $12.50 ÷ 10
	e. Time: The piano recital began at 7:15 p.m. It ended 75 minutes later. At what time did the recital end?
	f. Algebra: The perimeter of this rectangle is 14 in. What is *y*?
	g. Statistics: What is the mean number of letters in the words HE, HER, and HERE?
	h. Calculation: 6×6, $- 6$, $÷ 6$, $- 5$, $\times 2$, $+ 1$

5 in. *y*

problem solving

Choose an appropriate problem-solving strategy to solve this problem. The sum of the digits of a five-digit number is 25. What is the five digit number if the fifth digit is two less than the fourth, the fourth digit is two less than the third, the third is two less than the second, and the second digit is two less than the first digit?

Often fractional-parts statements contain more information than what is directly stated. Consider this fractional-parts statement:

Three fourths of the 28 students in the class are boys.

This sentence directly states information about the number of boys in the class. It also *indirectly* states information about the number of girls in the class. In this lesson, we will practice finding several pieces of information from fractional-parts statements.

Example 1

Diagram this statement. Then answer the questions that follow.

Three fourths of the 28 students in the class are boys.

a. **Into how many parts is the class divided?**

b. **How many students are in each part?**

c. **How many parts are boys?**

d. **How many boys are in the class?**

e. **How many parts are girls?**

f. **How many girls are in the class?**

We draw a rectangle to represent the whole class. Since the statement uses fourths to describe a part of the class, we divide the rectangle into four parts. Dividing the total number of students by four, we find there are seven students in each part. We identify three of the four parts as boys and one of the four parts as girls. Now we answer the questions.

28 students

$\frac{1}{4}$ are girls. { 7 students

7 students

$\frac{3}{4}$ are boys. { 7 students

7 students

a. The denominator of the fraction indicates that the class is divided into **four parts** for the purpose of this statement. It is important to distinguish between the number of *parts* (as indicated by the denominator) and the number of *categories*. There are two categories of students implied by the statement—boys and girls.

b. In each of the four parts there are **seven students.**

c. The numerator of the fraction indicates that **three parts** are boys.

d. Since three parts are boys and since there are seven students in each part, we find that there are **21 boys** in the class.

e. Three of the four parts are boys, so only **one part** is girls.

f. There are seven students in each part. One part is girls, so there are **seven girls.**

Example 2

There are thirty marbles in a bag. If one marble is drawn from the bag, the probability of drawing red is $\frac{2}{5}$.

a. **How many marbles are red?**

b. **How many marbles are not red?**

c. **The complement of drawing a red marble is drawing a not red marble. What is the probability of drawing a not red marble?**

d. **What is the sum of the probabilities of drawing red and drawing not red?**

a. The probability of drawing red is $\frac{2}{5}$, so $\frac{2}{5}$ of the 30 marbles are red.

$$\frac{2}{5} \cdot 30 = 12$$

There are **12 red marbles.**

b. Twelve of the 30 marbles are red, so **18 marbles are not red.**

c. The probability of drawing a not red marble is $\frac{18}{30} = \frac{3}{5}$.

d. The sum of the probabilities of an event and its complement is **1.**

$$\frac{2}{5} + \frac{3}{5} = 1$$

Lesson Practice

Model Diagram this statement. Then answer the questions that follow.

Three eighths of the 40 little engines could climb the hill.

a. Into how many parts was the group divided?

b. How many engines were in each part?

c. How many parts could climb the hill?

d. How many engines could climb the hill?

e. How many parts could not climb the hill?

f. How many engines could not climb the hill?

Read the statement and then answer the questions that follow.

The face of a spinner is divided into 12 equal sectors. The probability of spinning red on one spin is $\frac{1}{4}$.

g. How many sectors are red?

h. How many sectors are not red?

i. What is the probability of spinning not red in one spin?

j. What is the sum of the probabilities of spinning red and not red? How are the events related?

Written Practice

Distributed and Integrated

1. The weight of an object on the Moon is about $\frac{1}{6}$ of its weight on Earth.
   (24) A person weighing 114 pounds on Earth would weigh about how much on the Moon?

* 2. **Estimate** Estimate the weight of an object in your classroom such as a table
   (21, 24) or desk. Use the information in problem 1 to calculate what the approximate weight of the object would be on the moon. Round your answer to the nearest pound.

* 3. Mekhi was at bat 24 times and got 6 hits.
(25, 73)

    a. What fraction of the times at bat did Mekhi get a hit?

    b. What percent of the times at bat did Mekhi get a hit?

* 4. (**Model**) Diagram this statement. Then answer the questions that follow.
(75)

        There are 30 students in the class. Three fifths of them are boys.

    a. Into how many parts is the class divided?

    b. How many students are in each part?

    c. How many boys are in the class?

    d. How many girls are in the class?

* 5. a. In the figure below, what fraction of the group is shaded?
(69, 73)

    b. Convert the fraction in part **a** to a decimal number.

    c. What percent of the group is shaded?

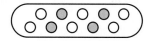

* 6. Write the decimal number 3.6 as a mixed number.
(68)

Find each unknown number:

7. $3.6 + a = 4.15$
(39)

8. $\frac{2}{5}x = 1$
(36)

9. ✏ (**Explain**) If the chance of rain is 60%, is it more likely to rain or not to rain? Why?
(54)

* 10. Three fifths of a circle is what percent of a circle?
(73)

11. A temperature of $-3°F$ is how many degrees below the freezing temperature of water?
(9)

* 12. Compare:
(71, 74)

    a. $0.35 \bigcirc \frac{7}{20}$ **b.** $3^2 \bigcirc 2^3$

13. $\frac{1}{2} + \frac{2}{3}$
(55)

14. $3\frac{1}{5} - 1\frac{3}{5}$
(58)

15. $\frac{1}{2} + \frac{3}{4} + \frac{7}{8}$
(56)

16. $3 \times 1\frac{1}{3}$
(62)

17. $3 \div 1\frac{1}{3}$
(64)

18. $1\frac{1}{3} \div 2$
(64)

19. What is the perimeter of this rectangle?
(23)

20. What is the area of this rectangle?
(23)

    1.5 cm

    0.9 cm

* 21. Write the prime factorization of 1000 using exponents.
(71)

22. Coats were on sale for 40% off. One coat was regularly priced at $80.
$^{(35)}$

    a. How much money would be taken off the regular price of the coat during the sale?

    b. What would be the sale price of the coat?

23. Patricia bought a coat that cost $38.80. The sales-tax rate was 7%.
$^{(35)}$

    a. What was the tax on the purchase?

    b. What was the total purchase price including tax?

24. (**Classify**) Is every quadrilateral a polygon?
$^{(59)}$

25. What time is one hour fourteen minutes before noon?
$^{(RF11, 7)}$

26. **Multiple Choice** What percent of this rectangle appears to be shaded?
$^{(Inv. 1)}$

    A 20%   **C** 60%

    B 40%   **D** 80%

* 27. (**Represent**) Sketch a coordinate plane on graph paper.
$^{(RF27)}$ Graph point W (2, 3), point X (1, 0), point Y (−3, 0), and point Z (−2, 3). Then draw $\overline{WX}$, $\overline{XY}$, $\overline{YZ}$, and $\overline{ZW}$.

* 28. **a.** (**Conclude**) Which segment in problem 27 is parallel to $\overline{WX}$?
$^{(66)}$

    b. Which segment in problem 27 is parallel to $\overline{XY}$?

29. Write the prime factorization of both the numerator and the denominator of this fraction. Then simplify the fraction.
$^{(63)}$

$$\frac{210}{350}$$

30. Dequon rolls a number cube. What is the probability that an even number is rolled?
$^{(70)}$

Real-World Connection

Simplify each prime factorization below. Then identify which prime factorization does not belong in the group. Explain your reasoning.

$3^2 \times 2^3$ $2^2 \times 3^3 \times 5$ $2^2 \times 3^2 \times 5^2$ 3^5 $2^4 \times 7^2$

California Mathematics Content Standards

AF 2.0, 2.1 Convert one unit of measurement to another (e.g., from feet to miles, from centimeters to inches).

MR 2.0, 2.4 Use a variety of methods, such as words, numbers, symbols, charts, graphs, tables, diagrams, and models, to explain mathematical reasoning.

• Metric System

Power Up

facts   Power Up M

mental math

    a. Number Sense: $380 + 155$

    b. Money: Each candle costs $1.99. How much would 4 candles cost?

    c. Money: 0.95×100

    d. Fractions: $\frac{1}{5}$ of $4.50

    e. Powers/Roots: $\sqrt{100} - 2$

    f. Statistics: The four sprinters finished the race with these times (in seconds): 14.9, 15.0, 15.4, and 16.9. What was the range?

    g. Algebra: If $2c = 10$, what is c?

    h. Calculation: $8 \times 8, - 4, \div 2, + 2, \div 4, \times 3, + 1, \div 5$

problem solving

Choose an appropriate problem-solving strategy to solve this problem. In the 4×200 m relay, Sarang ran first, then Sarai, then Joyce, and finally Karla. Each girl ran her 200 meters 2 seconds faster than the previous runner. The team finished the race in exactly 1 minute and 50 seconds. How fast did each runner run her 200 meters?

New Concept

The system of measurement used throughout most of the world is the **metric system,** also called the **International System.** The metric system has two primary advantages over the U.S. Customary System: it is a decimal system, and the units of one category of measurement are linked to units of other categories of measurement.

The metric system is a decimal system in that units within a category of measurement differ by a factor, or power, of 10. The U.S. Customary System is not a decimal system, so converting between units is more difficult. Here we show some equivalent measures of length in the metric system:

Units of Length
10 millimeters (mm) = 1 centimeter (cm)
1000 millimeters (mm) = 1 meter (m)
100 centimeters (cm) = 1 meter (m)
1000 meters (m) = 1 kilometer (km)

The basic unit of length in the metric system is the **meter.** Units larger than a meter or smaller than a meter are indicated by prefixes that are used across the categories of measurement. These prefixes, shown in the table below, indicate the multiplier of the basic unit.

Examples of Metric Prefixes

Prefix	Unit		Relationship
kilo-	kilometer	(km)	1000 meters
hecto-	hectometer	(hm)	100 meters
deka-	dekameter	(dkm)	10 meters
	meter	(m)	
deci-	decimeter	(dm)	0.1 meter
centi-	centimeter	(cm)	0.01 meter
milli-	millimeter	(mm)	0.001 meter

As we move up the table, the units become larger and the number of units needed to describe a length decreases:

$$1000 \text{ mm} = 100 \text{ cm} = 10 \text{ dm} = 1 \text{ m}$$

As we move down the table, the units become smaller and the number of units required to describe a length increases:

$$1 \text{ km} = 10 \text{ hm} = 100 \text{ dkm} = 1000 \text{ m}$$

To change lengths between the metric system and the U.S. Customary System, we may use these conversions:

$$1 \text{ kilometer} \approx 0.6 \text{ mile}$$

$$1 \text{ meter} \approx 1.1 \text{ yard}$$

$$2.54 \text{ cm} = 1 \text{ inch}$$

Note: The symbol ≈ means "approximately equal to."

(**Analyze**) Look at a meter stick. It is numbered according to what unit of measure? One meter equals how many centimeters? One centimeter equals how many millimeters?

Example 1

a. **Five kilometers is how many meters?**

b. **Three hundred centimeters is how many meters?**

a. One kilometer is 1000 meters, so 5 kilometers is **5000 meters.**

b. A centimeter is 0.01 $(\frac{1}{100})$ of a meter (just as a cent is 0.01 $(\frac{1}{100})$ of a dollar). One hundred centimeters equals 1 meter, so 300 centimeters equals **3 meters.**

The **liter** is the basic unit of capacity in the metric system. We are familiar with 2-liter bottles of soft drink. Each 2-liter bottle can hold a little more than a half gallon, so a liter is a little more than a quart. A milliliter is 0.001 ($\frac{1}{1000}$) of a liter.

Units of Capacity
1000 milliliters (mL) = 1 liter (L)

Example 2

A 2-liter bottle can hold how many milliliters of beverage?

One liter is 1000 mL, so 2 L is **2000 mL.**

Generalize Look around your classroom and find two objects: one that you would measure using metric units of length, and one that you would measure using metric units of capacity.

The basic unit of mass in the metric system is the **kilogram.** For scientific purposes, we distinguish between *mass* and *weight.* The weight of an object varies with the gravitational force, while its mass does not vary. The weight of an object on the Moon is about $\frac{1}{6}$ of its weight on Earth, yet its mass is the same. The mass of this book is about one kilogram. A gram is 0.001 ($\frac{1}{1000}$) of a kilogram—about the mass of a paperclip. A milligram, 0.001 ($\frac{1}{1000}$) of a gram and 0.000001 ($\frac{1}{1,000,000}$) of a kilogram, is a unit used for measuring the mass of smaller quantities of matter, such as the amount of vitamins in the food we eat.

Units of Mass/Weight
1000 grams (g) = 1 kilogram (kg)
1000 milligrams (mg) = 1 gram

Although a kilogram is a unit for measuring mass anywhere in the universe, here on Earth store clerks use kilograms to measure the weights of the goods they buy and sell. A kilogram mass weighs about 2.2 pounds.

Example 3

How many 250-mg tablets of vitamin C equal one gram of vitamin C?

A gram is 1000 mg, so **four** 250-mg tablets of vitamin C total one gram of vitamin C.

Discuss Does this weigh more or less than a pound? Why?

The Celsius and Kelvin scales are used by scientists to measure temperature. Both are **centigrade** scales because there are 100 gradations, or degrees, between the freezing and boiling temperatures of water. The Celsius scale places 0°C at the freezing point of water. The Kelvin scale places 0 K at **absolute zero,** which is 273 centigrade degrees below the freezing temperature of water (−273°C). The Celsius

Thinking Skills

Justify

A room's temperature is 22 degrees and it feels comfortably warm. Is this measurement in °C or °F? Explain.

scale is more commonly used by the general population. Below, we show frequently referenced temperatures on the Celsius scale, along with the Fahrenheit equivalents.

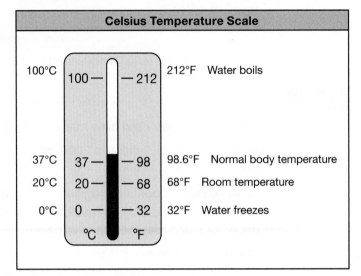

Celsius Temperature Scale

100°C	100 — — 212	212°F Water boils
37°C	37 — — 98	98.6°F Normal body temperature
20°C	20 — — 68	68°F Room temperature
0°C	0 — — 32	32°F Water freezes
	°C °F	

Example 4

A temperature increase of 100° on the Celsius scale is an increase of how many degrees on the Fahrenheit scale?

The Celsius and Fahrenheit scales are different scales. An increase of 1°C is not equivalent to an increase of 1°F. On the Celsius scale there are 100° between the freezing point of water and the boiling point of water. On the Fahrenheit scale water freezes at 32° and boils at 212°, a difference of 180°. So an increase of 100°C is an increase of **180°F**. Thus, a change of one degree on the Celsius scale is equivalent to a change of 1.8 degrees on the Fahrenheit scale.

Reading Math

The symbol ° means *degrees*. Read 32°F as "thirty-two degrees Fahrenheit."

Lesson Practice

a. The closet door is about 2 meters tall. How many centimeters is 2 meters?

b. A 1-gallon plastic jug can hold about how many liters of milk?

c. A metric ton is 1000 kilograms, so a metric ton is about how many pounds?

d. A temperature increase of 10° on the Celsius scale is equivalent to an increase of how many degrees on the Fahrenheit scale? (See Example 4.)

e. After running 800 meters of a 3-kilometer race, Michelle still had how many meters to run?

f. A 30-cm ruler broke into two pieces. One piece was 120 mm long. How long was the other piece? Express your answer in millimeters.

g. **Conclude** About how many inches long was the ruler in problem **f** before it broke?

1. What is the difference when the product of $\frac{1}{2}$ and $\frac{1}{2}$ is subtracted from the sum of
(4, 53) $\frac{1}{2}$ and $\frac{1}{2}$?

2. The claws of a Siberian tiger are 10 centimeters long. How many millimeters is
(76) that?

* 3. (**Analyze**) Quanda was thinking of a number between 40 and 50 that is a multiple
(19) of 3 and 4. Of what number was she thinking?

* 4. (**Model**) Diagram this statement. Then answer the questions that follow.
(75)
Four fifths of the 60 lights were on.

a. Into how many parts have the 60 lights been divided?

b. How many lights are in each part?

c. How many lights were on?

d. How many lights were off?

5. (**Classify**) Which counting number is neither a prime number nor a composite
(61) number?

Find each unknown number:

6. $\frac{4}{5}m = 1$
(36)

7. $\frac{4}{5} + w = 1$
(39)

8. $\frac{4}{5} \div x = 1$
(39)

9. $\frac{3}{4} = \frac{n}{100}$
(37)

* 10. **a.** What fraction of the rectangle below is shaded?
(69, 73)

b. Write the answer to part **a** as a decimal number.

c. What percent of the rectangle is shaded?

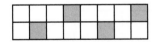

* 11. Convert the decimal number 1.15 to a mixed number.
(68)

* 12. Compare:
(71, 74)
a. $\frac{3}{5}$ ◯ 0.35

b. $\sqrt{100}$ ◯ $1^4 + 2^3$

13. $\frac{5}{6} - \frac{1}{2}$
(55)

14. $4\frac{1}{4} - 3\frac{1}{3}$
(58)

15. $\frac{1}{2} + \frac{2}{3} + \frac{5}{6}$
(56)

16. $1\frac{1}{2} \times 2\frac{2}{3}$
(62)

17. $1\frac{1}{2} \div 2\frac{2}{3}$
(64)

18. $2\frac{2}{3} \div 1\frac{1}{2}$
(64)

19. **a.** What is the perimeter of this square?
(32)

b. What is the area of this square?

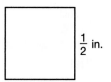

$\frac{1}{2}$ in.

20. **Verify** "The opposite sides of a rectangle are parallel." True or false?
(59)

* 21. What is the mean of 3^3 and 5^2?
(10, 71)

* 22. The diameter of the small wheel was 7 inches. The circumference was about
(46, 69) 22 inches. Write the ratio of the circumference to the diameter of the circle as a decimal number rounded to the nearest hundredth.

23. How many inches is $2\frac{1}{2}$ feet?
(62)

24. **Connect** Which arrow below could be pointing to 0.1?
(28)

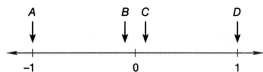

25. **Represent** Draw a quadrilateral that is not a rectangle.
(59)

26. **Represent** Find the prime factorization of 900 by using a factor tree. Then
(61, 71) write the prime factorization using exponents.

27. Diego randomly selects a letter from the word *turtle*. Find the probability that
(70) he selects a "t".

28. Find the length of this segment.
(76)

a. in centimeters.

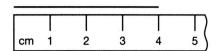

b. in millimeters?

* 29. **Multiple Choice** A liter is closest in size to which of the following?
(76)

    A pint   **B** quart   **C** $\frac{1}{2}$ gallon   **D** gallon

* 30. **a.** How many mL of liquid are in the container?
(76)

b. How many mL are in a liter?

c. The amount of liquid in this container is how much less than a liter?

California Mathematics Content Standards

AF 3.0, 3.1 Use variables in expressions describing geometric quantities (e.g., $P = 2w + 2l$, $A = \frac{1}{2}bh$, $C = \pi d$ - the formulas for the perimeter of a rectangle, the area of a triangle, and the circumference of a circle, respectively).

MR 2.0, 2.6 Indicate the relative advantages of exact and approximate solutions to problems and give answers to a specified degree of accuracy

• Area of a Triangle

Power Up

facts   Power Up J

mental math

    a. Number Sense: 311×5

    b. Number Sense: $565 - 250$

    c. Money: Each carton of milk costs $1.99. What would 5 cartons cost?

    d. Money: $7.50 + $1.99

    e. Decimals: $6.5 \div 100$

    f. Statistics: If the mean score for three games is 6 points, what is the total number of points scored in all three games?

    g. Geometry: The area of this rectangle is 12 cm². How many centimeters is the width w?

    h. Calculation: 10×10, $\times 10$, -1, $\div 9$, -11, $\div 10$

4 cm

problem solving

Choose an appropriate problem-solving strategy to solve this problem. Kathleen read a mean of 45 pages per day for four days. If she read a total of 123 pages during the first three days, how many pages did she read on the fourth day?

New Concept

Math Language

A small square at the vertex of an angle indicates that the angle is a right angle.

A triangle has a **base** and a **height** (or **altitude**).

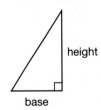

height

base

The base is one of the sides of the triangle. The height (or altitude) is the perpendicular distance between the base (or baseline) and the opposite vertex of the triangle. Since a triangle has three sides and any side can be the base, a triangle can have three base-height orientations, as we show by rotating this triangle.

One Right Triangle Rotated to Three Positions

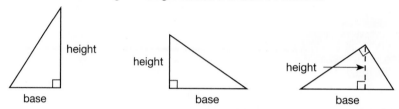

If one angle of a triangle is a right angle, the height may be a side of the triangle, as we see above. If none of the angles of a triangle are right angles, then the height will not be a side of the triangle. When the height is not a side of the triangle, a dashed line segment will represent it, as in the right-hand figure above. If one angle of a triangle is an obtuse angle, then the height is shown outside the triangle in two of the three orientations, as shown below.

One Obtuse Triangle Rotated to three Positions

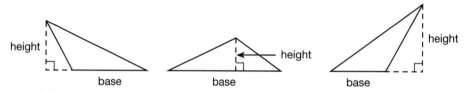

Model Using a ruler, draw a right triangle, an obtuse triangle and an acute triangle. Using the method described above, label the base and height of each triangle, adding a dashed line segment as needed.

The area of a triangle is half the area of a rectangle with the same base and height, as the following activity illustrates.

Activity

Area of a Triangle

Materials needed:
- Paper
- Ruler or straightedge
- Protractor
- Scissors

Use a ruler or straightedge to draw a triangle. Determine which side of the triangle is the longest side. The longest side will be the base of the triangle for this activity. To represent the height (altitude) of the triangle, draw a series of dashes from the topmost vertex of the triangle to the base. Make sure the dashes are perpendicular to the base, as in the figure below.

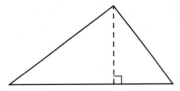

Now we draw a rectangle that contains the triangle. The base of the triangle is one side of the rectangle. The height of the triangle equals the height (width) of the rectangle.

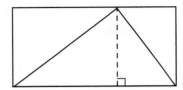

When you finish drawing the rectangle, consider this question, what fraction of the rectangle is the original triangle? The rest of this activity will answer this question.

Cut out the rectangle and set the scraps aside. Next, carefully cut out the triangle you drew from the rectangle. Save all three pieces.

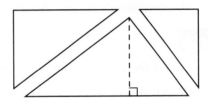

Rotate the two smaller pieces, and fit them together to make a triangle identical to the triangle you drew. The two triangles are **congruent** to each other because they are the same size and shape.

Since the two congruent triangles have equal areas, the original triangle must be half the area of the rectangle. Recall that the area of a rectangle can be found by this formula:

$$\text{Area} = \text{Length} \times \text{Width}$$

$$A = LW$$

Suggest a formula for finding the area of a triangle. Use base b and height h in place of length and width.

When we multiply two perpendicular dimensions, the product is the area of a rectangle with those dimensions.

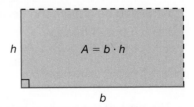

To find the area of a triangle with *a* base of *b* and *a* height of *h,* we find half of the product of *b* and *h.*

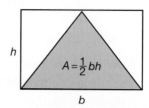

We show two formulas for finding the area of a triangle.

(**Discuss**) How are the formulas different? Why do both formulas yield the same result?

> Area of a triangle $= \frac{1}{2}bh$
>
> Area of a triangle $= \frac{bh}{2}$

Example 1

Find the area of this triangle.

(Use $A = \frac{bh}{2}$.)

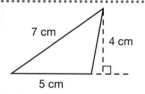

We find the area of the triangle by multiplying the base by the height then dividing the product by 2. The base and height are perpendicular dimensions. In this figure the base is 5 cm, and the height is 4 cm.

$$Area = \frac{5 \text{ cm} \times 4 \text{ cm}}{2}$$
$$= \frac{20 \text{ cm}^2}{2}$$
$$= \textbf{10 cm}^2$$

Example 2

High on the wall near the slanted ceiling was a triangular window with the dimensions shown. What is the area of the window? (Use $A = \frac{1}{2}bh$.)

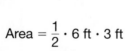

The base and height are perpendicular dimensions. Since one angle of this triangle is a right angle, the base and height are the perpendicular sides, which are 6 ft and 3 ft long.

$$Area = \frac{1}{2} \cdot 6 \text{ ft} \cdot 3 \text{ ft}$$
$$= \textbf{9 ft}^2$$

The area of the window is **9 ft²**.

Example 3

Jerry has a rectangular lawn that measures 39.75 ft by 48.5 ft. A triangular portion (shown shaded in the diagram) needs to be reseeded. Jerry wants to buy as little grass seed as necessary. Explain how you would estimate to decide how much grass seed he should buy?

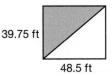

39.75 ft

48.5 ft

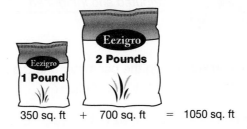

| covers | 350 sq. ft | 700 sq. ft | 1750 sq. ft | 3500 sq. ft |

The amount of grass seed Jerry needs depends on the area that will be reseeded. We can estimate the area of the triangular region by first rounding the base and height to 50 ft and 40 ft.

$$A = \frac{50 \text{ ft} \cdot 40 \text{ ft}}{2}$$

$$A = \frac{2000}{2} \text{ sq. ft}$$

$$A = 1000 \text{ sq. ft}$$

Jerry needs enough seed to cover 1000 sq. ft. He could buy a 2 lb bag and a 1 lb bag, which would cover 1050 sq. ft.

350 sq. ft   +   700 sq. ft   =   1050 sq. ft

Or, he could buy three 1 pound bags which would cover the same area.

350 sq. ft   +   350 sq. ft   +   350 sq. ft   =   1050 sq. ft

The 5 lb and 10 lb bags would cover the area, but would be much more than he needs.

Example 4 ·

This triangle has unknown height *a*. Write an equation that expresses the area of the triangle in terms of *a*.

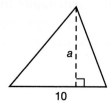

We use the formula for the area of a triangle.

$$A = \frac{1}{2}bh$$

We write 10 for the base and *a* for the height.

$$A = \frac{1}{2} \times 10a$$

Multiply $\frac{1}{2}$ and 10 to simplify.

$$A = 5a$$

Lesson Practice   Find the area of each triangle. Dimensions are in centimeters

a.

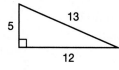

b.

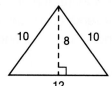

c.

d. Write two formulas for finding the area of a triangle.

Written Practice   *Distributed and Integrated*

1. **Explain** If you know both the perimeter and the length of a rectangle, how
(23) can you determine the width of the rectangle?

*2. A 2-liter bottle contained 2 qt 3.6 oz of beverage. Use this information to compare
(76) a liter and a quart:

1 liter ◯ 1 quart

3. Mr. Johnson was 38 years old when he started his job. He worked for 33 years.
(7) How old was he when he retired?

4. **Verify** Answer "true" or "false" for each statement:
(59)
a. "Every rectangle is a square."

b. "Every rectangle is a parallelogram."

5. Ninety percent of 30 trees are birch trees.
(16, 35)
a. How many trees are birch trees?

b. What is the ratio of birch trees to non-birch trees?

* 6. Eighteen of the twenty-four runners finished the race.
(73, 75)

    a. What fraction of the runners finished the race?

    b. What fraction of the runners did not finish the race?

    c. What percent of the runners did not finish the race?

* 7. **Analyze** This parallelogram is divided into two congruent triangles. What is the
(77) area of each triangle?

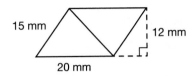

15 mm   12 mm

20 mm

* 8. $10^3 \div 10^2$
(71)

9. $6.42 + 12.7 + 8$
(41)

10. $1.2(0.12)$
(33)

11. $64 \div 0.08$
(44)

12. $3\frac{1}{3} \times \frac{1}{5} \times \frac{3}{4}$
(67)

13. $2\frac{1}{2} \div 3$
(64)

Find each unknown number:

14. $10 - q = 9.87$
(39)

15. $24m = 0.288$
(39, 41)

16. $n - 2\frac{3}{4} = 3\frac{1}{3}$
(39, 55)

17. $w + \frac{1}{4} = \frac{5}{6}$
(39, 55)

18. The perimeter of a square is 80 cm. What is its area?
(32)

19. April selects a marble at random from a can containing 4 red, 3 green, and 2 blue
(70) marbles. Find the probability that she does not pick green.

* 20. **Connect** Juana set the radius on the compass to 10 cm and drew a circle.
(42) What was the circumference of the circle? (Use 3.14 for π.)

21. **Multiple Choice** Which of these numbers is closest to zero?
(28)
    A -2 **B** 0.2 **C** 1 **D** $\frac{1}{2}$

22. **Estimate** Find the product of 6.7 and 7.3 by rounding each number to the
(46) nearest whole number before multiplying. Explain how you arrived at your
answer.

* 23. **Analyze** The expression 2^4 (two to the fourth power) is the prime factorization
(71) of 16. The expression 3^4 is the prime factorization of what number?

24. What number is halfway between 0.2 and 0.3?
(10, 41)

25. **Connect** To what decimal number is the arrow pointing on the number line
(28) below?

10   11

26. Which quadrilateral has only one pair of parallel sides?
(59)

* 27. **(Analyze)** The coordinates of the vertices of a quadrilateral are (−5, 5),
(RF27, 59) (1, 5), (3, 1), and (−3, 1). What is the name for this kind of quadrilateral?

(Analyze) In the figure below, a square and a regular hexagon share a common side.
The area of the square is 100 sq. cm. Use this information to answer problems **28** and **29.**

* 28. **a.** What is the length of each side of the square?
(32)
    b. What is the perimeter of the square?

29. **a.** What is the length of each side of the hexagon?
(RF7)
    b. What is the perimeter of the hexagon?

30. Write the prime factorization of both the numerator and the denominator of this
(63) fraction. Then simplify the fraction.

$$\frac{32}{48}$$

Mrs. Singh takes care of eight children. Half of the children drink four
cups of milk a day. The other half drink two cups of milk a day. How
many gallons of milk would Mrs. Singh have to purchase to have enough
milk for the children for four days?

*Real-World
Connection*

LESSON
78

California Mathematics Content Standards

NS **1.0**, **1.2** Interpret and use ratios in different contexts (e.g., batting averages, miles per hour) to show the relative sizes of two quantities, using appropriate notations ($\frac{a}{b}$, a to b, a:b).

NS **1.0**, **1.3** Use proportions to solve problems (e.g., determine the value of N if $\frac{4}{7} = \frac{N}{21}$, find the length of a side of a polygon similar to a known polygon). Use cross-multiplication as a method for solving such problems, understanding it as the multiplication of both sides of an equation by a multiplicative inverse.

• Using a Constant Factor to Solve Ratio Problems

facts   Power Up E

mental
math

    a. Number Sense: 4×325

    b. Money: The shampoo cost $2.99. Desiree paid for it with a $20 bill. How much change should she receive?

    c. Fractions: $\frac{1}{3} \times \$2.40$

    d. Measurement: 1500 grams + 275 grams

    e. Measurement: How many centimeters is 1.75 meters?

    f. Statistics: Find the mean of 1.0, 1.2, 1.2, 1.4

    g. Algebra: If $j - 9 = 9$, what is j?

    h. Calculation: $9 \times 11, + 1, \div 2, - 1, \div 7, - 2, \times 5$

problem
solving

Choose an appropriate problem-solving strategy to solve this problem. Raul's PE class built a training circuit on a circular path behind their school. There are six light poles spaced evenly around the circuit, and it takes Raul 64 seconds to mow the path from the first pole to the third pole. At this rate, how long will it take Raul to mow once completely around the path?

New Concept

Consider the following ratio problem:

    To make green paint, the ratio of blue paint to yellow paint is 3 to 2. For 6 ounces of yellow paint, how much blue paint is needed?

We see two uses for numbers in ratio problems. One use is to express a ratio. The other use is to express an actual count. A ratio box can help us sort the two uses by placing the ratio numbers in one column and the actual counts in another column. We write the items being compared along the left side of the rows.

	Ratio	Actual Count
Blue Paint	3	
Yellow Paint	2	6

We are told that the ratio of blue to yellow was 3 to 2. We place these numbers in the ratio column, assigning 3 to the blue paint row and 2 to the yellow paint row. We are given an actual count of 6 ounces of yellow paint, which we record in the box. We are asked to find the actual count of blue paint, so that portion of the ratio box is empty.

Ratio numbers and actual counts are related by a **constant factor.** If we multiply the terms of a ratio by the constant factor, we can find the actual count. Recall that a ratio is a reduced form of an actual count. If we can determine the factor by which the actual count was reduced to form the ratio, then we can recreate the actual count.

	Ratio	Actual Count
Blue Paint	3 × constant factor	?
Yellow Paint	2 × constant factor	6

We see that 2 can be multiplied by 3 to get 6. So 3 is the constant factor in this problem. That means we can multiply each ratio term by 3. Multiplying the ratio term 3 by the factor 3 gives us an actual count of 9 ounces of blue paint.

Example 1

Sadly, the ratio of flowers to weeds in the garden is 2 to 5. There are 30 flowers in the garden, how many weeds are there?

We will begin by drawing a ratio box.

	Ratio	Actual Count
Flowers	2	30
Weeds	5	

To determine the constant factor, we study the row that has two numbers. In the "flowers" row we see the ratio number 2 and the actual count 30. If we divide 30 by 2, we find the factor, which is 15. Now we will use the factor to find the prediction of weeds in the garden.

$$\text{Ratio} \times \text{constant factor} = \text{actual count}$$
$$5 \quad \times \quad 15 \quad = \quad 75$$

There are **75 weeds** in the garden.

Example 2

Derek drove 55 miles per hour on the highway. If he drove at this rate for 4 hours, how far did he travel?

A rate in miles per hour is a ratio of distance to time. Fifty-five miles per hour can be written as the ratio 55 miles to 1 hour. We can draw a ratio box.

	Ratio	Actual Count
Distance	55 mi	
Time	1 hr	4 hr

The constant factor is 4.

$$\text{Ratio} \times \text{constant factor} = \text{actual count}$$
$$55 \quad \times \quad 4 \quad = \quad 220$$

Derek traveled **220 miles.**

Lesson Practice

Model Draw a ratio box and use a constant factor to solve each ratio problem:

a. The ratio of boys to girls in the cafeteria was 6 to 5. If there were 60 girls, how many boys were there?

b. The ratio of ants to flies at the picnic was 8 to 3. If there were 24 flies, predict how many ants were there?

Written Practice *Distributed and Integrated*

1. What is the mean of 96, 49, 68, and 75? What is the range?
(Inv. 4)

2. The average depth of the ocean beyond the edges of the continents is $2\frac{1}{2}$ miles. How many feet is that? (1 mile = 5280 ft)
(62)

* 3. **Formulate** The 168 girls who signed up for soccer were divided equally into 12 teams. How many players were on each team? Write an equation and solve the problem.
(8)

Analyze Parallelogram *ABCD* is divided into two congruent triangles. Segments *BA* and *CD* measure 3 in. Segments *AD* and *BC* measure 5 in. Segment *BD* measures 4 in. Refer to this figure to answer problems **4** and **5**.

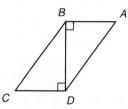

4. **a.** What is the perimeter of the parallelogram?
(66)
    b. What is the area of the parallelogram?

* 5. **a.** What is the perimeter of each triangle?
(77)
    b. What is the area of each triangle?

6. **Conclude** This quadrilateral has one pair of parallel sides. What is the name of
(59) this kind of quadrilateral?

7. **Verify** "All squares are rectangles." True or false?
(59)

* 8. **Represent** If four fifths of the 30 students in the class were present, then how
(75) many students were absent?

* 9. The ratio of dogs to cats in the neighborhood was 2 to 5. If there were 10 dogs,
(78) predict how many cats were there.

* 10. Write as a percent:
(73)
    a. $\dfrac{19}{20}$ **b.** 0.6

* 11. **Connect** **a.** What percent of the perimeter of a square is the length of one side?
(69, 73)
    b. What is the ratio of the side length of a square to its perimeter? Express the
    ratio as a fraction and as a decimal.

* 12. Compare:
(74)
    a. 0.5 $\bigcirc$ $\dfrac{3}{4}$ **b.** 3 qts $\bigcirc$ 1 gal

* 13. Write 4.4 as a reduced mixed number.
(68)

14. Write $\dfrac{1}{8}$ as a decimal number.
(69)

15. $\dfrac{5}{6} + \dfrac{1}{2}$ **16.** $\dfrac{5}{8} - \dfrac{1}{4}$ **17.** $2\dfrac{1}{2} \times 1\dfrac{1}{3} \times \dfrac{3}{5}$
(55)   (55)   (67)

Find each unknown number:

18. $4 - a = 2.6$ **19.** $3n = 1\dfrac{1}{2}$
(39)   (39, 64)

20. $5x = 0.36$ **21.** $0.9y = 63$
(39, 41)   (39, 44)

22. Round 0.4287 to the hundredths place.
(46)

Refer to the bar graph below to answer problems **23–25.**

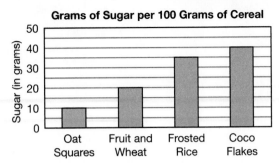

23. **Estimate** Frosted Rice contains about how many grams of sugar per 100
(Inv. 4) grams of cereal?

24. **Estimate** Fifty grams of CocoFlakes would contain about how many grams of
(Inv. 4) sugar?

*25. **Formulate** Write a problem about comparing that refers to the bar graph,
(Inv. 4) and then answer the problem.

26. The eyedropper held 2 mL of liquid.
(76)

    a. How many mL make a Liter?

    b. How many eyedroppers of liquid would it take to fill a 1-liter container?

27. Aaliyah rolls a number cube and gets a 5. If she rolls the number cube again, what
(70) is the probability that she will not roll another 5?

28. **Multiple Choice** **Conclude** Which of these angles could be the complement of
(60) a 30° angle?

29. If $A = \frac{1}{2}bh$, and if $b = 6$ and $h = 8$, then what does A equal?
(6)

30. **Model** Draw a pair of parallel segments that are the same length.
(59) Form a quadrilateral by drawing two segments between the endpoints of the
parallel segments. Is the quadrilateral a parallelogram?

✎ *California Mathematics Content Standards*

AF 2.0, 2.1 Convert one unit of measurement to another (e.g., from feet to miles, from centimeters to inches).

AF 2.0, 2.2 Demonstrate an understanding that *rate* is a measure of one quantity per unit value of another quantity.

• Arithmetic with Units of Measure

Power Up

facts	Power Up M
mental math	**a. Number Sense:** $1000 - 420$
	b. Money: Each greeting card costs $2.99. What is the total cost of 4 cards?
	c. Number Sense: Double $24.
	d. Decimals: 0.125×100
	e. Measurement: How many liters are in a kiloliter?
	f. Geometry: The height of this triangle is 1 cm. What is the area of the triangle?
	g. Statistics: At three different stores, the picture frame was priced at $7.49, $8.32, and $9.99. What was the range of prices?
	h. Calculation: $2 \times 2, \times 2, \times 2, -1, \times 2, +2, \div 2, \div 2$

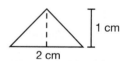
1 cm
2 cm

problem solving	Choose an appropriate problem-solving strategy to solve this problem. The restaurant serves four different soups and three different salads. How many different soup-and-salad combinations can diners order? Draw a diagram to support your answer.

New Concept

Recall that the operations of arithmetic are addition, subtraction, multiplication, and division. In this lesson we will practice adding, subtracting, multiplying, and dividing units of measure.

We may add or subtract measurements that have the same units. If the units are not the same, we first convert one or more measurements so that the units are the same. Then we add or subtract.

Example 1

Add: 2 ft + 12 in.

The units are not the same. Before we add, we either convert 2 feet to 24 inches or we convert 12 inches to 1 foot.

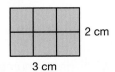

Convert to Inches	Convert to Feet
2 ft + 12 in.	2 ft + 12 in.
24 in. + 12 in. = **36 in.**	2 ft + 1 ft = **3 ft**

Either answer is correct, because 3 feet equals 36 inches.

Notice that in each equation in Example 1, the units of the sum are the same as the units of the addends. The units do not change when we add or subtract measurements. However, the units *do* change when we multiply or divide measurements.

When we find the area of a rectangle, we multiply the lengths. Notice how the units change when we multiply.

2 cm

3 cm

To find the area of this rectangle, we multiply 2 cm by 3 cm. The product has a different unit of measure than the factors.

$$2 \text{ cm} \cdot 3 \text{ cm} = 6 \text{ sq. cm}$$

A centimeter and a square centimeter are two different kinds of units. A centimeter is used to measure length. It can be represented by a line segment.

─────────

1 cm

A square centimeter is used to measure area. It can be represented by a square that is 1 centimeter on each side.

1 sq. cm

The unit of the product is a different unit because we multiplied the units of the factors. When we multiply 2 cm by 3 cm, we multiply both the numbers and the units.

$$2 \text{ cm} \cdot 3 \text{ cm} = \underbrace{2 \cdot 3}_{6} \ \underbrace{\text{cm} \cdot \text{cm}}_{\text{sq. cm}}$$

Instead of writing "sq. cm," we may use exponents to write "cm $\cdot$ cm" as "cm^2." Recall that we read cm^2 as "square centimeters."

$$2 \text{ cm} \cdot 3 \text{ cm} = \underbrace{2 \cdot 3}_{6} \ \underbrace{\text{cm} \cdot \text{cm}}_{\text{cm}^2}$$

Example 2

Multiply: 6 ft · 4 ft

We multiply the numbers. We also multiply the units.

$$6 \text{ ft} \cdot 4 \text{ ft} = \underbrace{6 \cdot 4}_{24} \; \underbrace{\text{ft} \cdot \text{ft}}_{\text{ft}^2}$$

The product is **24 ft²,** which can also be written as "24 sq. ft."

Units also change when we divide measurements. For example, if we know both the area and the length of a rectangle, we can find the width of the rectangle by dividing.

Area = 21 cm²

7 cm

To find the width of this rectangle, we divide 21 cm² by 7 cm.

$$\frac{21 \text{ cm}^2}{7 \text{ cm}} = \frac{\overset{3}{\cancel{21}} }{\underset{1}{\cancel{7}}} \frac{\cancel{\text{cm}} \cdot \text{cm}}{\cancel{\text{cm}}}$$

We divide the numbers and write "cm²" as "cm × cm" in order to reduce the units. The quotient is 3 cm, which is the width of the rectangle.

Example 3

Divide: $\dfrac{25 \text{ mi}^2}{5 \text{ mi}}$

To divide the units, we write "mi²" as "mi · mi" and reduce.

$$\frac{\overset{5}{\cancel{25}}}{\underset{1}{\cancel{5}}} \frac{\cancel{\text{mi}} \cdot \text{mi}}{\cancel{\text{mi}}}$$

The quotient is **5 mi.**

Sometimes when we divide measurements, the units will not reduce. When units will not reduce, we leave the units in fraction form. For example, if a car travels 300 miles in 6 hours, we can find the average speed of the car by dividing.

$$\frac{300 \text{ mi}}{6 \text{ hr}} = \frac{\overset{50}{\cancel{300}}}{\underset{1}{\cancel{6}}} \frac{\text{mi}}{\text{hr}}$$

The quotient is $50\frac{\text{mi}}{\text{hr}}$, which is 50 miles per hour (50 mph).

Notice that speed is a quotient of distance divided by time.

Speed is a rate because it is a measure of distance per unit value of time.

Reading Math

The word *per* means "for each" and is used in place of the division bar.

Example 4

Divide: $\dfrac{300 \text{ mi}}{10 \text{ gal}}$

We divide the numbers. The units do not reduce.

$$\frac{300 \text{ mi}}{10 \text{ gal}} = \frac{\overset{30}{\cancel{300}} \text{ mi}}{\underset{1}{\cancel{10}} \text{ gal}}$$

The quotient is $30\,\frac{\text{mi}}{\text{gal}}$, which is 30 miles per gallon. This is a rate because it is a measure of distance per unit value of capacity.

Lesson Practice   Simplify:

    a. 2 ft − 12 in. (Write the difference in inches.)

    b. 2 ft · 4 ft   **c.** $\dfrac{12 \text{ cm}^2}{3 \text{ cm}}$ **d.** $\dfrac{300 \text{ mi}}{5 \text{ hr}}$

    e. The train traveled 300 miles in 4 hours. What is the rate of travel for the train?

Written Practice   *Distributed and Integrated*

---

* 1.
(77)
  a. Find the perimeter of the triangle.

    b. Find the area of the triangle.

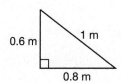

* 2. **Connect** One quart of milk is about 945 milliliters of milk. Use this information
(76) to compare:

$$1 \text{ gallon} \bigcirc 4 \text{ liters}$$

3. **Analyze** Carol cut $2\frac{1}{2}$ inches off her hair three times last year. How much
(62) longer would her hair have been at the end of the year if she had not cut it?

* 4. The plane flew 1200 miles in 3 hours. Divide the distance by the time to find the
(79) average speed of the plane.

5. Write the prime factorization of both the numerator and the denominator
(63) of this fraction. Then simplify the fraction.

$$\frac{54}{135}$$

6. The basketball team scored 60% of its 80 points in the second half.
(24, 27) Write 60% as a reduced fraction. Then find the number of points the
team scored in the second half.

7. What is the area of this parallelogram?
(66)

8. What is the perimeter of this parallelogram?
(66)

9. 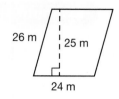 **Verify** "Some rectangles are trapezoids." True or false?
(59) Why?

Figure: parallelogram with 26 m slant side, 25 m height, 24 m base.

* 10. **Predict** The ratio of red marbles to blue marbles in the
(78) bag was 3 to 4. If 24 marbles were blue, how many were red?

* 11. Arrange these numbers in order from least to greatest:
(74)

$$\frac{1}{2}, \frac{1}{5}, 0.4$$

* 12. a. What decimal number is equivalent to $\frac{4}{25}$?
(69, 73)

b. What percent is equivalent to $\frac{4}{25}$?

13. $(10 - 0.1) \times 0.1$
(41)

14. $(0.4 + 3) \div 2$
(41)

15. $\frac{5}{8} + \frac{3}{4}$
(55)

16. $3 - 1\frac{1}{8}$
(58)

17. $4\frac{1}{2} - 1\frac{3}{4}$
(58)

18. $\frac{5}{6} \cdot \frac{4}{5} \cdot \frac{3}{8}$
(67)

19. $4\frac{1}{2} \times 1\frac{1}{3}$
(62)

20. $3\frac{1}{3} \div 1\frac{2}{3}$
(64)

Analyze The perimeter of this square is two meters. Refer to this figure to answer problems **21** and **22.**

21. How many centimeters long is each side of the square
(32, 76) (1 meter = 100 centimeters)?

* 22. a. What is the diameter of the circle in centimeters?
(42)

b. What is the circumference of the circle in centimeters? (Use 3.14 for π.)

23. If the sales-tax rate is 6%, what is the tax on a $12.80 purchase?
(35)

24. What time is two-and-one-half hours after 10:40 a.m.?
(RF11, 7)

25. Use a ruler to find the length of this line segment to the nearest sixteenth of an
(14) inch.

_____

26. **Connect** What is the area of a quadrilateral with the vertices (0, 0),
(RF27, 66) (4, 0), (6, 3), and (2, 3)?

27. June spins a spinner with 8 equal sections numbered 1–8. Find the probability
(70) that she spins a number less than 1.

28. If the area of a square is one square foot, what is the perimeter?
(32)

* 29. Simplify:
(79)

    a. 2 yd + 3 ft   **b.** 5 m · 3 m

    (Write the sum in yards.)

    c. $\dfrac{36 \text{ ft}^2}{6 \text{ ft}}$ **d.** $\dfrac{400 \text{ miles}}{20 \text{ gallons}}$

* 30. (**Model**) Draw a pair of parallel segments that are not the same length. Form
(59) a quadrilateral by drawing two segments between the endpoints of the parallel
segments. What is the name of this type of quadrilateral?

California Mathematics Content Standards

SDAP 3.0, 3.4 Understand that the probability of either of two disjoint events occurring is the sum of the two individual probabilities and that the probability of one event following another, in independent trials, is the product of the two probabilities.

SDAP 3.0, 3.5 Understand the difference between independent and dependent events.

• Independent and Dependent Events

Power Up

facts   Power Up J

mental math

   a. Number Sense: 4×315

   b. Money: The package of markers cost $2.99. What would be the cost of 5 packages?

   c. Money: The puzzle book cost $7.99. Aditi paid for it with a $10 bill. How much change should she receive?

   d. Fractions: $\frac{1}{4}$ of $4.80

   e. Decimals: $37.5 \div 100$

   f. Probability: Raul placed these pieces of paper in a hat. If he draws one number without looking, what is the probability it will be a prime number?

4	5
6	7

   g. Algebra: If $20 + b = 60$, what is b?

   h. Calculation: $5 \times 5, \times 5, - 25, \div 4, \div 5, -5$

problem solving

Choose an appropriate problem-solving strategy to solve this problem. A seven digit phone number consists of a three-digit prefix followed by four digits. How many different phone numbers are possible for a particular prefix?

New Concept

Events are **independent events** when the probability of an event occurring is not affected by other events. For example, if a coin is tossed several times, the likelihood of each toss is constant—it does not depend on the result of the previous toss.

Thinking Skills

Explain

In your own words explain the difference between independent and dependent events?

Events are **dependent events** if the probability that one event occurs is influenced by the occurrence of another event. For example, if a marble is drawn from a bag of mixed marbles and is not replaced, then the probability for the next draw differs from the probability of the first draw because the number and mix of marbles remaining in the bag has changed.

Independent and dependent events are examples of **compound events**. A compound event is composed of two or more simple events. For example, "getting heads twice" with two flips of a coin is a compound event.

To find the probability of a compound event we multiply the probabilities of each part of the event. If events are dependent, we must take into consideration that the probabilities change.

The probability of dependent events occurring in a specific order is a product of the first event and the recalculated probabilities of each subsequent event.

Example 1

Math Language

Sometimes dependent events are referred to as "not independent events".

Two red marbles, three white marbles, and four blue marbles are in a bag. If one marble is drawn and not replaced, and a second marble is drawn, what is the probability that both marbles will be red?

Two of the nine marbles are red, so the probability of red on the first draw is $\frac{2}{9}$.

$$\text{1st Draw: probability of "red"} = \frac{2}{9}$$

If a red marble is removed from the bag on the first draw, then only one of the remaining eight marbles is red. Thus, the probability of red on the second draw is $\frac{1}{8}$.

$$\text{2nd Draw: probability of "red"} = \frac{1}{8}$$

To find the probability of red on both the first and second draw, we multiply the probabilities of the two events.

$$\text{probability of "red" and "red"} = \frac{2}{9} \times \frac{1}{8} = \frac{1}{36}$$

Explain Why is the product of $\frac{2}{9} \times \frac{1}{8}$ shown as $\frac{1}{36}$?

Example 2

Two cards are drawn from a regular shuffled deck of 52 cards without replacement. What is the probability of drawing two aces?

Although two cards are drawn simultaneously we calculate the probability of each card separately and in sequence. Four of the 52 cards are aces, so the probability of the first card being an ace is $\frac{4}{52}$ which reduces to $\frac{1}{13}$.

$$\text{First card: probability of "Ace"} \ \frac{4}{52} = \frac{1}{13}$$

Since the first card must be an ace in order for both cases to be aces, only three of the remaining 51 cards are aces. Thus the probability of the second card being an ace is $\frac{3}{51}$ which reduces to $\frac{1}{17}$.

$$\text{Second card: probability of "Ace"} \quad \frac{3}{51} = \frac{1}{17}$$

The probability that both cards drawn are aces is the product of the probabilities of each event.

$$\text{probability of "Ace" and "Ace"} = \frac{1}{13} \times \frac{1}{17} = \frac{1}{221}$$

Example 3

Two cards are drawn from a regular shuffled deck of 52 cards. After the first card is drawn, it is replaced, the deck is reshuffled, and the second card is drawn. What is the probability of drawing two aces?

In each trial, the probability of drawing an ace is $\frac{4}{52}$, which is $\frac{1}{13}$.

$$\text{probability of "Ace" and "Ace"} = \frac{1}{13} \times \frac{1}{13}$$

$$= \frac{1}{169}$$

Explain Which probability is greater, when the cards are drawn with replacement or without replacement? Explain why your answer is reasonable.

Explain In the experiment from Example 2, in which two cards are drawn without replacement, are the events independent or dependent? In the experiment from Example 3, in which two cards are drawn with replacement, are the events independent or dependent? Explain your answer.

Lesson Practice

Use the following information to answer problems **a** and **b**.

Analyze A bag contains two red marbles, three white marbles, and four blue marbles.

a. If one marble is drawn from the bag and not replaced and then a second marble is drawn from the bag, what is the probability of drawing two blue marbles?

b. If one marble is drawn from the bag, then replaced, and a second marble is drawn, what is the probability of drawing two blue marbles?

c. In which experiment (a or b) are the events independent?

d. If two cards are drawn from a regular shuffled deck of 52 cards without replacement, what is the probability that both cards will be diamonds?

1. Sam placed the five alphabet cards shown to the right face
(80) down on the table. Nadia will turn over any two cards.

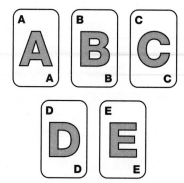

   a. What is the probability that Nadia will turn over
   two vowels?

   b. What is the probability that Nadia will turn over
   two consonants?

2. Tennis balls are sold in cans containing 3 balls. What would be
(8) the total cost of one dozen tennis balls if the price per can was
$2.49?

* 3. (**Analyze**) A cubit is about 18 inches. If Ruben was 4 cubits tall, about how many
(8) feet tall was he?

* 4. **a.** Write $\frac{7}{100}$ as a percent.
(73)
   b. Write $\frac{7}{10}$ as a percent.

5. Write 90% as a reduced fraction. Then write the fraction as a decimal number.
(27, 69)

* 6. Of the 50 students who went on a trip, 23 wore a hat. What percent of the
(73) students wore a hat?

7. Write $\frac{9}{25}$ as a percent.
(73)

8. (**Connect**) A box of cereal has the shape of what geometric solid?
(RF25)

Find each unknown number:

9. $w - 3\frac{5}{6} = 2\frac{1}{3}$
(39, 55)

10. $3\frac{1}{4} - y = 1\frac{5}{8}$
(39, 58)

11. $6n = 0.12$
(39, 41)

12. $0.12m = 6$
(39, 44)

13. $5n = 10^2$
(5, 32)

14. $1\frac{1}{2}w = 6$
(39, 64)

* 15. **a.** What fraction of this group is shaded?
(73)
   b. What percent of this group is shaded?

16. $0.5 + (0.5 \div 0.5) + (0.5 \times 0.5)$
(41)

17. $\frac{1}{2} + \frac{1}{5} + \frac{1}{10}$
(56)

18. $1\frac{4}{5} \times 1\frac{2}{3}$
(62)

19. Which digit in 6.3457 has the same place value as the 8 in 128.90?
(28)

20. Estimate the product of 39 and 41.
(21)

21. In a bag there are 12 red marbles and 36 blue marbles.
(16, 54)

    a. What is the ratio of red marbles to blue marbles?

    b. (**Predict**) If one marble is taken from the bag, what is the probability that the marble will be red? Express the probability ratio as a fraction and as a decimal.

* 22. (**Analyze**) What is the area of this parallelogram?
(66)

23. What is the perimeter of this parallelogram?
(66)

24. Write the prime factorization of 252 using exponents.
(71)

25. Sakura selects a letter at random from the word *fall*. Find the probability that she does *not* choose "a."
(70)

* 26. (**Model**) A quadrilateral has vertices with the coordinates (−2, −1), (1, −1), (3, 3), and (−3, 3). Graph the quadrilateral on a coordinate plane. The figure is what type of quadrilateral?
(RF27, 59)

* 27. Two liters of water has a mass of 2 kg. How many grams is that?
(76)

* 28. Simplify:
(79)

    a. $\dfrac{49 \text{ m}^2}{7 \text{ m}}$ **b.** $\dfrac{400 \text{ miles}}{8 \text{ hours}}$

* 29. Three of the dozen eggs were cracked. What percent of the eggs were cracked?
(73)

30. Find the area of the triangle below.
(77)

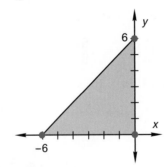

California Mathematics Content Standards

SDAP 3.0, 3.4 Understand that the probability of either of two disjoint events occurring is the sum of the two individual probabilities and that the probability of one event following another, in independent trials, is the product of the two probabilities.

MR 2.0, 2.4 Use a variety of methods, such as words, numbers, symbols, charts, graphs, tables, diagrams, and models, to explain mathematical reasoning.

• Thinking About the Probability of More than One Outcome

Probability refers to how great a chance an event will occur. Probability ranges from a 0% chance all the way up to a 100% chance.

Remember that to find the probability of an event, we divide the number of favorable outcomes for the event by the total number of outcomes. Finding the probability of more than one outcome at the same time involves knowing whether you need to add or multiply the probabilities.

> For these examples and the problems that follow, we will work with a coin and a bag containing 4 yellow marbles, 3 green marbles, and 5 blue marbles.

Example 1

What is the probability that a yellow or a blue marble will be chosen when one marble is removed?

In this case, either outcome, yellow or blue, satisfies the event, so the likelihood of the event is greater than either outcome by itself. When the likelihood of an event increases, then the probabilities are ADDED.

$$P(\text{yellow}) + P(\text{blue}) = \frac{4}{12} + \frac{5}{12} = \frac{9}{12} \text{ or } 75\%$$

Explain Why would it not make sense to multiply these probabilities?

Example 2

What is the probability that a green marble is chosen and heads is flipped on a coin?

In this case, both outcomes, a green marble and heads, are needed to satisfy the event, so the likelihood of the event decreases because the chance of both of these outcomes happening at the same time is smaller than either one by itself. When the likelihood of an event decreases, then the probabilities need to be MULTIPLIED.

$$P(\text{green}) \times P(\text{heads}) = \frac{3}{12} \times \frac{1}{2} = \frac{3}{24} \text{ or } 12.5\%$$

Explain Why would it not make sense to add these probabilities?

Apply For the following situations, determine whether addition or multiplication is needed to find the probability. Then find the probability.

   a. What is the probability that a yellow or a green marble will be chosen in one draw?

   b. What is the probability that a blue marble will be chosen in one draw and heads will be tossed in one flip?

   c. What is the probability that tails will be tossed in one flip and a blue or green marble will be chosen after one draw?

California Mathematics Content Standards

SDAP 3.0, 3.1 Represent all possible outcomes for compound events in an organized way (e.g., tables, grids, tree diagrams) and express the theoretical probability of each outcome.

SDAP 3.0, 3.2 Use data to estimate the probability of future events (e.g., batting averages or number of accidents per mile driven).

SDAP 3.0, 3.3 Represent probabilities as ratios, proportions, decimals between 0 and 1, and percentages between 0 and 100 and verify that the probabilities computed are reasonable; know that if P is the probability of an event, 1-P is the probability of an event not occurring.

SDAP 3.0, 3.5 Understand the difference between independent and dependent events.

Focus on
Probability and Predicting

In this investigation, we will look at the relationship between real events and probability. We will use data from real events to **estimate** probabilities, and we will use probabilities to predict events.

In Investigation **7,** we collected and used data to estimate probabilities. This is commonly done in everyday life.

1. In baseball, batting average is the ratio of hits to at bats, and it is expressed as a decimal to three places. (However, when we say a batting average, we ignore the decimal point.) In the first half of the season, Fernando had 8 hits and 25 at bats. Estimate the probability that Fernando has a hit at his next at bat.

2. Researchers concerned with traffic safety compute accidents per mile ratios for stretches of highway. This is the ratio of accidents in a year on a portion of highway to the number of miles on that stretch. In one year, on a 5 mile stretch of highway, there were 226 accidents. Determine the accidents per mile.

3. A truckload of DVDs was inspected for quality. A selection of 50 were inspected, and 8 DVDs were found to be damaged. Estimate the probability that a DVD selected from the truckload is damaged. Write the probability as a fraction and as a proportion using words.

Explain Is this the true proportion of damaged DVDs in the truckload? Explain.

In this activity, you will act out a scenario similar to problem 3:

An inspector will check a crate of apples, some of which have worms. The inspector will estimate the probability that a customer, picking an apple from the crate, chooses an apple with a worm.

4. With a partner, decide who will be the inspector and who will create the crate of apple. The crate-maker will follow these instructions:

Cut a piece of paper into 64 rectangles. You can do this by folding it in half lengthwise two times and then folding it in half widthwise two times. This gives you a guide for cutting 16 rectangles. By cutting each of these into 4 rectangles, you will have a total of 64. Each rectangle is an apple.

The crate-maker will then select a certain number (10 to 40) of apples to have worms. Mark the rectangles to indicate that they have worms; then turn them over to hide the mark so that the apples with worms and without worms are indistinguishable. Secretly write down the number of apples with worms, and compute the probability that a customer, selecting an apple from the crate, picks an apple with a worm.

5. The inspector will now inspect the crate. Be sure the rectangles are well mixed. The inspector will select 20 apples (choose 20 rectangles), cut open the apples (turn the rectangles over), and tally the number of apples with worms. What proportion of apples selected had worms? Estimate the probability that a customer, selecting an apple from the crate, chooses an apple with a worm?

6. Compare the theoretical probability computed in problem **4** to the estimated probability from problem **5**. If they are different, explain why. In some cases, why do we need to rely on estimated probabilities?

Just as what has happened can give us an idea about probabilities, probabilities can also give us an idea of what may happen.

When we flip a coin, the probability it lands on heads is $\frac{1}{2}$. We *expect* the coin to land on heads about $\frac{1}{2}$ of the time. If we flip a coin 100 times, we expect it to land on heads about 50 times.

$$\frac{1}{2} \text{ of } 100 = \frac{1}{2} \times 100 = 50$$

7. If this spinner is spun once, what is the probability it stops on an even number? If the spinner is spun 30 times, predict the number of times it stops on an even number.

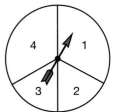

8. Asia knew that for every 1 adult theater ticket that was sold, 3 children's tickets were sold. In the balcony, Asia counted 20 adults. About how many children were in the balcony?

9. Two number cubes are rolled. Complete the grid to list all possible outcomes.

Second Number Cube

	1	2	3	4	5	6
1	1,1	1,2	1,3	1,4	1,5	1,6
2	2,1	2,2				
3						
4						
5						
6						

First Number Cube

What is the probability of rolling "1,1"? Are the events independent or dependent?

What is the probability of rolling doubles ("1,1" or "2,2" or "3,3" and so on)?

10. Cheryl rolls two number cubes 30 times. Predict the number of times she rolls doubles.

California Mathematics Content Standards
AF 2.0, 2.1 Convert one unit of measurement to
another (e.g., from feet to miles, from centimeters
to inches).
MR 2.0, 2.1 Use estimation to verify the
reasonableness of calculated results.
MR 2.0, 2.3 Estimate unknown quantities graphically
and solve for them by using logical reasoning and
arithmetic and algebraic techniques.
MR 2.0, 2.4 Use a variety of methods, such as
words, numbers, symbols, charts, graphs, tables,
diagrams, and models, to explain mathematical
reasoning.

• U.S. Customary System

Power Up

facts   Power Up M

**mental
math**

a. Number Sense: $(1 \times 100{,}000) + (8 \times 10{,}000) + (6 \times 1000)$

b. Number Sense: 311×7

c. Number Sense: $2000 - 1250$

d. Money: Each T-shirt costs $9.99. What is the cost for 4 T-shirts?

e. Decimals: 0.075×100

f. Estimation: The perimeter of this rectangle is 20 cm.
Estimate the width of the rectangle to the nearest
centimeter.

5.95 cm

g. Measurement: How many milliliters are in
3 liters?

h. Calculation: 8×8, $+ 6$, $\div 2$, $+1$, $\div 6$, $\times 3$, $\div 2$

**problem
solving**

Choose an appropriate problem-solving strategy to solve this problem.
Alexis has 6 coins that total exactly $1.00. Name one coin she must have
and one coin she cannot have.

New Concept

In this lesson we will consider units of the **U.S. Customary System.** We
can measure an object's dimensions, weight, volume, or temperature.
Each type of measurement has a set of units. We should remember
common equivalent measures and have a "feel" for the units so that we
can estimate measurements reasonably.

The following table shows the common weight equivalences in the U.S.
Customary System:

Units of Weight	
16 ounces (oz)	= 1 pound (lb)
2000 pounds	= 1 ton (tn)

Example 1

Suppose a pickup truck can carry a load of $\frac{1}{2}$ of a ton. How many pounds can the pickup truck carry?

One ton is 2000 pounds, so $\frac{1}{2}$ of a ton is **1000 pounds.**

The following table shows the common length equivalences in the U.S. Customary System:

Units of Length		
12 inches (in.)	=	1 foot (ft)
3 feet	=	1 yard (yd)
1760 yards	=	1 mile (mi)
5280 feet	=	1 mile

Example 2

One yard is equal to how many inches?

One yard equals 3 feet. One foot equals 12 inches. Thus 1 yard is equal to 36 inches.

$$1 \text{ yard} = 3 \times 12 \text{ inches} = \textbf{36 inches}$$

Example 3

A mountain bicycle is about how many feet long?

We should develop a feel for various units of measure. Most mountain bicycles are about $5\frac{1}{2}$ feet long, so a good estimate would be **about 5 or 6 feet.**

Verify Without measuring a mountain bicycle, how could we determine that this estimate is reasonable?

Just as an inch ruler is divided successively in half, so units of liquid measure are divided successively in half. Half of a gallon is a half gallon. Half of a half gallon is a quart. Half of a quart is a pint. Half of a pint is a cup.

| 1 gallon | $\frac{1}{2}$ gallon | 1 quart | 1 pint | 1 cup |

Common container sizes based on the U.S. Customary System are illustrated above. These containers are named by their **capacity**, that is, by the amount of liquid they can contain.

The following chart shows some equivalent liquid measures in the U.S. Customary System.

Units of Liquid Measure		
8 ounces (oz)	=	1 cup (c)
2 cups	=	1 pint (pt)
2 pints	=	1 quart (qt)
4 quarts	=	1 gallon (gal)

Example 4

Below is a function table showing the number of cups in a given number of pints.

Pints	Cups
1	2
2	4
3	6

Steve drinks at least 4 pints of water every day. Determine the number of cups he drinks in a day in the following two ways:

a. Graph the order pairs in the function table shown above and estimate the number of cups of water Steve drinks in a day.

b. Write a rule for the function and use it to determine the number of cups of water Steve drinks in a day.

a.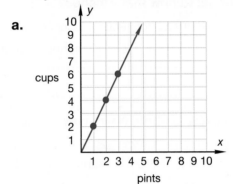

The next ordered pair on the ray would be (4,8) so Steve drinks **8 cups** of water in a day.

b. The rule to find the number of cups is to **multiply the number of pints by 2.** We will take 4 pints and multiply by 2 to find that Steve drinks **8 cups** of water a day.

The following diagram of a **Fahrenheit scale** shows important benchmark temperatures in the U.S. Customary System.

Fahrenheit Temperature Scale

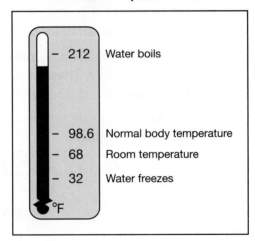

Example 5

How many Fahrenheit degrees are between the freezing and boiling temperatures of water?

$212°F - 32°F = \mathbf{180°F}$

A **function** is a mathematical rule that identifies the relationship between two sets of numbers. The rule uses an input number to generate an output number. For each input number, there is one and only one output number.

The rule for a function may be described with words or in an equation, and the relationship between the sets of numbers may be illustrated in tables or graphs.

In this lesson we will use function tables to help us solve problems. We will also study function tables to discover the rules of functions.

Example 6

Thinking Skills

Connect

How is the rule for a function similar to a rule for a number pattern?

This table shows the weight in ounces for a given weight in pounds.

a. Describe the rule of this function.

b. Mattie weighed 7 pounds when she was born. Use the function rule to find how many ounces Mattie weighed when she was born.

INPUT Pounds	OUTPUT Ounces
1	16
2	32
3	48
4	64
5	80

a. To find the number of ounces (the output), **multiply the number of pounds (the input) by 16.**

b. To find Mattie's birth weight in ounces, we multiply 7 (her birth weight in pounds) by 16. Mattie weighed **112 ounces** at birth.

Example 7

Hana bought four yards of fabric to make a costume for a school play. Make a function table that shows the number of feet (output) for a given number of yards (input). Use the function table to find the number of feet of fabric Hana bought.

The number of feet Hana bought depends on the numbers of yards. We can say that the number of feet is a *function* of the number of yards. To solve the problem, we can set up a table to record input and output numbers for this function. Each yard equals three feet. So the input is yards and the output is feet.

INPUT Yards	OUTPUT Feet
1	3
2	6
3	9
4	12

In the function table we see 12 feet paired with 4 yards. Hana bought **12 feet** of fabric.

Conclude What is the rule for this function?

Lesson Practice

a. A typical door may be about how many feet tall?

b. How many quarts are in a half-gallon?

c. (**Estimate**) When Alberto was born, he weighed 8 lb. 7 oz. Is that weight closer to 8 lb or 9 lb?

d. How many ounces are in a 2-cup measure?

e. Both pots are filled with water. What is the temperature difference, in degrees Fahrenheit, between the two pots of water?

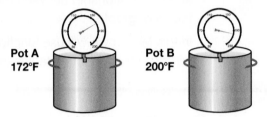

Pot A
172°F

Pot B
200°F

Simplify.

f. $\frac{3}{8}$ in. + $\frac{5}{8}$ in.

g. 32°F + 180°F

h. 2(3 ft + 4 ft)

i. 1 ton − 1000 pounds

j. (**Analyze**) A sheet of plywood is 4 feet wide. Copy and complete the function table to determine the width, in inches, of a sheet of plywood.

INPUT Feet	OUTPUT Inches
1	12
2	
3	
4	

How wide is a sheet of plywood?

What is the rule for this function?

Written Practice — *Distributed and Integrated*

1. What is the product when the sum of 0.2 and 0.2 is multiplied by the difference of 0.2 and 0.2?
(4, 41)

2. (**Analyze**) Arabian camels travel about 3 times as fast as Bactrian camels. If Bactrian camels travel at $1\frac{1}{2}$ miles per hour, at how many miles per hour do Arabian camels travel?
(62)

3. (**Connect**) Mark was paid at a rate of $4 per hour for cleaning up a neighbor's yard. If he worked from 1:45 p.m. to 4:45 p.m., how much was he paid?
(RF11, 7)

4. Write 55% as a reduced fraction.
(27)

5. **a.** Write $\frac{9}{100}$ as a percent.
(73)

    b. Write $\frac{9}{10}$ as a percent.

6. The whole class was present. What percent of the class was present?
(73)

* 7. **Connect** A century is 100 years. A decade is 10 years.
(25, 73)

    a. What fraction of a century is a decade?

    b. What percent of a century is a decade?

8. **a.** Write 0.48 as a reduced fraction.
(68, 73)

    b. Write 0.48 as a percent.

9. Write $\frac{7}{8}$ as a decimal number.
(69)

10. $\left(1\frac{1}{3} + 1\frac{1}{6}\right) - 1\frac{2}{3}$
(58)

11. $1\frac{1}{2} \times 3 \times 1\frac{1}{9}$
(67)

12. $4\frac{2}{3} \div 1\frac{1}{6}$
(64)

13. $0.1 + (1 - 0.01)$
(41)

* 14. A pint of milk weighs about one pound. About how many pounds does a gallon of milk weigh?
(81)

15. Draw a ratio box to solve: Two workers painted 5 rooms in a day. If 10 workers painted, how many rooms could be painted?
(78)

16. Write the standard numeral for the following:
(RF18)

$$(8 \times 10,000) + (4 \times 100) + (2 \times 10)$$

17. **a.** Compare: $2^4 \bigcirc 4^2$ **b.** 1 km $\bigcirc$ 1 mi
(71, 76)

18. Write the prime factorization of both the numerator and the denominator of this fraction. Then simplify the fraction.
(63)

$$\frac{24}{32}$$

* 19. Miyau scored 21 goals in 10 soccer games. Find the number of goals per game. In the next three games, about how many goals might Miyau make?
(Inv. 8)

20. **Multiple Choice** **Estimate** If the diameter of a ceiling fan is 4 ft, then the tip of one of the blades on the fan moves about how far during one full turn? Choose the closest answer.
(42)

    A 8 ft   **B** 12 ft   **C** $12\frac{1}{2}$ ft   **D** 13 ft

21. What is the perimeter of this trapezoid?
(RF7)

22. **a.** Estimate the length of $\overline{AB}$ in centimeters.
(76)

    b. Use a centimeter scale to find the length of $\overline{AB}$ to the nearest centimeter.

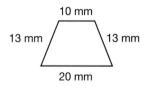

* 23. **a.** What is the area of this parallelogram?
(66, 77)

b. What is the area of the shaded triangle?

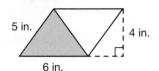

5 in.   6 in.   4 in.

* 24. One fourth of the 120 students took wood shop. How many students
(75) did not take wood shop?

25. How many millimeters is 2.5 centimeters?
(76)

* 26. **a.** What is the name of this
(RF25) geometric solid?

b. Sketch a net of this solid.

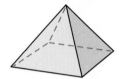

* 27. Simplify:
(79)
    a. 3 quarts + 2 pints (Write the sum in pints.)

    b. $\dfrac{64 \text{ cm}^2}{8 \text{ cm}}$ **c.** $\dfrac{60 \text{ students}}{3 \text{ teachers}}$

* 28. DeShawn delivers newspapers to 20 of the 25 houses on North Street. What
(73) percent of the houses on North Street does DeShawn deliver papers to?

29. **Represent** Draw a triangle that has two perpendicular sides.
(RF24,
20)

30. Gloria had two sets of alphabet cards (A–Z). She mixed the two sets together to
(80) form a single stack of cards. Then Gloria drew 3 cards from
the stack without replacing them. What is the probability that she drew
either a card with an L or a card with an R all 3 times?

Early Finishers
Real-World
Connection

Use a protractor to measure each angle listed. Then classify each angle
as acute, right, obtuse, or straight.

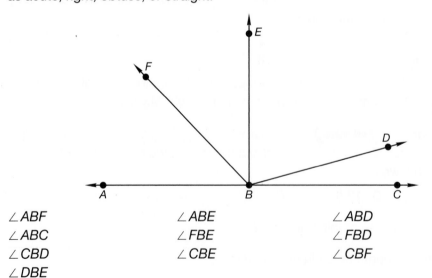

∠ABF	∠ABE	∠ABD
∠ABC	∠FBE	∠FBD
∠CBD	∠CBE	∠CBF
∠DBE		

California Mathematics Content Standards

NS **1.0**, **1.3** Use proportions to solve problems (e.g., determine the value of N if $\frac{4}{7} = \frac{N}{21}$, find the length of a side of a polygon similar to a known polygon). Use cross-multiplication as a method for solving such problems, understanding it as the multiplication of both sides of an equation by a multiplicative inverse.

MR 3.0, 3.3 Develop generalizations of the results obtained and the strategies used and apply them in new problem situations.

• Proportions

Power Up

facts　　　　　Power Up N

mental math

   a. **Probability:** What is the probability of rolling an even number with one roll of a number cube?

   b. **Money:** $1.99 + $2.99

   c. **Fractions:** $\frac{1}{3}$ of $7.50

   d. **Decimals:** 2.5×10

   e. **Number Sense:** $\frac{800}{40}$

   f. **Measurement:** How many ounces are in a cup?

   g. **Algebra:** If $n = 0$, what is $14 + n$?

   h. **Calculation:** 10×10, -1, $\div 3$, -1, $\div 4$, $+1$, $\div 3$

problem solving

Choose an appropriate problem-solving strategy to solve this problem. One-fifth of Ronnie's number is $\frac{1}{3}$. What is $\frac{3}{5}$ of Ronnie's number?

New Concept

If peaches are on sale for 3 pounds for 4 dollars then the ratio $\frac{3}{4}$ expresses the relationship between the quantity and the price of peaches. Since the ratio is constant, we can buy 6 pounds for 8 dollars, 9 pounds for 12 dollars and so on. With two equal ratios we can write a proportion.

PEACHES
3 lbs.
for $4

A **proportion** is a true statement that two ratios are equal. Here is an example of a proportion:

$$\frac{3}{4} = \frac{6}{8}$$

We read this proportion as "Three is to four as six is to eight." Two ratios that are not equivalent are not proportional.

Math Language

A **ratio** is a comparison of two numbers by division.

Example 1

Which ratio forms a proportion with $\frac{2}{3}$?

A $\frac{2}{4}$ B $\frac{3}{4}$ C $\frac{4}{6}$ D $\frac{3}{2}$

Equivalent ratios form a proportion. Equivalent ratios also reduce to the same rate. Notice that $\frac{2}{4}$ reduces to $\frac{1}{2}$; that $\frac{3}{4}$ and $\frac{2}{3}$ are reduced, and that $\frac{4}{6}$ reduces to $\frac{2}{3}$. Thus the ratio equivalent to $\frac{2}{3}$ is **C.**

Verify How can we verify that $\frac{2}{3}$ and $\frac{4}{6}$ form a proportion?

Example 2

Write this proportion with digits: Four is to six as six is to nine.

We write "four is to six" as one ratio and "six is to nine" as the equivalent ratio. We are careful to write the numbers in the order stated.

$$\frac{4}{6} = \frac{6}{9}$$

We can use proportions to solve a variety of problems. Proportion problems often involve finding an unknown term. The letter *a* represents an unknown term in this proportion:

$$\frac{3}{5} = \frac{6}{a}$$

Math Language

A **scale factor** is a number that relates corresponding sides of similar figures and corresponding terms in equivalent ratios.

One way to find an unknown term in a proportion is to determine the fractional name for 1 that can be multiplied by one ratio to form the equivalent ratio. The first terms in these ratios are 3 and 6. Since 3 times 2 equals 6, we find that the scale factor is 2. So we multiply $\frac{3}{5}$ by $\frac{2}{2}$ to form the equivalent ratio.

$$\frac{3}{5} \cdot \frac{2}{2} = \frac{6}{10}$$

We find that *a* represents the number 10.

Example 3

Complete this proportion: Two is to six as what number is to 30?

We write the terms of the proportion in the stated order, using a letter to represent the unknown number.

$$\frac{2}{6} = \frac{n}{30}$$

We are not given both first terms, but we are given both second terms, 6 and 30. The scale factor is 5, since 6 times 5 equals 30. We multiply $\frac{2}{6}$ by $\frac{5}{5}$ to complete the proportion.

$$\frac{2}{6} \times \frac{5}{5} = \frac{10}{30}$$

The unknown term of the proportion is **10.**

$$\frac{2}{6} = \frac{10}{30}$$

Evaluate How can we check the answer?

Example 4

On the scale drawing of a new gymnasium, the length measures 2 in. and the width measures 4 in. If the actual width of the gymnasium will be 200 ft, what will be the length? Use a proportion to solve.

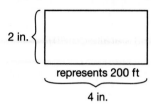

We can state the problem, "two is to four as what is to 200?"

$$\frac{2}{4} = \frac{w}{200}$$

The scale factor is 50, since 4 times 50 is 200.

$$\frac{2}{4} \times \frac{50}{50} = \frac{100}{200}$$

The length of the new gymnasium will be **100 ft.**

Lesson Practice

a. Multiple Choice Which ratio forms a proportion with $\frac{5}{2}$?

A $\frac{3}{2}$ **B** $\frac{4}{10}$ **C** $\frac{15}{6}$ **D** $\frac{5}{20}$

b. Write this proportion with digits: Six is to eight as nine is to twelve.

c. Write and complete this proportion: Four is to three as twelve is to what number? How did you find your answer?

d. **Explain** Write and complete this proportion: Six is to nine as what number is to thirty-six? How can you check your answer?

Written Practice

Distributed and Integrated

1. **Classify** What is the ratio of prime numbers to composite numbers in
(16, 61) this list?

$$2, 3, 4, 5, 6, 7, 8, 9, 10$$

* 2. Bianca poured four cups of milk from a full half-gallon container. How many cups
(81) of milk were left in the container?

* 3. **Analyze** $6 + 6 \times 6 - 6 \div 6$
(72)

4. Write 30% as a reduced fraction. Then write the fraction as a decimal number.
(27, 69)

Find the area of each triangle:

* 5.
(77)

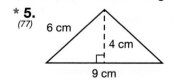

* 6.
(77)

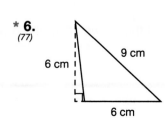

* 7. **a.** Write $\frac{1}{20}$ as a decimal number.
(69, 73)

b. Write $\frac{1}{20}$ as a percent.

8. **Verify** "Some parallelograms are rectangles." True or false? Why?
(59)

9. What is the area of this parallelogram?
(66)

10. What is the perimeter of this parallelogram?
(66)

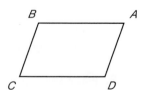
24 cm   25 cm

16 cm

11. $\left(3\frac{1}{8} + 2\frac{1}{4}\right) - 1\frac{1}{2}$ **12.** $\frac{5}{6} \times 2\frac{2}{3} \times 3$
(58)   (67)

13. $8\frac{1}{3} \div 100$ **14.** $(4 - 3.2) \div 10$
(64)   (41)

15. $0.5 \times 0.5 + 0.5 \div 0.5$ **16.** $8 \div 0.04$
(72)   (44)

17. Which digit is in the hundredths place in 12.345678?
(28)

18. **Explain** How do you round $5\frac{1}{8}$ to the nearest whole number?
(21)

19. **Analyze** Write the prime factorization of 700 using exponents.
(71)

20. Two ratios form a proportion if the ratios reduce to the same fraction. Which
(82) two ratios below form a proportion?

$$\frac{15}{12} \qquad \frac{15}{9} \qquad \frac{25}{10} \qquad \frac{35}{21}$$

21. **Connect** The perimeter of a square is 1 meter. How many centimeters long is
(32, 76) each side?

* 22. Fong scored 9 of the team's 45 points.
(25, 73)
a. What fraction of the team's points did Fong score?

b. What percent of the team's points did she score?

23. What time is 5 hours 30 minutes after 9:30 p.m.?
(RF11, 7)

* 24. **Analyze** Write and complete this proportion: Six is to four as
(82) what number is to eight?

25. **Conclude** Figure *ABCD* is a parallelogram. Its opposite angles
(60) ($\angle A$ and $\angle C$, $\angle B$ and $\angle D$) are congruent. Its adjacent angles (such
as $\angle A$ and $\angle B$) are supplementary. If $\angle A$ measures 70°, what are the
measures of $\angle B$, $\angle C$, and $\angle D$?

B   A

C   D

26. **a.** There are two red, three white, and four blue marbles in a bag. If
(80) Cordell pulls two marbles from the bag, what is the probability that will be
white and the other blue?

b. **Verify** Will the order of the draws affect the probability? Support your
answer.

Simplify:

* 27. 2 ft + 24 in. (Write the sum in inches.)
(79)

* 28. **a.** $\dfrac{100 \text{ cm}^2}{10 \text{ cm}}$ **b.** $\dfrac{180 \text{ pages}}{4 \text{ days}}$
(79)

29. (**Model**) A triangle has vertices at the coordinates (4, 4) and (4, 0) and at
(RF27) the origin. Draw the triangle on graph paper. Notice that inside the triangle
are some full squares and some half squares.

a. How many full squares are in the triangle?

b. How many half squares are in the triangle?

30. (**Analyze**) This year Moises has read 24 books. Sixteen of the books were
(16) non-fiction and the rest were fiction. What is the ratio of fiction to non-fiction
books Moises has read this year?

LESSON
83

California Mathematics Content Standards

NS 1.0, 1.3 Use proportions to solve problems (e.g., determine the value of N if $\frac{4}{7} = \frac{N}{21}$, find the length of a side of a polygon similar to a known polygon). Use cross-multiplication as a method for solving such problems, understanding it as the multiplication of both sides of an equation by a multiplicative inverse.

MR 3.0, 3.3 Develop generalizations of the results obtained and the strategies used and apply them in new problem situations.

• Using Cross Products to Solve Proportions

facts   Power Up I

mental math

    a. Number Sense: 20×34

    b. Decimals: $2.5 \div 100$

    c. Probability: What is the probability of rolling a number less than 3 with one roll of a number cube?

    d. Measurement: The bottle contains 1 liter of water. If Chad pours out 125 milliliters, how much water remains in the bottle?

    e. Money: One baseball costs $3.99. What would be the total price of 3 baseballs?

    f. Statistics: Find the median of 2, 3, 4, and 7.

    g. Geometry: These two angles are supplementary angles. What is m∠A?

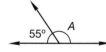

    h. Calculation: $9 \times 9, -1, \div 2, +2, \div 6, +2, \div 3$

problem solving

Choose an appropriate problem-solving strategy to solve this problem. Compare the following two separate quantities:

$$1\tfrac{7}{8} + 2\tfrac{5}{6} \bigcirc 3\tfrac{4}{5} \qquad 6.142 \times 9.065 \bigcirc 54$$

Describe how you performed the comparisons.

We have compared fractions by writing the fractions with common denominators. In Lesson 52, we saw that a variation of this method is to determine whether two fractions have equal **cross products.** If the cross products are equal, then the fractions are equal. The cross products of two fractions are found by cross multiplication, as we show on the next page.

$$8 \times 3 = 24 \qquad 4 \times 6 = 24$$

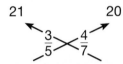

$$\frac{3}{4} \times \frac{6}{8}$$

Both cross products are 24. Since the cross products are equal, we can conclude that the fractions are equal.

> **Equal fractions have equal cross products.**

Example 1

Use cross products to determine whether $\frac{3}{5}$ and $\frac{4}{7}$ are equal.

To find the cross products, we multiply the numerator of each fraction by the denominator of the other fraction. We write the cross product above the numerator that is multiplied.

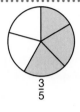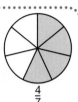

$$\frac{3}{5} \qquad \frac{4}{7}$$

$$21 \qquad\qquad 20$$

$$\frac{3}{5} \times \frac{4}{7}$$

The cross products are not equal, so **the fractions are not equal.** The greater cross product is above the greater fraction. So $\frac{3}{5}$ is greater than $\frac{4}{7}$.

When we find the cross products of two fractions, we are simply renaming the fractions with common denominators. The common denominator is the product of the two denominators and is usually not written. Look again at the two fractions we compared:

$$\frac{3}{5} \qquad \frac{4}{7}$$

The denominators are 5 and 7.

If we multiply $\frac{3}{5}$ by $\frac{7}{7}$ and multiply $\frac{4}{7}$ by $\frac{5}{5}$, we form two fractions that have common denominators.

$$\frac{3}{5} \times \frac{7}{7} = \frac{21}{35} \qquad \frac{4}{7} \times \frac{5}{5} = \frac{20}{35}$$

The numerators of the renamed fractions are 21 and 20, which are the cross products of the fractions. So when we compare cross products, we are actually comparing the numerators of the renamed fractions.

Example 2

Math Language

A **proportion** is a statement that shows two ratios are equal.

Do these two ratios form a proportion?

$$\frac{8}{12}, \frac{12}{18}$$

If the cross products of two ratios are equal, then the ratios are equal and therefore form a proportion. To find the cross products of the ratios above, we multiply 8 by 18 and 12 by 12.

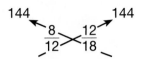

The cross products are 144 and 144, so **the ratios form a proportion.**

$$\frac{8}{12} = \frac{12}{18}$$

Since equivalent ratios have equal cross products, we can use cross products to find an unknown term in a proportion. By cross multiplying, we form an equation. Then we solve the equation to find the unknown term of the proportion.

Example 3

Use cross products to complete this proportion: $\frac{6}{9} = \frac{10}{m}$

The cross products of a proportion are equal. So 6 times m equals 9 times 10, which is 90.

$$\frac{6}{9} = \frac{10}{m}$$

$$6m = 9 \cdot 10$$

We solve this equation:

$$6m = 90$$

$$m = 15$$

The unknown term is 15. We complete the proportion.

$$\frac{6}{9} = \frac{10}{15}$$

Example 4

Use cross products to find the unknown term in this proportion: Fifteen is to twenty-one as what number is to seventy?

We write the ratios in the order stated.

$$\frac{15}{21} = \frac{w}{70}$$

The cross products of a proportion are equal.

$$15 \cdot 70 = 21w$$

To find the unknown term, we divide $15 \cdot 70$ by 21. Notice that we can simplify as follows:

$$\frac{\overset{5}{\cancel{15}} \cdot \overset{10}{\cancel{70}}}{\underset{1}{\cancel{\underset{7}{\cancel{21}}}}} = w$$

The unknown term is **50.**

Lesson Practice

Use cross products to determine whether each pair of ratios forms a proportion:

a. $\dfrac{6}{9}, \dfrac{7}{11}$

b. $\dfrac{6}{8}, \dfrac{9}{12}$

Use cross products to complete each proportion:

c. $\dfrac{6}{10} = \dfrac{9}{x}$

d. $\dfrac{12}{16} = \dfrac{y}{20}$

e. Use cross products to find the unknown term in this proportion: 10 is to 15 as 30 is to what number?

Written Practice — *Distributed and Integrated*

*** 1.** Twenty-one of the 25 books Aretha has are about crafts. What percent of the
(73) books are about crafts?

2. By the time the blizzard was over, the temperature had dropped from 17°F to
(9) − 6°F. This was a drop of how many degrees?

3. The cost to place a collect call was $1.50 for the first minute plus $1.00 for each
(4) additional minute. What was the cost of a 5-minute phone call?

*** 4.** The ratio of runners to walkers at the 10K fund-raiser was 5 to 7. If there were 350
(78) runners, how many walkers were there?

*** 5.** (Conclude) The two acute angles in △ABC are complementary. If
(60) the measure of ∠B is 55°, what is the measure of ∠A?

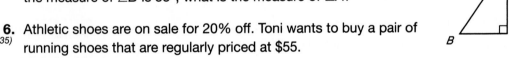

6. Athletic shoes are on sale for 20% off. Toni wants to buy a pair of
(35) running shoes that are regularly priced at $55.

a. How much money will be subtracted from the regular price if she buys the shoes on sale?

b. What will be the sale price of the shoes?

7. Freddy bought a pair of shoes for a sale price of $39.60. The sales-tax
(35) rate was 8%.

a. What was the sales tax on the purchase?

b. What was the total price including tax?

*** 8.** **a.** Write $\dfrac{1}{25}$ as a decimal number.
(69, 73)
b. Write $\dfrac{1}{25}$ as a percent.

*** 9.** Use cross products to determine whether this pair of ratios forms a
(83) proportion:

$$\dfrac{5}{11}, \dfrac{6}{13}$$

* 10. *(83)* ✏ **Explain** Use cross products to find the unknown term in this proportion: 4 is to 6 as 10 is to what number? Describe how you found your answer.

11. *(64)* $10 \div 2\frac{1}{2}$

12. *(41)* $6.5 - (4 - 0.32)$

13. *(33)* $(6.25)(1.6)$

14. *(41)* $0.06 \div 12$

Find each unknown number:

15. *(39, 58)* $2\frac{1}{2} + x = 3\frac{1}{4}$

16. *(39, 58)* $4\frac{1}{8} - y = 1\frac{1}{2}$

* 17. *(83)* $\frac{9}{12} = \frac{n}{20}$

* 18. *(74)* Arrange in order from least to greatest:

$$\frac{1}{2}, 0.4, 30\%$$

19. *(16, 25)* In a school with 300 students and 15 teachers, what is the student-teacher ratio?

20. *(54, 61)* If a number cube is rolled once, what is the probability that it will stop with a composite number on top?

21. *(75)* One fourth of 32 students have pets. How many students do not have pets?

22. *(RF27, 66)* **Connect** What is the area of a parallelogram that has vertices with the coordinates (0, 0), (4, 0), (5, 3), and (1, 3)?

* 23. *(72)* $2 + 2^2 - 2 \div 2$

* 24. *(RF11, 7)* Alejandro started the 10-kilometer race at 8:22 a.m. He finished the race at 9:09 a.m. How long did it take him to run the race?

25. *(21)* **Estimate** Refer to the table below to answer this question: Ten kilometers is about how many miles? Round the answer to the nearest mile.

1 meter ≈ 1.093 yards
1 kilometer ≈ 0.621 mile

> **Reading Math**
>
> The symbol ≈ means "is approximately equal to."

* 26. *(80)* In a bag are two red marbles, three white marbles, and four blue marbles.

    a. One marble is drawn and put back in the bag. Then a marble is drawn again. What is the probability of drawing a white marble on both draws? Are the events dependent or independent?

    b. One marble is drawn and not replaced. Then a second marble is drawn. What is the probability of drawing a white marble on both draws? Are the events independent or dependent?

* 27. *(79)* Simplify:

    a. 2 ft + 24 in. (Write the sum in feet.)

    b. 3 yd · 3 yd

28. **Connect** A quart is what percent of a gallon?
(73, 81)

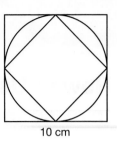

10 cm

Analyze This figure shows a square inside a circle, which is itself inside a larger square. Refer to this figure to answer problems **29** and **30**.

* 29. The area of the smaller square is half the area of the larger square.
(32)
    a. What is the area of the larger square?

    b. What is the area of the smaller square?

30. **Predict** Based on your answers to the questions in problem 29, make an
(32)
educated guess as to the area of the circle. Explain your reasoning.

Real-World Connection

One day Mr. Holmes brought home a $50\frac{1}{4}$ ounce tin of maple syrup. Mrs. Holmes knew they could never finish that much maple syrup. Mr. Holmes explained to her that he was going to share the maple syrup with two neighbors. Mr. Holmes brought back $26\frac{1}{4}$ ounces of maple syrup after sharing with both neighbors. How much maple syrup did each of Mr. Holmes' neighbors receive if they received equal amounts of syrup?

• How to Use the Multiplicative Inverse to Solve Proportions

A proportion is a statement that two ratios are equal. We have used a scale factor to find an unknown number in a proportion.

$$\frac{a}{3} = \frac{6}{9}$$ Since $3 \times 3 = 9$, then $a \times 3$ must be 6, and $a = 2$.

We can also solve a proportion by multiplying both sides of the proportion by the multiplicative inverse of the ratio without the unknown number. The multiplicative inverse is another name for the reciprocal. So for the proportion above, we multiply both sides by $\frac{9}{6}$, the multiplicative inverse of $\frac{6}{9}$.

$$\frac{a}{3} \times \frac{9}{6} = \frac{6}{9} \times \frac{9}{6}$$

Remember that a number multiplied by its multiplicative inverse is 1.

$$\frac{a \times 9}{3 \times 6} = 1$$

When a fraction is equal to 1, the numerator is equal to the denominator.

$$a \times 9 = 3 \times 6 \qquad \text{So } 9a = 18, \text{ and } a = 2.$$

Now look at the original proportion. In this proportion, $a \times 9$ and 3×6 are cross products. The loops below show why they have this name.

$$\frac{a}{3} ⨯ \frac{6}{9} \qquad a \times 9 = 3 \times 6$$

We have shown that the cross products of this proportion are equal. Will other proportions also have equal cross products?

Verify Compute the following cross products.

a. $\frac{1}{4} = \frac{7}{28}$ **b.** $\frac{5}{6} = \frac{15}{18}$ **c.** $\frac{12}{14} = \frac{6}{7}$ **d.** $\frac{30}{48} = \frac{5}{8}$

We can use the steps above to show that the cross products of any proportion are equal. We will use the proportion $\frac{a}{b} = \frac{c}{d}$ (where $b \neq 0$, $d \neq 0$)

$$\frac{a}{b} = \frac{c}{d}$$

$$\frac{a}{b} \times \frac{d}{c} = \frac{c}{d} \times \frac{d}{c} \qquad \text{Multiplication Property of Equality}$$

$$\frac{a \times d}{b \times c} = 1 \qquad \text{The product of a number and its reciprocal is 1.}$$

$$a \times d = b \times c \qquad \text{When a fraction is equal to 1, the numerator equals the denominator.}$$

The cross products are equal.

Summarize Why is it helpful to know two ways to solve proportions?

LESSON
84

• Area of a Circle

\ **California Mathematics Content Standards**

AF 3.0, 3.2 Express in symbolic form simple relationships arising from geometry.

MG 1.0, 1.1 Understand the concept of a constant such as pi; know the formulas for the circumference and area of a circle.

MG 1.0, 1.2 Know common estimates of π (3.14; $\frac{22}{7}$) and use these values to estimate and calculate the circumference and the area of circles; compare with actual measurements.

facts   Power Up N

mental math

   a. Number Sense: 980 − 136

   b. Fractions: $\frac{1}{4}$ of $10.00

   c. Decimals: 7.5 × 100

   d. Money: Juan bought a package of pushpins for $2.99 and a poster for $5.99. What was the total cost?

   e. Probability: What is the probability of rolling a number greater than 2 on a number cube?

   f. Measurement: How many ounces are in a pint?

   g. Algebra: If $k = 0.9$, what is $4.1 + k$?

   h. Calculation: 8 × 8, − 4, ÷ 3, + 4, ÷ 4, + 2, ÷ 4

problem solving

Choose an appropriate problem-solving strategy to solve this problem. Sally's hourglass sand timer runs for exactly three minutes. Jessi's timer runs for exactly four minutes. The two girls want to play a game where they each get one five-minute turn. Explain how the girls can use their timers to mark off exactly five minutes.

(**Understand**) We need to time a 5-minute turn. We have one 3-minute timer and one 4-minute timer.

(**Plan**) We will *work backwards* and *use logical reasoning* to find the answer. We can use the 4-minute timer by itself to mark off four minutes, so we will look for a way to mark off one minute using both timers.

(**Solve**) We see that we can mark off one minute by turning both timers over at the same time. When the 3-minute timer is empty, there is exactly one minute left in the 4-minute timer. At this point the player should begin her turn. When the minute left in the 4-minute timer runs out, we can immediately turn it back over to time the remaining four minutes.

(**Check**) The minute left in Jessi's timer plus the four minutes when it is turned back over totals five minutes.

We can estimate the area of a circle drawn on a grid by counting the number of square units enclosed by the figure.

Example 1

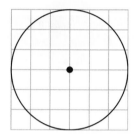

This circle is drawn on a grid.

   a. How many units is the radius of the circle?

   b. Estimate the area of the circle.

  a. To find the **radius** of the circle, we may either find the diameter of the circle and divide by 2, or we may locate the center of the circle and count units to the circle. We find that the radius is **3 units.**

Thinking Skills

Discuss

What is another way we could count the squares and find the area of the circle?

  b. To estimate the area of the circle, we count the square units enclosed by the circle. We show the circle again, this time shading the squares that lie completely or mostly within the circle. We have also marked with dots the squares that have about half their area inside the circle.

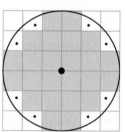

We count 24 squares that lie completely or mostly within the circle. We count 8 "half squares." Since $\frac{1}{2}$ of 8 is 4, we add 4 square units to 24 square units to get an estimate of **28 square units** for the area of the circle.

Finding the exact area of a circle involves the number π. To find the area of a circle, we first find the area of a square built on the radius of the circle. The circle below has a radius of 10 mm, so the area of the square is 100 mm². Notice that four of these squares would cover more than the area of the circle. However, the area of three of these squares is less than the area of the circle.

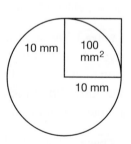

The area of the circle is exactly equal to π times the area of one of these squares. To find the area of this circle, we multiply the area of the square by π. We will continue to use 3.14 for the approximation of π.

$$3.14 \times 100 \text{ mm}^2 = 314 \text{ mm}^2$$

The area of the circle is approximately 314 mm².

Example 2

The radius of a circle is 3 cm. What is the area of the circle? (Use 3.14 for π. Round the answer to the nearest square centimeter.)

We will find the area of a square whose sides equal the radius. Then we multiply that area by 3.14.

Area of square: 3 cm $\times$ 3 cm = 9 cm²

Area of circle: (3.14)(9 cm²) = 28.26 cm²

We round 28.26 cm² to the nearest whole number of square centimeters and find that the area of the circle is approximately **28 cm²**.

The area of any circle is π times the area of a square built on a radius of the circle. The following formula uses *A* for the area of a circle and *r* for the radius of the circle to relate the area of a circle to its radius:

$$A = \pi r^2$$

Example 3

This circle has unknown radius b. Which of the following is a correct way of expressing the area of the circle? Explain why it is correct.

A $A = 2\pi b$ **B** $A = \pi b^2$ **C** $A = bh$

The correct way to express the area of this circle is **B** $A = \pi b^2$.
The formula for the are of a circle is $A = \pi r^2$. Since the radius of this circle is *b*, we replace *r* with *b*.

Lesson Practice

a. The radius of this circle is 4 units. Estimate the area of the circle by counting the squares within the circle.

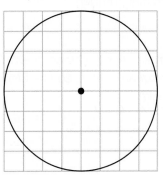

In problems **b–e,** use 3.14 for π.

b. Calculate the area of a circle with a radius of 4 cm.

c. Calculate the area of a circle with a radius of 2 feet.

d. Calculate the area of a circle with a diameter of 2 feet.

e. Calculate the area of a circle with a diameter of 10 inches.

Written Practice

Distributed and Integrated

1. What is the quotient when the decimal number ten and six tenths is divided by
(44) four hundredths?

2. The time in Los Angeles is 3 hours earlier than the time in New York. If it is
(32) 1:15 p.m. in New York, what time is it in Los Angeles?

3. Geraldine paid with a $10 bill for 1 dozen keychains that cost 75¢ each. How
(4, 15) much should she get back in change?

* 4. (**Analyze**) $32 + 1.8(50)$
(72)

* 5. In a can are 100 marbles: 10 yellow, 20 red, 30 green and 40 blue.
(80)
    a. If a marble is drawn from the can, what is the chance
    that the marble will not be red?

    b. If the first marble is not replaced and a second marble
    is drawn from the can, what is the probability that both
    marbles will be yellow?

6. (**Predict**) The ratio of hardbacks to paperbacks in the school library was 5 to 2.
(78) If there were 600 hardbacks, how many paperbacks were there?

7. Nate missed three of the 20 questions on the test. What percent of the questions
(73) did he miss?

8. (**Analyze**) The credit card company charges 1.5% (0.015) interest on the unpaid
(35) balance each month. If Mr. Jones has an unpaid balance of $2000, how much
interest does he need to pay this month?

9. **a.** Write $\frac{4}{5}$ as a decimal number.
(69, 73)
    b. Write $\frac{4}{5}$ as a percent.

10. Serena is stuck on a multiple-choice question that has four choices. She has no
(54) idea what the correct answer is, so she just guesses. What is the probability that
her guess is correct?

11. $5\frac{1}{2} + 3\frac{7}{8}$
(55)

12. $3\frac{1}{4} - \frac{5}{8}$
(58)

13. $\left(4\frac{1}{2}\right)\left(\frac{2}{3}\right)$
(62)

14. $12\frac{1}{2} \div 100$
(64)

15. $5 \div 1\frac{1}{2}$
(64)

16. $\frac{5}{6}$ of $30
(24)

Find each unknown number:

17. 4.72 + 12 + n = 50.4
(39)

18. $10 − m = $9.87
(5)

19. 3n = 0.48
(39, 41)

20. $\dfrac{w}{8} = \dfrac{25}{20}$
(83)

21. (Predict) What are the next three terms in this sequence of perfect squares?
(32)

1, 4, 9, 16, 25, 36, 49, 64, 81, 100, _____, _____, _____, . . .

* 22. (Analyze) This parallelogram is divided into two congruent triangles.
(66, 77)
  a. What is the area of the parallelogram?

  b. What is the area of each triangle?

* 23. Sidell drew a circle with a radius of 10 cm. What was the
(84) approximate area of the circle? (Use 3.14 for π.)

24. **Multiple Choice** Choose the appropriate unit for the area of a
(23) garage.

  A square inches   **B** square feet   **C** square miles

* 25. (Connect) Which two ratios form a proportion? How do you know?
(82, 83)

$$\frac{9}{12} \qquad \frac{8}{14} \qquad \frac{12}{21} \qquad \frac{20}{36}$$

26. **Multiple Choice** The wheel of the covered wagon turned around once in
(42) about 12 feet. The diameter of the wheel was about

  A 6 feet   **B** 4 feet   **C** 3 feet   **D** 24 feet

27. Anabel drove her car 348 miles in 6 hours. Divide the distance by the time to find
(79) the average speed of the car.

28. (Conclude) The opposite angles of a parallelogram are congruent.
(60, 66) The adjacent angles are supplementary. If ∠X measures 110°, then
what are the measures of ∠Y and ∠Z?

29. (Estimate) The diameter of each wheel on the lawn mower is 10 inches. How far
(42) must the lawn mower be pushed in order for each
wheel to complete one full turn? Round the answer to the nearest inch.
(Use 3.14 for π.)

* 30. The ratio of long stems to short stems was 1 to 3. If there were 12 short stems,
(78) how many long stems were there?

Real-World Connection

The Great Frigates are large birds with long, slender wings. Frigates are great flyers and have one of the greatest wingspan to weight ratios of all birds. If a 3-pound Great Frigate bird has a wingspan of 6 feet, what would be the approximate wingspan of a 4-pound Great Frigate bird? Assume that the ratio of wingspan to weight is fairly constant.

⬛ *California Mathematics Content Standards*

MG 1.0, (1.1) Understand the concept of a constant such as π; know the formulas for the circumference and area of a circle.
MR 2.0, 2.6 Indicate the relative advantages of exact and approximate solutions to problems and give answers to a specified degree of accuracy.

• What is Pi?

We use the symbol π to represent pi. The exact value of pi cannot be written as a fraction, as a terminating decimal, or as repeating decimal. Therefore, pi is an irrational number.

> The Greek letter for "P" is written π and pronounced pī. In 1706, William Jones, a mathematician, first used the symbol π to represents the *periphery* of a circular region because of the sound of the word's first letter. *Periphery* means the boundary of a region or its perimeter.

One way to think about pi is as the area of a unit circle.

Look at this circle. This circle has a radius of 1 unit. The area of the entire grid is 1 × 1 or 1 square unit, and part of the grid covers one fourth of the circle.

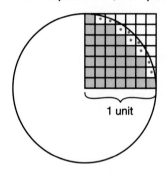

1 unit

The grid is divided into 7 × 7 or 49 unit squares. The area of each unit square is $\frac{1}{49}$ of the grid.

The sum of the number of whole shaded squares and the number of dotted partial squares is a way to represent the area of one-fourth of the circle.

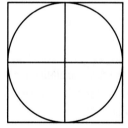

Shaded Whole Squares	Dotted Partial Squares
In the portion of the grid that covers one fourth of the circle, there are 32 shaded unit squares. We can multiply 32 by 4 to find the number of whole squares in the entire circle.	A reasonable estimate of the combined dotted squares covering one fourth of a circle is $6\frac{1}{2}$ unit squares. We can multiply $6\frac{1}{2}$ by 4 to find the number of squares in the entire circle.
$4 \times 32 = 128$ unit squares. Since each unit square is $\frac{1}{49}$ of the grid, the entire circle contains $\frac{128}{49}$ unit squares.	$4 \times 6\frac{1}{2} = 26$ unit squares. Since each unit square is $\frac{1}{49}$ of the grid, the entire circle contains $\frac{26}{49}$ unit squares.

Now, we can estimate the area of the circle to be about $\frac{128}{49} + \frac{26}{49}$ or $\frac{154}{49}$ square units, which reduces to $\frac{22}{7}$.

Discuss Why do you think we use 3.14 as the decimal approximation for pi?

California Mathematics Content Standards

AF 1.0, 1.1 Write and solve one-step linear equations in one variable.

MR 3.0, 3.3 Develop generalizations of the results obtained and the strategies used and apply them in new problem situations.

• Finding Unknown Factors

facts	Power Up M
mental math	**a.** **Number Sense:** 20×35
	b. **Measurement:** Inez will run 1 kilometer and then stop. If she has already run 625 meters, how many more meters will she run?
	c. **Money:** One binder costs $3.99. How much would four binders cost?
	d. **Money:** Double $1.25.
	e. **Decimals:** $7.5 \div 10$
	f. **Measurement:** 2.5 liters is how many milliliters
	g. **Algebra:** If $y = 6$, what does $6y$ equal?
	h. **Calculation:** $7 \times 7, + 1, \div 2, - 1, \div 2, \times 5, \div 2$
problem solving	Choose an appropriate problem-solving strategy to solve this problem. Copy the problem below and fill in the missing digits. No two digits in the problem may be alike.

$$\begin{array}{r} \_\,\_\,\_ \\ \times \quad 7 \\ \hline 9\_\,\_ \end{array}$$

New Concept

Since Lesson 5 we have practiced solving unknown factor problems. In this lesson we will solve problems in which the unknown factor is a mixed number or a decimal number. Remember that we can find an unknown factor by dividing the product by the known factor.

Example 1

Solve: $5n = 21$

To find an unknown factor, we divide the product by the known factor.

Note: We will write the answer as a mixed number unless there are decimal numbers in the problem.

Since there are no decimal numbers in the problem, we write our answer as a mixed number.

$$\begin{array}{r} 4\frac{1}{5} \\ 5\overline{)21} \\ \underline{20} \\ 1 \end{array}$$

$$n = 4\frac{1}{5}$$

Example 2

Solve: $0.6m = 0.048$

Again, we find the unknown factor by dividing the product by the known factor. Since there are decimal numbers in the problem, we write our answer as a decimal number.

$$\begin{array}{r} 0.08 \\ 0\underset{\frown}{6.}\overline{)0\underset{\frown}{0.48}} \\ \underline{48} \\ 0 \end{array}$$

$$m = 0.08$$

Example 3

Solve: $45 = 4x$

This problem might seem "backward" because the multiplication is on the right-hand side. However, an equal sign is not directional. It simply states that the quantities on either side of the sign are equal. In this case, the product is 45 and the known factor is 4. We divide 45 by 4 to find the unknown factor.

$$\begin{array}{r} 11\frac{1}{4} \\ 4\overline{)45} \\ \underline{4} \\ 05 \\ \underline{4} \\ 1 \end{array}$$

$$x = 11\frac{1}{4}$$

Lesson Practice

a. $6w = 21$ **b.** $50 = 3f$ **c.** $5n = 36$

d. $0.3t = 0.24$ **e.** $8m = 3.2$ **f.** $0.8 = 0.5x$

Written Practice

Distributed and Integrated

1. If the divisor is 12 and the quotient is 24, what is the dividend?
(5)

2. The brachiosaurus, one of the largest dinosaurs, weighed only $\frac{1}{4}$ as much as a blue whale. A blue whale can weigh 140 tons. How much could a brachiosaurus have weighed?
(24)

3. **Analyze** Fourteen of the 32 students in the class are boys. What is the ratio of boys to girls in the class?
(16, 25)

Find each unknown number:

* 4. $0.3m = 0.27$
(85)

* 5. $31 = 5n$
(85)

* 6. (Analyze) $3n = 6^2$
(32, 85)

* 7. $4n = 0.35$
(85)

8. Write 0.25 as a fraction and add it to $3\frac{1}{4}$. What is the sum?
(22, 68)

9. Write $\frac{3}{5}$ as a decimal and add it to 6.5. What is the sum?
(41, 69)

10. Write $\frac{1}{50}$ as a decimal number and as a percent.
(69, 73)

11. $12\frac{1}{5} - 3\frac{4}{5}$ **12.** $6\frac{2}{3} \times 1\frac{1}{5}$ **13.** $11\frac{1}{9} \div 100$
(58)   (62)   (64)

14. $4.75 + 12.6 + 10$ *** 15.** $35 - (0.35 \times 100)$
(41)   (72)

* 16. $4 + 4 \times 4 - 4 \div 4$
(72)

17. Write the decimal numeral twelve and
(RF19) five hundredths.

* 18. Dimensions are in mm.
(77)
    a. Find the perimeter of the triangle.

    b. Find the area of the triangle.

19. (Evaluate) If a equals 15, then what number does $2a - 5$ equal?
(6)

20. What is the area of this parallelogram?
(66)

21. What is the perimeter of this parallelogram?
(RF7)

22. (Verify) "All rectangles are parallelograms."
(59) True or false?

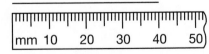

* 23. Charles spent $\frac{1}{10}$ of his 100 shillings. How many shillings does he still have?
(75)

24. The temperature rose from $-18°F$ to $19°F$. How many degrees did the
(9) temperature increase?

25. How many **centimeters** long is the line below?
(76)

26. Johann poured 500 mL of water from a full 2-liter container. How many milliliters
(76) of water were left in the container?

27. Carmela has a deck of 26 alphabet cards. She mixes the cards, places them face
(80) down on the table, and turns them over one at a time. What is the probability that
the first two cards she turns over are vowel cards (A, E, I, O, U)?

* 28. Simplify:
(79)

    a. 2 meters + 100 centimeters (Write the answer in meters.)

    b. 2 m · 4 m

* 29. (**Analyze**) Solve this proportion: $\dfrac{12}{m} = \dfrac{18}{9}$
(83)

30. (**Connect**) What is the perimeter of a rectangle with vertices at $(-4, -4)$, $(-4, 4)$,
(RF27) $(4, 4)$, and $(4, -4)$?

Real-World Connection

Frida and her family went on a summer vacation to Chicago, Illinois from Boston, Massachusetts. Her family drove 986 miles in 17 hours.

    a. How many miles per hour did Frida's family average on their drive to Chicago?

    b. If the family's car averages 29 miles per gallon, how many gallons of gas did they use on their trip?

LESSON

86

• Using Proportions to Solve Ratio Word Problems

California Mathematics Content Standards

NS **1.0, 1.2** Write and evaluate an algebraic expression for a given situation, using up to three variables.

NS **1.0, 1.3** Apply algebraic order of operations and the commutative, associative, and distributive properties to evaluate expressions; and justify each step in the process.

MR **2.0, 2.4** Use a variety of methods, such as words, numbers, symbols, charts, graphs, tables, diagrams, and models, to explain mathematical reasoning.

Power Up

facts	Power Up K
mental math	**a. Powers/Roots:** 60^2
	b. Number Sense: $850 - 170$
	c. Fractions: $\frac{1}{5}$ of $2.50
	d. Decimals: 0.08×100
	e. Geometry: These angles are complementary angles. What is m $\angle D$?
	f. Measurement: How many cups are in 1 pint?
	g. Algebra: If $10x = 100$, what is x?
	h. Calculation: $6 \times 6, -6, \div 2, -1, \div 2, \times 8, -1, \div 5$

66°

D

problem solving

Choose an appropriate problem-solving strategy to solve this problem. Thomasita was thinking of a number less than 90 that she says when counting by sixes and when counting by fives, but not when counting by fours. Of what number was she thinking?

New Concept

Proportions can be used to solve many types of word problems. In this lesson we will use proportions to solve ratio word problems such as those in the following examples.

Example 1

The ratio of salamanders to frogs was 5 to 7. If there were 20 salamanders, how many frogs were there?

In this problem there are two kinds of numbers: ratio numbers and actual-count numbers. The ratio numbers are 5 and 7. The number 20 is an actual count of the salamanders. We will arrange these numbers in two columns and two rows to form a ratio box.

Lesson 86   533

We were not given the actual count of frogs, so we use the letter *f* to stand for the actual number of frogs.

Instead of using scale factors in this lesson, we will practice using proportions. We use the positions of numbers in the ratio box to write a proportion. By solving the proportion, we find the actual number of frogs.

	Ratio	Actual Count
Salamanders	5	20
Frogs	7	f

	Ratio	Actual Count
Salamanders	5	20
Frogs	7	f

$$\frac{5}{7} = \frac{20}{f}$$

We can solve the proportion in two ways. We can multiply $\frac{5}{7}$ by $\frac{4}{4}$, or we can use cross products. Here we show the solution using cross products:

$$\frac{5}{7} = \frac{20}{f}$$

$$5f = 7 \cdot 20$$

$$f = \frac{7 \cdot 20}{5}$$

$$f = 28$$

We find that there were **28 frogs.**

Example 2

Thinking Skills

Justify

Two methods for solving proportions are using cross products and using a constant factor. Under what circumstance is one method preferable to the other?

If 3 sacks of concrete will make 12 square feet of sidewalk, predict how many sacks of concrete are needed to make 40 square feet of sidewalk?

We are given the ratio 3 sacks to 12 square feet and the actual count of square feet of sidewalk needed.

	Ratio	Actual Count
Sacks	3	n
Sq. ft	12	40

$$\frac{3}{12} = \frac{n}{40}$$

The constant factor from the first ratio to the second is not obvious, so we use cross products.

$$\frac{3}{12} = \frac{n}{40}$$

$$12 \cdot n = 3 \cdot 40$$

$$n = \frac{3 \cdot 40}{12}$$

$$n = 10$$

We find that **10 sacks** of concrete are needed.

Lesson Practice **Model** For each problem, draw a ratio box. Then solve the problem using proportions.

a. The ratio of DVDs to CDs was 5 to 4. If there were 60 CDs, how many DVDs were there?

b. **Explain** At the softball game, the ratio of fans for the home team to the fans for the away team is 5 to 3. If there are 30 fans for the home team, how many fans for the away team are there? How can you check your answer?

Written Practice *Distributed and Integrated*

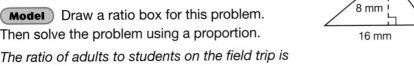

1. Mavis scored 12 of the team's 20 points. What percent of the team's points did
(73) Mavis score?

* 2. One fourth of an inch of snow fell every hour during the storm. How many hours
(45) did the storm last if the total accumulation of snow was 4 inches?

* 3. Eamon wants to buy a new baseball glove that costs $50. He has $14 and he
(4, 85) earns $6 per hour cleaning yards. How many hours must he work to have enough money to buy the glove?

* 4. **Analyze** Find the area of this triangle.
(77)

* 5. **Model** Draw a ratio box for this problem.
(86) Then solve the problem using a proportion.

*The ratio of adults to students on the field trip is
3 to 5. If there
are 15 students on the field trip, how many adults are there?*

```
        /\
       /  \  10 mm
  8 mm/  | \
     /___|__\
      16 mm
```

6. If two marbles are pulled one at a time without replacement from a bag
(69, 80) containing 8 blue marbles, 7 green marbles, and 6 yellow marbles, what is the probability that both marbles will be green? Express the probability as a decimal number.

Find each unknown factor:

* 7. $10w = 25$
(85)

* 8. $20 = 9m$
(85)

* 9. a. What is the perimeter of this triangle?
(RF7, 77)
b. What is the area of this triangle?

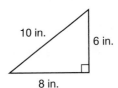

10. Write 5% as a
(27)
a. decimal number.   **b.** fraction.

11. Write $\frac{2}{5}$ as a decimal, and multiply it by 2.5. What is the product?
(69)

12. Compare: $\frac{2}{3} + \frac{3}{2} \bigcirc \frac{2}{3} \cdot \frac{3}{2}$
(24, 55)

13. $\frac{1}{3} \times \frac{100}{1}$
(24, 51)

14. $6 \div 1\frac{1}{2}$
(64)

15. $12 \div 0.25$
(44)

16. 0.025×100
(RF22)

17. If the tax rate is 7%, what is the tax on a $24.90 purchase?
(35)

18. The prime factorization of what number is $2^2 \cdot 3^2 \cdot 5^2$?
(71)

* 19. **Multiple Choice** Which of these is a composite number?
(61)

    A 61   **B** 71   **C** 81   **D** 101

20. Round the decimal number one and twenty-three hundredths to the nearest tenth.
(46)

21. Albert baked 5 dozen muffins and gave away $\frac{7}{12}$ of them. How many muffins were left?
(75)

* 22. $6 \times 3 - 6 \div 3$
(72)

23. How many milliliters is 4 liters?
(76)

24. **Model** Draw a line segment $2\frac{1}{4}$ inches long. Label the endpoints A and C. Then make a dot at the midpoint of $\overline{AC}$ (the point halfway between points A and C), and label the dot B. What are the lengths of $\overline{AB}$ and $\overline{BC}$?
(RF9)

25. On a coordinate plane draw a rectangle with vertices at $(-2, -2)$, $(4, -2)$, $(4, 2)$, and $(-2, 2)$. What is the area of the rectangle?
(RF9)

26. What is the ratio of the length to the width of the rectangle in problem 25?
(16)

* 27. **Explain** How do you calculate the area of a triangle?
(77)

28. In the figure at right, angles ADB and BDC are supplementary.
(25, 60)

    a. What is m$\angle ADB$?

    b. What is the ratio of m$\angle BDC$ to m$\angle ADB$? Write the answer as a reduced fraction.

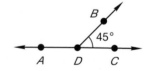

* 29. **Analyze** Nathan drew a circle with a radius of 10 cm. Then he drew a square around the circle.
(32, 84)

    a. What was the area of the square?

    b. What was the area of the circle? (Use 3.14 for π.)

* 30. **Formulate** Write a word problem that can be solved using the proportion $\frac{6}{8} = \frac{w}{100}$. Solve the problem.
(83)

LESSON 87

California Mathematics Content Standards

AF 3.0, 3.1 Use variables in expressions describing geometric quantities (e.g., $P = 2w + 2l$, $A = \frac{1}{2}bh$, $C = \pi d$ - the formulas for the perimeter of a rectangle, the area of a triangle, and the circumference of a circle, respectively).

MR 2.0, 2.5 Express the solution clearly and logically by using the appropriate mathematical notation and terms and clear language; support solutions with evidence in both verbal and symbolic work.

• Geometric Formulas

Power Up

facts   Power Up L

mental math

   a. Number Sense: $70 \cdot 70$

   b. Number Sense: 20×45

   c. Decimals: $62.5 \div 100$

   d. Geometry: How many degrees are in one full rotation (turn)?

   e. Estimation: Estimate 19×21 by rounding the factors and then multiplying.

   f. Statistics: The daily high temperatures for Monday through Friday were 84°, 85°, 85°, 87°, and 87°. What was the median high temperature?

   g. Algebra: If $n = 2$, what does $2n$ equal?

   h. Calculation: $5 \times 5, -5, \times 5, \div 2, -1, \div 7, \times 3, -1, \div 2$

problem solving

Choose an appropriate problem-solving strategy to solve this problem. In his 1859 autobiography, Abraham Lincoln wrote, "Of course when I came of age I did not know much. Still somehow, I could read, write, and cipher to the Rule of Three." When Lincoln wrote these words, "ciphering to the Rule of Three" is what students called setting up a proportion. For example, "ciphering to the Rule of Three the numbers 2, 5, and 4" means "2 is to 5 as 4 is to ____," which is the same as solving the proportion $\frac{2}{5} = \frac{4}{x}$. We find that $x = 10$, so we can say that "2 is to 5 as 4 is to 10."

Cipher to the Rule of Three the numbers 2, 6, and 7.

New Concept

We have found the area of a rectangle by multiplying the length of the rectangle by its width. This procedure can be described with the following formula:

$$A = lw$$

The letter *A* stands for the area of the rectangle. The letters *l* and *w* stand for the length and width of the rectangle. Written side by side, *lw* means that we multiply the length by the width. The table below lists formulas for the perimeter and area of squares, rectangles, parallelograms, and triangles.

Figure	Perimeter	Area
Square	$P = 4s$	$A = s^2$
Rectangle	$P = 2l + 2w$	$A = lw$
Parallelogram	$P = 2b + 2s$	$A = bh$
Triangle	$P = s_1 + s_2 + s_3$	$A = \frac{1}{2}bh$

The letters *P* and *A* are abbreviations for **perimeter** and **area.** Other abbreviations are illustrated below:

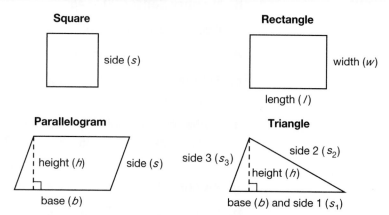

Since squares and rectangles are also parallelograms, the formulas for the perimeter and area of parallelograms may also be used for squares and rectangles.

To use a formula, we substitute each known measure in place of the appropriate letter in the formula. When substituting a number in place of a letter, it is a good practice to write the number in parentheses.

Example 1

Write the formula for the perimeter of a rectangle. Then substitute 8 cm for the length and 5 cm for the width. Solve the equation to find *P*.

The formula for the perimeter of a rectangle is

$$P = 2l + 2w$$

We rewrite the equation, substituting 8 cm for *l* and 5 cm for *w*. We write these measurements in parentheses.

$$P = 2(8 \text{ cm}) + 2(5 \text{ cm})$$

We multiply 2 by 8 cm and 2 by 5 cm.

$$P = 16 \text{ cm} + 10 \text{ cm}$$

Now we add 16 cm and 10 cm.

$$P = 26 \text{ cm}$$

The perimeter of the rectangle is **26 cm.**

We summarize the steps below to show how your work should look.

$$P = 2l + 2w$$

$$P = 2(8 \text{ cm}) + 2(5 \text{ cm})$$

$$P = 16 \text{ cm} + 10 \text{ cm}$$

$$P = 26 \text{ cm}$$

Example 2

Write the formula area of a parallelogram. Then substitute 7 cm for the base and 4 cm for the height. Solve the equation for A.

$$A = bh$$

We rewrite the equation, substituting 7 cm for b and 4 cm for h.

$$A = (7 \text{ cm}) (4 \text{ cm})$$

$$A = 28 \text{ cm}^2$$

The area of the parallelogram is **28 cm².**

We summarize the steps below to show you how your work should look.

$$A = bh$$

$$A = (7 \text{ cm}) (4 \text{ cm})$$

$$A = 28 \text{ cm}^2$$

Lesson Practice

a. Write the formula for the area of a rectangle. Then substitute 8 cm for the length and 5 cm for the width. Solve the equation to find the area of the rectangle.

b. Write the formula for the perimeter of a parallelogram. Then substitute 10 cm for the base and 6 cm for the side. Solve the equation to find the perimeter of the parallelogram.

c. **Estimate** Look around the room for a rectangular or triangular shape. Estimate its dimensions and write a word problem about its perimeter or area. Then solve the problem.

Written Practice   *Distributed and Integrated*

1. What is the difference when the product of $\frac{1}{2}$ and $\frac{1}{2}$ is subtracted from the sum of
(4, 17) $\frac{1}{4}$ and $\frac{1}{4}$?

2. A dairy cow can give 4 gallons of milk per day. How many cups of milk is that? (1
(81) gallon = 4 quarts; 1 quart = 4 cups)

* 3. The recipe called for $\frac{3}{4}$ cup of sugar. If the recipe is doubled, how much sugar
(24) should be used?

* 4. (Model) Draw a ratio box for this problem. Then solve the problem using a
(86) proportion.

> The recipe called for sugar and flour in the ratio of 2 to 9. If the chef
> used 18 pounds of flour, how many pounds of sugar were needed?

* 5. Write the formula for finding the area of a parallelogram. Then substitute 12 for the
(87) height and 10 for the base. Find the area of the parallelogram.

* 6. Express the unknown factor as a mixed number:
(85)
$$7n = 30$$

7. Marissa used a compass to draw a circle with a radius of 4 inches.
(42)
   a. What is the diameter of the circle?

   b. What is the circumference of the circle?

* 8. In problem **7** what is the area of the circle Marissa drew?
(84)

Use 3.14 for π.

* 9. What is the area of the triangle at right?
(77)

10. **a.** What is the area of this parallelogram?
(66)
   b. What is the perimeter of this parallelogram?

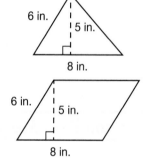

11. Write 0.5 as a fraction and subtract it from $3\frac{1}{4}$.
(58, 68) What is the difference?

12. Write $\frac{3}{4}$ as a decimal, and multiply it by 0.6. What is the product?
(33, 69)

* 13. (Analyze) $2 \times 15 + 2 \times 12$
(72)

* 14. In a bag were six red marbles and 4 blue marbles. If Latrina pulls a marble out of
(80) the bag with her left hand and then pulls a marble out with her right hand, what is
the probability that the marble in each hand will be blue?

15. $6 \div 8
(RF2)

16. $1\frac{3}{5} \times 10 \times \frac{1}{4}$
(67)

17. $37\frac{1}{2} \div 100$
(64)

18. $3 \div 7\frac{1}{2}$
(64)

19. What is the place value of the 7 in 987,654.321?
(28)

20. The ratio of tidy rooms to unkempt rooms was 3 to 2. If 33 rooms were tidy, how
(78) many were unkempt?

21. $30 + 60 + m = 180$
(5)

22. (**Analyze**) Half of the students are girls. Half of the girls have brown hair. Half of
(67) the brown-haired girls wear their hair long. Of the 32 students, how many are girls with long, brown hair?

Refer to the pictograph below to answer problems **23–25.**

Books Read This Year

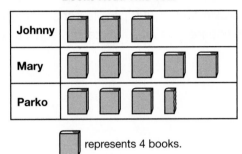

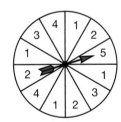 represents 4 books.

23. How many books has Johnny read?
(Inv. 4)

24. Mary has read how many more books than Parko?
(Inv. 4)

25. (**Formulate**) Write a question that relates to this graph and answer the question.
(Inv. 4)

* 26. (**Analyze**) Solve this proportion: $\dfrac{12}{8} = \dfrac{21}{m}$
(83)

27. The face of this spinner is divided into 12 congruent regions. If the spinner is spun
(54, 69) once, what is the probability that it will stop on a 3? Express the probability ratio as a fraction and as a decimal number rounded to the nearest hundredth.

28. Multiple Choice (**Conclude**) If two angles are complementary, and if one angle
(20, 60) is acute, then the other angle is what kind of angle?

   A acute   **B** right   **C** obtuse

* 29. Simplify:
(79)
   a. 100 cm + 100 cm (Write the answer in meters.)
   b. $\dfrac{(5 \text{ in.})(8 \text{ in.})}{2}$

* **30.** In the Fahrenheit scale, the difference between the boiling and
(76) freezing temperatures of water is 212°F − 32°F = 180°F. What is
the difference between the boiling and freezing temperatures of
water in the Celsius scale?

*Real-World
Connection*

There are close to 4 million people living on the island of Puerto Rico,
making it one of the most densely populated islands in the world. If the
population density is approximately 1,000 people per square mile, how
many people live in a 3.5 square mile area? Write and solve a proportion
to answer this question.

California Mathematics Content Standards

AF 1.0, 1.2 Write and evaluate an algebraic expression for a given situation, using up to three variables.

AF 1.0, 1.3 Apply algebraic order of operations and the commutative, associative, and distributive properties to evaluate expressions; and justify each step in the process.

AF 3.0, 3.1 Use variables in expressions describing geometric quantities (e.g., $P = 2w + 2l$, $A = \frac{1}{2}bh$, $C = \pi d$ - the formulas for the perimeter of a rectangle, the area of a triangle, and the circumference of a circle, respectively).

• Distributive Property

facts	Power Up N
mental math	**a. Powers/Roots:** 80^2
	b. Money: One juice box costs $1.98. What is the price for two juice boxes?
	c. Fractions: $\frac{1}{10}$ of $5.00
	d. Decimals: 0.15×100
	e. Number Sense: $\frac{750}{250}$
	f. Measurement: How many milliliters are in 5 liters?
	g. Algebra: If $10.5 - y = 8.5$, what is y?
	h. Calculation: 4×4, $- 1$, $\times 2$, $+ 3$, $\div 3$, $- 1$, $\times 10$, $- 1$, $\div 9$

problem solving

Choose an appropriate problem-solving strategy to solve this problem. Kim hit a target like the one shown 6 times, earning a total score of 20. Find two sets of scores Kim could have earned.

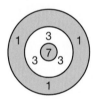

New Concept

In Lesson 23 it was stated that the area (*A*) of a rectangle is related to the length (*l*) and width (*w*) of the rectangle by this formula:

$$A = lw$$

This formula means "the area of a rectangle equals the product of its length and width." If we are given measures for *l* and *w*, we can replace the letters in the formula with numbers and calculate the area.

Find _A_ in _A = lw_ when _l_ is 8 ft and _w_ is 4 ft.

We replace _l_ and _w_ in the formula with 8 ft and 4 ft respectively. Then we simplify.

$$A = lw$$

$$A = (8 \text{ ft})(4 \text{ ft})$$

$$A = \textbf{32 ft}^2$$

Notice the effect on the units when the calculation is performed. Multiplying two units of length results in a unit of area.

Evaluate 2(_l + w_) when _l_ is 8 cm and _w_ is 4 cm.

In place of _l_ and _w_ we substitute 8 cm and 4 cm. Then we simplify.

$$2(l + w)$$

$$2(8 \text{ cm} + 4 \text{ cm})$$

$$2(12 \text{ cm})$$

$$\textbf{24 cm}$$

Activity

Perimeter Formulas

Lamar finds the perimeter of a rectangle by doubling the length, doubling the width, and then adding the two numbers.

Molly finds the perimeter by adding the length and width, then doubling that number.

Write a formula for Lamar's method and another formula for Molly's method.

There are two formulas commonly used to relate the perimeter (_P_) of a rectangle to its length and width.

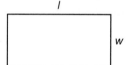

$$P = 2(l + w)$$

$$P = 2l + 2w$$

Both formulas describe how to find the perimeter of a rectangle if we are given its length and width. The first formula means "add the length and width and then double this sum." The second formula means "double the length and double the width and then add."

Example 3

Use the two perimeter formulas to find the perimeter of this rectangle.

30 in.

20 in.

In both formulas we replace *l* with 30 in. and *w* with 20 in. Then we simplify.

$P = 2(l + w)$	$P = 2l + 2w$
$P = 2(30 \text{ in.} + 20 \text{ in.})$	$P = 2(30 \text{ in.}) + 2(20 \text{ in.})$
$P = 2(50 \text{ in.})$	$P = 60 \text{ in.} + 40 \text{ in.}$
$P = \mathbf{100 \text{ in.}}$	$P = \mathbf{100 \text{ in.}}$

Both formulas in Example 3 yield the same result because the two formulas are equivalent.

$$2(l + w) = 2l + 2w$$

(**Conclude**) Evaluate $2(l + w)$ and $2l + 2w$ when *l* is 9 cm and *w* is 5 cm. Are the results equal?

Thinking Skills

(**Formulate**)

Write an equivalent formula for the perimeter of a rectangle using only variables.

These equivalent expressions illustrate the **Distributive Property of Multiplication Over Addition,** often called simply the **Distributive Property.** Applying the Distributive Property, we distribute, or "spread," the multiplication over the terms that are being added (or subtracted) within the parentheses. In this case we multiply *l* by 2, giving us 2*l*, and we multiply *w* by 2, giving us 2*w*.

$$2(l + w) = 2l + 2w$$

The Distributive Property is often expressed in equation form using variables:

$$a(b + c) = ab + ac$$

The Distributive Property also applies over subtraction.

$$a(b - c) = ab - ac$$

Example 4

Show two ways to simplify this expression:

$$6(20 + 5)$$

One way is to add 20 and 5 and then multiply the sum by 6.

$$6(20 + 5)$$

$$6(25)$$

$$150$$

Another way is to multiply 20 by 6 and multiply 5 by 6. Then add the products.

$$6(20 + 5)$$

$$(6 \cdot 20) + (6 \cdot 5)$$

$$120 + 30$$

$$150$$

Example 5

Simplify: $2(3 + n) + 4$

Justify each step.

We show and justify each step.

$2(3 + n) + 4$	Given
$2 \cdot 3 + 2n + 4$	Distributive Property
$6 + 2n + 4$	$2 \cdot 3 = 6$
$2n + 6 + 4$	Commutative Property
$2n + (6 + 4)$	Associative Property
$2n + 10$	$6 + 4 = 10$

Lesson Practice

a. (**Connect**) Find A in $A = bh$ when b is 15 in. and h is 8 in.

b. Evaluate $\frac{ab}{2}$ when a is 6 ft and b is 8 ft.

c. (**Formulate**) Write an equation using the letters x, y, and z that illustrates the Distributive Property of Multiplication Over Addition.

d. (**Analyze**) Show two ways to simplify this expression:

$$6(20 - 5)$$

e. Write two formulas for finding the perimeter of a rectangle.

f. (**Explain**) Describe two ways to simplify this expression:

$$2(6 + 4)$$

g. Simplify: $2(n + 5)$. Show and justify the steps.

Written Practice *Distributed and Integrated*

1. What is the mean of 4.2, 4.8, and 5.1?
(Inv. 4)

2. Which property is illustrated by the equation $a(b + c) = ab + ac$?
(88)

3. Fifteen of the 25 students in Room 20 are boys. What percent of the students in
(73) Room 20 are boys?

4. This triangular prism has how many more edges than vertices?
(RF25)

* 5. The teacher cut a 12-inch diameter circle from a sheet of
(84) construction paper.

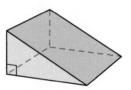

　　a. What was the radius of the circle?

　　b. What was the area of the circle? (Use 3.14 for π.)

6. *(59)* **Explain** Write a description of a trapezoid.

7. *(9, 14)* Arrange these numbers in order from least to greatest:

$$1, -2, 0, -4, \frac{1}{2}$$

* 8. *(85)* **Analyze** Express the unknown factor as a mixed number:

$$25n = 70$$

Refer to the triangle to answer questions **9–11.**

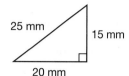

9. *(77)* What is the area of this triangle?

10. *(RF7)* What is the perimeter of this triangle?

* 11. *(16, 69)* What is the ratio of the length of the shortest side to the length of the longest side? Express the ratio as a fraction and as a decimal.

12. *(58, 68)* **Connect** Write 6.25 as a mixed number. Then subtract $\frac{5}{8}$ from the mixed number. What is the difference?

13. *(88)* Evaluate $2\pi r$ when π is 3.14 and r is 10.

14. *(27)* Write 28% as a reduced fraction.

* 15. *(83)* **Analyze** $\dfrac{n}{12} = \dfrac{20}{30}$

16. *(41)* $0.625 \div 10$

17. *(44)* $\dfrac{25}{0.8}$

18. *(55)* $3\frac{3}{8} + 3\frac{3}{4}$

19. *(43)* $5\frac{1}{8} - 1\frac{7}{8}$

20. *(67)* $6\frac{2}{3} \times \dfrac{3}{10} \times 4$

21. *(75)* One third of the two dozen knights were on horseback. How many knights were not on horseback?

* 22. *(10)* **Evaluate** Weights totaling 38 ounces were placed on the left side of this scale, while weights totaling 26 ounces were placed on the right side of the scale. How many ounces of weights should be moved from the left side to the right side to balance the scale? (*Hint:* Find the mean of the weights on the two sides of the scale.)

23. *(87)* Write the formula for finding the area of a rectangle.

24. *(46)* Round forty-eight hundredths to the nearest tenth.

25. *(80)* Arianna writes the letters of her first name on index cards, one letter per card. She turns the cards over and mixes them up. Then she chooses a card. If this card is an *A*, what is the probability that the next card will also be an *A*? Tell how you found the answer.

26. The ratio of dogs to cats in the neighborhood is 6 to 5. What is the
(16) ratio of cats to dogs?

27. $10 + 10^2 - 10 \div 10$
(72)

28. ✎ **Analyze** The Thompsons drink a gallon of milk every two days. There
(81) are four people in the Thompson family. Each person drinks an average of
how many pints of milk per day? Explain your thinking.

* 29. Simplify:
(76, 79)
  a. 10 cm + 100 mm (Write the answer in millimeters.)

  b. 300 books ÷ 30 students

30. **Model** On a coordinate plane draw a segment from point A $(-3, -1)$ to
(RF27) point B $(5, -1)$. What are the coordinates of the point that is halfway
between points A and B?

Early Finishers
Real-World Connection

The students in Mrs. Fitzgerald's cooking class will make buttermilk
biscuits using the recipe below. If 56 students each work with a partner
to make one batch of biscuits, how much of each ingredient will Mrs.
Fitzgerald need to buy? Hint: Keep each item in the unit measure given.

$1\frac{3}{4}$ cups of all-purpose flour

1 teaspoon of baking soda

1 stick of butter

$1\frac{1}{4}$ cups of milk

LESSON 89

California Mathematics Content Standards

NS **2.0**, **2.4** Determine the least common multiple and the greatest common divisor of whole numbers; use them to solve problems with fractions (e.g., to find a common denominator to add two fractions or to find the reduced form for a fraction).

AF **2.0**, **2.2** Demonstrate an understanding that *rate* is a measure of one quantity per unit value of another quantity.

AF **2.0**, **2.3** Solve problems involving rates, average speed, distance, and time.

• Reducing Rates Before Multiplying

facts   Power Up L

mental math

    a. Number Sense: 90 · 90

    b. Measurement: Tessa drank 405 milliliters from the full 1-liter container of water. How many milliliters remained in the container?

    c. Money: Each set of blank CDs costs $7.99. What is the cost of 6 sets of CDs?

    d. Money: Double $27.00.

    e. Decimals: 87.5 ÷ 100

    f. Geometry: What is the area of a triangle that has a base of 4 cm and a height of 2.5 cm?

    g. Geometry: How many degrees is one half of a full rotation?

    h. Calculation: $3 \times 3, + 2, \times 5, - 5, \times 2, \div 10, + 5, \div 5$

problem solving

Choose an appropriate problem-solving strategy to solve this problem. A loop of string was arranged to form a square with sides 9 inches long. If the same loop of string is arranged to form an equilateral triangle, how long will each side be? If a regular hexagon is formed, how long will each side be?

Since Lesson 65 we have practiced reducing fractions before multiplying. This is sometimes called *canceling*.

$$\frac{\overset{1}{\cancel{3}}}{\underset{2}{\cancel{4}}} \cdot \frac{\overset{1}{\cancel{2}}}{\underset{1}{\cancel{5}}} \cdot \frac{\overset{1}{\cancel{5}}}{\underset{2}{\cancel{6}}} = \frac{1}{4}$$

We can cancel **units** before multiplying just as we cancel numbers.

$$\frac{4 \text{ miles}}{1 \text{ hour}} \times \frac{2 \text{ hours}}{1} = \frac{8 \text{ miles}}{1} = 8 \text{ miles}$$

Since **rates** are ratios of two measures, multiplying and dividing rates involves multiplying and dividing units.

Example 1

Multiply 55 miles per hour by six hours.

We write the rate 55 miles per hour as the ratio 55 miles over 1 hour, because "per" indicates division. We write six hours as the ratio 6 hours over 1.

$$\frac{55 \text{ miles}}{1 \text{ hour}} \times \frac{6 \text{ hours}}{1}$$

The unit "hour" appears above and below the division line, so we can cancel hours.

$$\frac{55 \text{ miles}}{1 \text{ hour}} \times \frac{6 \text{ hours}}{1} = \textbf{330 miles}$$

Connect Can you think of a word problem to fit this equation?

> **Math Language**
>
> Recall that a ratio is a comparison of two numbers by division.

Example 2

Antonia flew a plane for 3 hours, traveling at 500 miles per hour. How far did she go?

Each hour, Antonia traveled 500 miles. In three hours, she traveled 3×500 miles which is **1500 miles.**

We can also find this answer by multiplying 500 miles per hour by 3 hours.

$$\frac{500 \text{ miles}}{1 \text{ hour}} \times \frac{3 \text{ hours}}{1}$$

Cancel the "hour" unit.

$$\frac{500 \text{ miles}}{1 \text{ hour}} \times \frac{3 \text{ hours}}{1} = \textbf{1500 miles}$$

Example 3

Multiply 5 feet by 12 inches per foot.

We write ratios of 5 feet over 1 and 12 inches over 1 foot. We then cancel units and multiply.

$$\frac{5 \text{ feet}}{1} \cdot \frac{12 \text{ inches}}{1 \text{ foot}} = \textbf{60 inches}$$

Lesson Practice

When possible, cancel numbers and units before multiplying:

a. $\dfrac{3 \text{ dollars}}{1 \text{ hour}} \times \dfrac{8 \text{ hours}}{1}$

b. $\dfrac{6 \text{ baskets}}{10 \text{ shots}} \times \dfrac{100 \text{ shots}}{1}$

c. $\dfrac{10 \text{ cents}}{1 \text{ kwh}} \times \dfrac{26.3 \text{ kwh}}{1}$

Math Language

The abbreviation **"kwh"** stands for *kilowatt hours,* a rate used to measure energy.

d. $\dfrac{160 \text{ km}}{2 \text{ hours}} \cdot \dfrac{10 \text{ hours}}{1}$

e. Multiply 18 teachers by 29 students per teacher.

f. Multiply 2.3 meters by 100 centimeters per meter.

g. Solve this problem by multiplying two ratios: How far will the train travel in 6 hours at 45 miles per hour?

Written Practice

Distributed and Integrated

1. What is the ratio of prime numbers to composite numbers in this list?
(16, 61)

10, 11, 12, 13, 14, 15, 16, 17, 18, 19, 20, 21

2. Sunrise was at 6:15 a.m. and sunset was at 5:45 p.m. How many hours and
(RF11, 7) minutes were there from sunrise to sunset?

* 3. (**Model**) Draw a ratio box for this problem. Then solve the problem using
(86) a proportion.

> *When the good news was announced many leaped for joy and others just smiled. The ratio of leapers to smilers was 3 to 2. If 12 leaped for joy, how many just smiled?*

4. A rectangular prism has how many more faces than a triangular prism?
(RF25)

* 5. Write the formula for the area of a parallelogram. Then substitute 15 cm for
(87) the base and 4 cm for the height. Solve the equation to find the area of the parallelogram.

* 6. Latoya is asked to select and hold 3 marbles from a bag of 52 marbles. Four of the
(80) marbles are green. If the first two marbles she selects are green, what is the chance that the third marble she selects will be one of the two remaining green marbles?

7. A pyramid with a triangular base is shown at right.
(RF25)

    a. How many faces does it have?

    b. How many edges does it have?

    c. How many vertices does it have?

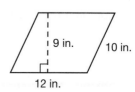

* 8. **a.** What is the perimeter of this parallelogram?
(66)

    b. What is the area of this parallelogram?

9. **a.** Write $\frac{7}{20}$ as a decimal number.
(69, 73)

    b. Write $\frac{7}{20}$ as a percent.

* **10.** **Connect** The distance between bases in a softball diamond is 60 feet.
(87) What is the shortest distance a player could run to score a home run?

11. $6\frac{2}{3} + 1\frac{3}{4}$ **12.** $5 - 1\frac{2}{5}$ **13.** $4\frac{1}{4} - 3\frac{5}{8}$
(55) (58) (58)

14. $3 \times \frac{3}{4} \times 2\frac{2}{3}$ **15.** $6\frac{2}{3} \div 100$ **16.** $2\frac{1}{2} \div 3\frac{3}{4}$
(67) (64) (64)

17. Compare: $\frac{9}{20} \bigcirc 50\%$
(73)

18. **a.** What fraction of this group is shaded?
(73)
 b. What percent of this group is shaded?

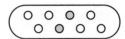

19. If $\frac{5}{6}$ of the 300 seeds sprouted, how many seeds did not sprout?
(75)

* **20.** $6y = 10$ * **21.** $\frac{w}{20} = \frac{12}{15}$
(85) (83)

22. What is the area of this triangle?
(77)

23. Multiply 150 feet by 1 yard per 3 feet.
(89)

$$\frac{150 \text{ ft}}{1} \times \frac{1 \text{ yd}}{3 \text{ ft}}$$

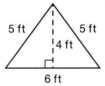

24. Write the prime factorization of 225 using exponents.
(71)

25. **Connect** The length of segment AC is 56 mm. The length of segment BC is
(RF9) 26 mm. How long is segment AB?

* **26.** Compare: $(12 \cdot 7) + (12 \cdot 13) \bigcirc 12(7 + 13)$.
(88)

27. Evaluate $\frac{1}{2}bh$ when $b = 12$ and $h = 10$.
(88)

28. **Model** A square has vertices at the coordinates (2, 0), (0, −2),
(RF27) (−2, 0), and (0, 2). Graph the points on graph paper, and draw segments
  from point to point in the order given. To complete the square, draw a
  fourth segment from (0, 2) to (2, 0).

* **29.** **Evaluate** The square in problem 28 encloses some whole squares
(RF27) and some half squares on the graph paper.

 a. How many whole squares are enclosed by the square?

 b. How many half squares are enclosed by the square?

 c. Counting two half squares as a whole square, calculate the area of the
  entire square.

* **30.** **Analyze** John will toss a coin three times. What is the probability that the coin
(Inv. 7) will land heads up all three times? Express the probability ratio as a fraction and
  as a decimal.

California Mathematics Content Standards

MG 2.0, 2.3 Draw quadrilaterals and triangles from given information about them (e.g., a quadrilateral having equal sides but no right angles, a right isosceles triangle).

MR 2.0, 2.4 Use a variety of methods, such as words, numbers, symbols, charts, graphs, tables, diagrams, and models, to explain mathematical reasoning.

LESSON

90

• Classifying Triangles

Power Up

facts   Power Up K

mental math

    a. Powers/Roots: $\sqrt{100}$

    b. Money: $\$1.98 + \$2.98$

    c. Fractions: The sweatshirt was on sale for $\frac{1}{3}$ off the regular price of $\$24.00$. What is $\frac{1}{3}$ of $\$24.00$?

    d. Decimals: 0.375×100

    e. Estimation: Estimate 25×25 by rounding one factor up and one factor down and then multiplying.

    f. Geometry: A cube has a height of 4 cm. What is the volume of the cube?

    g. Algebra: If $d = 120$, what is $\frac{d}{30}$?

    h. Calculation: $2 \times 2, \times 2, \times 2, -1, \times 2, +2, \div 4, \div 4$

problem solving

Choose an appropriate problem-solving strategy to solve this problem. One state used a license plate that included one letter followed by five digits. How many different license plates could be made that started with the letter A?

New Concept

Thinking Skills

Generalize

Explain in your own words how the number of equal sides of a triangle compares to the number of equal angles it has.

All three-sided polygons are triangles, but not all triangles are alike. We distinguish between different types of triangles by using the lengths of their sides and the measures of their angles. We will first classify triangles based on the lengths of their sides.

Triangles Classified by Their Sides

Name	Example	Description
Equilateral triangle		All three sides are equal in length.
Isosceles triangle		At least two of the three sides are equal in length.
Scalene triangle		All three sides have different lengths.

An **equilateral triangle** has three equal sides and three equal angles.

An **isosceles triangle** has at least two equal sides and two equal angles.

A **scalene triangle** has three unequal sides and three unequal angles.

Next, we consider triangles classified by their angles. In Lesson 20 we learned the names of three different kinds of angles: **acute, right,** and **obtuse.** We can also use these words to describe triangles.

Thinking Skills

Justify

Is an equilateral triangle also an isosceles triangle? Why or why not?

Triangles Classified by Their Angles

Name	Example	Description
Acute triangle		All three angles are acute.
Right triangle		One angle is a right angle.
Obtuse triangle		One angle is an obtuse angle.

Each angle of an equilateral triangle measures 60°, so an equilateral triangle is also an acute triangle. An isosceles triangle may be an acute triangle, a right triangle, or an obtuse triangle. A scalene triangle may also be an acute triangle, a right triangle, or an obtuse triangle.

Example 1

Draw a triangle that matches the description if it is possible. If it is not possible, explain why not.

a. right triangle that is scalene

b. right triangle that is isosceles

c. right triangle that is equilateral

a.

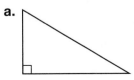

b.

c. On a plane, it is not possible to draw a right equilateral triangle because each angle of an equilateral triangle measures 60°.

Example 2

Draw a triangle that matches the description if it is possible. If it is not possible, explain why not.

    a. acute triangle that is scalene

    b. obtuse triangle that is equilateral

    c. acute triangle that is isosceles

    a.

    b. On a plane, it is not possible to draw an obtuse equilateral triangle because each angle of an equilateral triangle measures 60°.

    c.

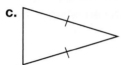

Lesson Practice

    a. One side of an equilateral triangle measures 15 cm. What is the perimeter of the triangle?

    b. (**Verify**) "An equilateral triangle is also an acute triangle." True or false?

    c. (**Verify**) "All acute triangles are equilateral triangles." True or false?

    d. Two sides of a triangle measure 3 inches and 4 inches. If the perimeter is 10 inches, what type of triangle is it?

    e. (**Verify**) "Every right triangle is a scalene triangle." True or false?

Written Practice   *Distributed and Integrated*

1. (**Explain**) The weather forecast stated that the chance of rain for
(54)   Wednesday is 40%. Does this forecast mean that it is more likely to rain or not to rain? Why?

2. A set of 36 shape cards contains an equal number of cards with hexagons,
(54, 69) squares, circles, and triangles. What is the probability of drawing a square
from this set of cards? Express the probability ratio as a fraction and as
a decimal.

3. (Connect) If the sum of three numbers is 144, what is the mean of the
(10) three numbers?

4. The ratio of clear days to cloudy days was 5 to 2. If there were 20 clear
(78) days, how many days were cloudy?

5. Classify this triangle by side lengths. Then classify it by angle measures.
(90)

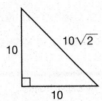

* 6. $2 \cdot 3^2 - 3 + (3 - 1)^3$
(72)

* 7. Write the formula for the perimeter of a rectangle. Then substitute
(87) 12 in. for the length and 6 in. for the width. Solve the equation to find the
perimeter of the rectangle.

8. Arrange these numbers in order from least to greatest:
(40)
$$1, 0, 0.1, -1$$

9. If $\frac{5}{6}$ of the 30 members were present, how many members were absent?
(75)

10. Reduce before multiplying or dividing: $\dfrac{(24)(36)}{48}$
(65)

11. Multiply 8 miles by 8 minutes per mile:
(89)
$$\frac{8 \text{ miles}}{1} \times \frac{8 \text{ min.}}{1 \text{ mile}} =$$

12. $12\frac{5}{6} + 15\frac{1}{3}$
(55)

13. $100 - 9.9$
(41)

14. $\frac{4}{7} \times 100$
(24)

15. $\dfrac{5}{8} = \dfrac{w}{48}$
(37)

16. $0.25 \times \$4.60$
(33)

* 17. (Estimate) The diameter of a circular saucepan is 6 inches. What is the area of
(84) the circular base of the pan? Round the answer to the nearest square inch. (Use
3.14 for π.)

18. Write $3\frac{3}{4}$ as a decimal number and subtract that number from 7.4.
(69)

* 19. What percent of the first ten letters of the alphabet are vowels?
(73)

20. **Connect** Show two ways to simplify this expression: 4 (5 + 6)
(88)

21. **Estimate** Find the product of 6.95 and 12.1 to the nearest whole number.
(33, 46)

* 22. **Analyze** Write and solve a proportion for this statement:
(83) 16 is to 10 as what number is to 25?

23. What is the area of the triangle below?
(77)

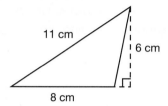

11 cm
6 cm
8 cm

24. This figure is a rectangular prism.
(RF25)
   a. How many faces does it have?

   b. How many edges does it have?

25. **Predict** Each term in this sequence is $\frac{1}{16}$ more than the previous term. What
(14) are the next four terms in the sequence?

$$\frac{1}{16}, \frac{1}{8}, \frac{3}{16}, \frac{1}{4}, \underline{\hspace{1cm}}, \underline{\hspace{1cm}}, \underline{\hspace{1cm}}, \underline{\hspace{1cm}}$$

Use a ruler to find the length and width of the rectangle below to the nearest quarter of an inch. Then refer to the rectangle to answer problems **26** and **27**.

* 26. What is the perimeter of the rectangle?
(87)

* 27. What is the area of the rectangle?
(87)

28. **Connect** The coordinates of the vertices of a parallelogram are (4, 3), (−2, 3),
(RF27) (0, −2), and (−6, −2). What is the area of the parallelogram?

* 29. Simplify:
(79)

    a. (12 cm)(8 cm)
                               b. $\dfrac{36\ \text{ft}^2}{4\ \text{ft}}$

30. Fernando poured water from one-pint bottles into a three-gallon bucket. How
(81) many pints of water could the bucket hold?

Real-World Connection

Alejandra owns a triangular plot of land. She hopes to buy another triangular section adjacent to the one she owns. Use the figure below to find the area of the land Alejandra owns and the area of the land she hopes to buy.

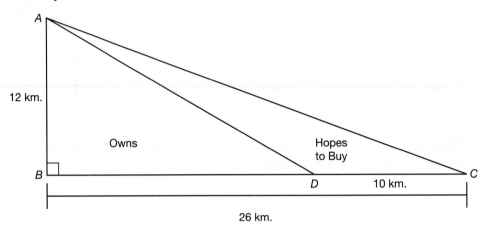

\\ California Mathematics Content Standards

AF 1.0, 1.1 Write and solve one-step linear equations in one variable.

MR 2.0, 2.7 Make precise calculations and check the validity of the results from the context of the problem.

Focus on
Balanced Equations

Equations are sometimes called **balanced equations** because the two sides of the equation "balance" each other. A balance scale can be used as a model of an equation. We replace the equal sign with a balanced scale. The left and right sides of the equation are placed on the left and right trays of the balance. For example, $x + 12 = 33$ becomes

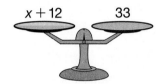

Using a balance-scale model we think of how to simplify the equation to get the unknown number, in this case x, alone on one side of the scale. Using our example, we could remove 12 (subtract 12) from the left side of the scale. However, if we did that, the scale would no longer be balanced. So we make this rule for ourselves.

> Whatever operation we perform on one side of an equation, we also perform on the other side of the equation to maintain a balanced equation.

We see that there are two steps to the process.

Step 1: Select the operation that will isolate the variable.

Step 2: Perform the selected operation on both sides of the equation.

We select "subtract 12" as the operation required to isolate x (to "get x alone"). Then we perform this operation on both sides of the equation.

Select operation:

To isolate x, *subtract 12.*

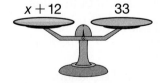

Perform operation:

To keep the scale balanced, *subtract* 12 *from both sides of the equation.*

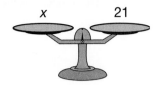

x   21

(**Summarize**) State what we do when we "isolate the variable."

After subtracting 12 from both sides of the equation, x is isolated on one side of the scale, and 21 balances x on the other side of the scale. This shows that x = 21. We check our solution by replacing x with 21 in the original equation.

$$x + 12 = 33$$ original equation

$$21 + 12 = 33$$ replaced x with 21

$$33 = 33$$ simplified left side

Both sides of the equation equal 33. This shows that the solution, x = 21, is correct.

Now we will illustrate a second equation 45 = x + 18.

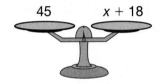

45   x + 18

This time the unknown number is on the right side of the balance scale, added to 18.

1. (**Analyze**) Select the operation that will isolate the variable, and write that operation on your paper.

2. Describe how to perform the operation and keep a balanced scale.

3. (**Discuss**) What will remain on the left and right side of the balance scale after the operation is performed.

We show the line-by-line solution of the equation below.

$$45 = x + 18$$ original equation

$$45 - 18 = x + 18 - 18$$ subtracted 18 from both sides

$$27 = x + 0$$ simplified both sides

$$27 = x$$ x + 0 = x

We check the solution by replacing x with 27 in the original equation.

$$45 = x + 18$$ original equation

$$45 = 27 + 18$$ replaced x with 27

$$45 = 45$$ simplified right side

By checking the solution in the original equation, we see that the solution is correct. Now we revisit the equation to illustrate one more idea.

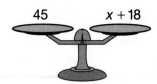

$$45 \qquad x + 18$$

4. **Predict** Suppose the contents of the two trays of the balance scale were switched. That is, $x + 18$ was moved to the left side, and 45 was moved to the right side. Would the scale still be balanced? Write what the equation would be.

Now we will consider an equation that involves multiplication rather than addition.

$$2x = 132$$

Since $2x$ means two x's $(x + x)$, we may show this equation on a balance scale two ways.

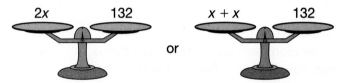

$$2x \qquad 132 \qquad \text{or} \qquad x + x \qquad 132$$

Our goal is to isolate x, that is, to have one x. We must perform the operations necessary to get one x alone on one side of the scale. We do not subtract 2, because 2 is not added to x. We do not subtract an x, because there is no x to subtract from the other side of the equation. To isolate x in this equation, we *divide by 2*. To keep the equation balanced, we *divide both sides by 2.*

Select operation:

To isolate x, divide by 2.

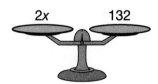

$$2x \qquad 132$$

Perform operation:

To keep the equation balanced, divide both sides by 2.

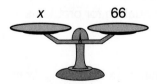

$$x \qquad 66$$

Here we show the line-by-line solution of this equation.

$$2x = 132 \qquad \text{original equation}$$
$$\frac{2x}{2} = \frac{132}{2} \qquad \text{divided both sides by 2}$$
$$1x = 66 \qquad \text{simplified both sides}$$
$$x = 66 \qquad 1x = x$$

Next we show the check of the solution.

$$2x = 132 \quad \text{original equation}$$
$$2(66) = 132 \quad \text{replaced } x \text{ with 66}$$
$$132 = 132 \quad \text{simplified left side}$$

This check shows that the solution, $x = 66$, is correct.

5. **Model** Draw a balance-scale model for the equation $3x = 132$.

6. **Analyze** Select the operation that will isolate the variable, and write that operation on your paper.

7. Describe how to perform the operation and keep a balanced scale.

8. **Model** Draw a balance scale and show what is on both sides of the scale after the operation is performed.

9. **Justify** Write the line-by-line solution of the equation.

10. Show the check of the solution.

Most students choose to solve the equation $3x = 132$ by dividing both sides of the equation by 3. There is another operation that could be selected that is often useful, which we will describe next. First note that the number multiplying the variable, in this case 3, is called the **coefficient** of x. Instead of dividing by the coefficient of x, we could choose to **multiply by the reciprocal** of the coefficient. In this case we could multiply by $\frac{1}{3}$. In either case the goal is to make the coefficient 1.

$$3x = 132$$
$$\frac{1}{3} \cdot 3x = \frac{1}{3} \cdot 132$$
$$1x = \frac{132}{3}$$
$$x = 44$$

Discuss Why is $\frac{1}{3}$ the reciprocal of 3?

When solving equations with whole number or decimal number coefficients, it is usually easier to think about dividing by the coefficient. However, when solving equations with fractional coefficients, it is usually easier to multiply by the reciprocal of the coefficient. Refer to the following equation for problems 11–14:

$$\frac{3}{4}x = \frac{9}{10}$$

11. **Analyze** Select the operation that will result in $\frac{3}{4}x$ becoming $1x$ in the equation. Tell why you chose the operation.

12. Describe how to perform the operation and keep the equation balanced.

13. Write a line-by-line solution of the equation.

14. Show the check of the solution.

We find that the solution to the equation is $\frac{6}{5}$ (or $1\frac{1}{5}$). In arithmetic we usually convert improper fractions to mixed numbers. In algebra we usually leave improper fractions in improper form unless the problem states or implies that a mixed number answer is preferable.

Evaluate For each of the following equations:

    a. State the operation you select to isolate the variable.

    b. Describe how to perform the operation and keep the equation balanced.

    c. Write a line-by-line solution of the equation.

    d. Show the check of the solution.

15. $x + 2.5 = 7$

16. $3.6 = y + 2$

17. $4w = 132$

18. $1.2m = 1.32$

19. $x + \dfrac{3}{4} = \dfrac{5}{6}$

20. $\dfrac{3}{4}x = \dfrac{5}{6}$

21. **Formulate** Make up your own equations. Solve and check each equation.

    a. Make up an addition equation with decimal numbers.

    b. Make up a multiplication equation with a fractional coefficient.

We have used some **properties of equality** to solve the equations in this investigation. The following table summarizes properties of equality involving the four operations of arithmetic.

1. Addition: If $a = b$, then $a + c = b + c$.

2. Subtraction: If $a = b$, then $a - c = b - c$.

3. Multiplication: If $a = b$, then $ac = bc$.

4. Division: If $a = b$, and if $c \neq 0$, then $\frac{a}{c} = \frac{b}{c}$.

The addition rule means this: If two quantities are equal and the same number is added to both quantities, then the sums are also equal.

22. Summarize Use your own words to describe the meaning of each of the other rules in the table.

Investigate Further

a. Justify Write an equation for each problem below. Then, choose between mental math, estimation, paper and pencil, or calculator to solve the equation. Justify your choice.

Marsha hiked a total of 12 miles on two trails in a national park. One trail was twice as long as the other. How long was each trail? Let s = the shorter trail.

Jake earned $50. He put half of his money in the bank. He spent half of the money he did not put in the bank. How much money does Jake have left? Let m the money Jake has left.

Mr. Wang's business averages $14,250 in expenses each month. Company salaries average $26,395 each month. The rent for the company is $2,825 each month. How much should Mr. Wang budget for salaries, expenses, and rent for the year? Let a = the total budget.

b. Justify Look at the four equations below. Which equation does not belong? Explain your reasoning.

$$4x = 8 \qquad 132 = x + 130 \qquad 22x = 66 \qquad x + 12 = 14$$

California Mathematics Content Standards
AF 1.0, 1.1 Write and solve one-step linear equations in one variable.
MR 2.0, 2.3 Estimate unknown quantities graphically and solve for them by using logical reasoning and arithmetic and algebraic techniques.
MR 2.0, 2.4 Use a variety of methods, such as words, numbers, symbols, charts, graphs, tables, diagrams, and models, to explain mathematical reasoning.
MR 2.0, 2.7 Make precise calculations and check the validity of the results from the context of the problem.

• Using Properties of Equality to Solve Equations

Power Up

facts	Power Up M
mental math	**a. Powers/Roots:** $5^2 + \sqrt{100}$
	b. Number Sense: $1000 - 875$
	c. Geometry: What is the perimeter of this rectangle? Dimensions are in centimeters.
	d. Money: Double $125.00.
	e. Statistics: The attendance figures for the first three games were 120, 134, and 188. What is the range of those figures?
	f. Number Sense: 20×42
	g. Measurement: How many pints are in a quart?
	h. Calculation: $3 \times 4, \div 2, \times 3, + 2, \times 2, + 2, \div 2, \div 3$

3.6
4.4

problem solving
Choose an appropriate problem-solving strategy to solve this problem. If Sam can read 20 pages in 30 minutes, how long will it take Sam to read 200 pages?

New Concept

We have practiced solving equations by inspection using logical reasoning. In this lesson we will use inverse operations to solve equations. Inverse operations "undo" each other.

- Addition and subtraction are inverse operations.
  A number plus five, minus five equals the number.

- Multiplication and division are inverse operations.
  A number times five, divided by five equals the number.

We use inverse operations to isolate the variable in an equation. We can illustrate the process with a balance-scale model:

The balanced scale shows that the weight on the left side equals the weight on the right side. We can represent the relationship with this equation:

$$x + 5 = 12$$

To isolate x we need to remove the 5. Since 5 is added to x, we remove the 5 by subtracting (which is the inverse operation of addition). However, if we subtract 5 from the left side, we must also subtract 5 from the right side to keep the scale balanced.

$$x + 5 - 5 = 12 - 5$$

$$x = 7$$

We find that x equals 7.

Isolating a variable means to get the variable, like x, "alone" on one side of the equal sign.

Notice these two aspects of isolating a variable.

1. We choose the operation that isolates the variable.

2. We perform the operation on both sides of the equals sign to keep the equation balanced.

Performing the same operation on both sides of an equation preserves the equality of the equation. Below are four **properties of equality** we can use to manipulate equations.

Operation Properties of Equality		
Addition:	If $a = b$, then	$a + c = b + c$.
Subtraction:	If $a = b$, then	$a - c = b - c$.
Multiplication:	If $a = b$, then	$ac = bc$.
Division:	If $a = b$, then	$\frac{a}{c} = \frac{b}{c}$ if c is not 0.

(Formulate) Use numbers to give an example of each property.

Example 1

After spending $2.30, Rondall had $5.70. How much money did he have before he spent $2.30? Solve this equation to find the answer. Then check your answer.

$$x - 2.30 = 5.70$$

We see that 2.30 is subtracted from x. To isolate x we add 2.30 to the left side to undo the subtraction, and we add 2.30 to the right side to keep the equation balanced. The result is a simpler equivalent equation.

Step:	Justification:
$x - 2.30 = 5.70$	Given equation
$x - 2.30 + 2.30 = 5.70 + 2.30$	Added 2.30 to both sides
$x = 8.00$	Simplified

We find that Rondall had **$8.00.** To check our answer, we substitute 8.00 for x in the equation.

$x - 2.30 = 5.70$	Equation
$8.00 - 2.30 = 5.70$	Replaced x with 8.00
$5.70 = 5.70$	Simplified

Since both sides of the equation equal 5.70, the solution is correct.

Example 2

If a 12-month subscription to a magazine costs $8.40, what is the average cost per month? Solve this equation to find the answer.

Then check your answer.

$$12m = 8.40$$

The variable is multiplied by 12. To isolate the variable we divide the left side by 12. To keep the equation balanced we divide the right side by 12.

Step:	Justification:
$12m = 8.40$	Given equation
$\dfrac{12m}{12} = \dfrac{8.4}{12}$	Divided both sides by 12
$1m = 0.70$	Simplified
$m = 0.70$	Simplified

(**Explain**) In the context of the problem, what does $m = 0.70$ mean?

We check the answer by substituting 0.70 for m.

$12m = 8.40$	Equation
$12(0.70) = 8.40$	Replaced m with 0.70
$8.40 = 8.40$	Simplified

Our answer is correct.

Example 3

Solve and check: $\frac{2}{3}w = \frac{3}{4}$

The variable is multiplied by $\frac{2}{3}$. We can divide by $\frac{2}{3}$, or multiply both sides by the multiplicative inverse (reciprocal) of $\frac{2}{3}$.

Thinking Skills

Connect

When the variable is multiplied by a fraction, why do we multiply by the reciprocal to help us isolate the variable?

Step:	Justification:
$\frac{2}{3}w = \frac{3}{4}$	Given equation
$\frac{3}{2} \cdot \frac{2}{3}w = \frac{3}{2} \cdot \frac{3}{4}$	Multiplied both sides by $\frac{3}{2}$
$1w = \frac{9}{8}$	Simplified
$w = \frac{9}{8}$	Simplified

We substitute $\frac{9}{8}$ for w in the equation to check.

$\frac{2}{3}w = \frac{3}{4}$	equation
$\frac{2}{3} \times \frac{9}{8} = \frac{3}{4}$	replaced w with $\frac{9}{8}$
$\frac{\cancel{2}^1}{\cancel{3}_1} \times \frac{\cancel{9}^3}{\cancel{8}_4} = \frac{3}{4}$	reduced
$\frac{3}{4} = \frac{3}{4}$	simplified

Our answer is correct.

It is customary to express a solution with the variable on the left side of the equal sign. Since the quantities on either side of an equal sign are equal, we can reverse everything on one side of an equation with everything on the other side of the equation. This property of equality is called the **symmetric property**.

Symmetric Property of Equality

If $a = b$, then $b = a$.

Example 4

Solve and check: $1.5 = x - 2.3$

We may apply the Symmetric Property at any step.

Step:	Justification:
$1.5 = x - 2.3$	Given equation
$1.5 + 2.3 = x - 2.3 + 2.3$	Added 2.3 to both sides
$3.8 = x$	Simplified
$x = 3.8$	Symmetric Property

We check our answer.

$$1.5 = x - 2.3 \qquad \text{equation}$$
$$1.5 = 3.8 - 2.3 \qquad \text{replaced } x \text{ with } 3.8$$
$$1.5 = 1.5 \qquad \text{simplified}$$

Our answer is correct.

We have used cross products to solve proportions.

$$\frac{15}{21} \bowtie \frac{w}{70}$$

$$70 \cdot 15 = w \cdot 21$$

Cross products are equal because they are numerators of equal fractions with a common denominator.

$$\frac{15}{21} = \frac{w}{70}$$

$$\frac{70 \cdot 15}{70 \cdot 21} = \frac{w \cdot 21}{70 \cdot 21}$$

$$70 \cdot 15 = w \cdot 21$$

Another way to understand why cross products are equal is to multiply both sides of the proportion by the denominators.

Consider the proportion again.

$$\frac{15}{21} = \frac{w}{70}$$

If we multiply both sides by $\frac{21}{1}$, the denominator on the left-hand side cancels. Since $\frac{21}{1}$ and $\frac{1}{21}$ are multiplicative inverses, their product is 1.

$$\frac{\cancel{21}}{1} \cdot \frac{15}{\cancel{21}} = \frac{w}{70} \cdot \frac{21}{1}$$

$$\frac{15}{1} = \frac{w \cdot 21}{70}$$

Then we multiply both sides by $\frac{70}{1}$, and the other denominator cancels.

$$\frac{70}{1} \cdot \frac{15}{1} = \frac{w \cdot 21}{\cancel{70}} \cdot \frac{\cancel{70}}{1}$$

We end up with equal cross products.

$$70 \cdot 15 = w \cdot 21$$

Discuss In the first step of this method, how do we know that the equality statement is still true once we have multiplied both sides of the equation by $\frac{21}{1}$?

Example 5

Solve this proportion using two different methods.

$$\frac{2}{3} = \frac{x}{4}$$

a. Multiply both sides of the equation by the denominators.

b. Use equal cross products.

a. We will multiply both sides of the equation by 3 and by 4.

$\frac{2}{3} = \frac{x}{4}$	Original equation
$\frac{3}{1} \cdot \frac{2}{3} = \frac{x}{4} \cdot \frac{3}{1}$	Multiplied both sides by 3
$\frac{2}{1} = \frac{x \cdot 3}{4}$	Canceled the denominator 3
$\frac{4}{1} \cdot \frac{2}{1} = \frac{x \cdot 3}{4} \cdot \frac{4}{1}$	Multiplied both sides by 4
$4 \cdot 2 = x \cdot 3$	Canceled the denominator 4
$8 = x \cdot 3$	Simplified
$\frac{8}{3} = \frac{x \cdot 3}{3}$	Divided both side by 3
$\frac{8}{3} = x$	Simplified

b. we will solve by using equal cross products

$\frac{2}{3} = \frac{x}{4}$	Original equation
$2 \cdot 4 = 3 \cdot x$	Cross products are equal
$8 = 3x$	Simplified
$\frac{8}{3} = \frac{3x}{3}$	Divided both sides by 3
$\frac{8}{3} = x$	Simplified

Analyze Explain why using cross-multiplication is a reasonable method for solving proportion problems.

Lesson Practice

a. What does "isolate the variable" mean?

b. Which operation is the inverse of subtraction?

c. Which operation is the inverse of multiplication?

Justify Solve each equation. Justify the steps. Then check your answer.

d. $x + \frac{1}{2} = \frac{3}{4}$

e. $x - 1.3 = 4.2$

f. $1.2 = \frac{x}{4}$

g. $\frac{3}{5}x = \frac{1}{4}$

h. Peggy bought 5 pounds of oranges for $4.50. Solve the equation $5x = \$4.50$ to find the price per pound.

i. From 8 a.m. to noon, the temperature rose 5 degrees to 2°C. Solve the equation $t + 5 = 2$ to find the temperature at 8 a.m.

* 1. The ratio of the length to the width of the rectangular lot was 5 to 2.
(86)   *If the lot was 60 ft wide, how long was the lot?*

* 2. Keenan does not know the correct answer to two multiple-choice
(Inv. 7)   questions. The choices are A, B, C, and D. If Keenan just guesses, what is
the probability that Keenan will guess both answers correctly?

3. If the sum of four numbers is 144, what is the mean of the four numbers?
(10)

4. The rectangular prism shown below is constructed
(RF29)   of 1-cubic-centimeter blocks. What is the volume of
the prism?

5. Write $\frac{9}{25}$ as a decimal number and as a percent.
(69, 73)

6. Write $3\frac{1}{5}$ as a decimal number and add it to 3.5. What is the sum?
(69)

7. What number is 45% of 80?
(35)

8. $(0.3)^3$
(41, 71)

9. $\left(2\frac{1}{2}\right)^2$
(62, 71)

10. With one toss of a pair of number cubes, what is the probability that the total
(80)   rolled will be a prime number? (Add the probability for each prime number
total.)

11. Twenty of the two dozen members voted yes. What fraction of the members
(25)   voted yes?

12. (**Analyze**) If the rest of the members in problem 11 voted no, then what was the
(16)   ratio of "no" votes to "yes" votes?

Find each unknown number:

13. $w + 4\frac{3}{4} = 9\frac{1}{3}$
(39, 58)

* 14. $\frac{6}{5} = \frac{m}{30}$
(83)

* 15. (**Conclude**) In what type of triangle are all three sides the same length?
(90)

16. What mixed number is $\frac{3}{8}$ of 100?
(24)

17. Which property is illustrated by the equation
(88)

$$\frac{5}{6}\left(\frac{6}{5} + \frac{6}{7}\right) = \frac{5}{6} \cdot \frac{6}{5} + \frac{5}{6} \cdot \frac{6}{7}$$

18. A triangular prism has how many faces?
(RF25)

19. How many pints of milk is 5 quarts of milk?
(81)
   a. Create a function table showing the number of pints in 1, 2, and
      3 quarts. Graph the ordered pairs in your table and estimate the
      answer to the question.

   b. Find the rule for the function and use it to find the number of pints in 5 quarts.

* 20. **Represent** Use a factor tree to find the prime factors of 800. Then write
(61, 71) the prime factorization of 800 using exponents.

21. Round the decimal number one hundred twenty-five thousandths to the
(46) nearest tenth.

* 22. $0.08n = \$1.20$
(85)

23. The diagonal segment through this rectangle divides the rectangle into two
(77) congruent right triangles. What is the area of one of the triangles?

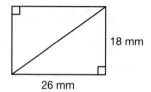

24. Write $\frac{17}{20}$ as a percent.
(73)

* 25. $\frac{2}{7}x = \frac{3}{4}$
(91)

26. Multiply 5 hours times 7 dollars per hour.
(89)

$$\frac{5 \text{ hours}}{1} \times \frac{\$7}{1 \text{ hour}} =$$

27. **a.** What is the probability of rolling a 6 with a single roll of a number cube?
(54)

　　b. What is the probability of rolling a number less
　　　　than 6 with a single roll of a number cube?

　　c. Name the event and its complement. Then describe
　　　　the relationship between the two probabilities.

28. **Represent** The coordinates of the four vertices of a quadrilateral are
(RF27, $(-3, -2)$, $(0, 2)$, $(3, 2)$, and $(5, -2)$. What is the name for this type of
59) quadrilateral?

* 29. **Explain** The formula for the area of a triangle is
(87)

$$A = \frac{bh}{2}$$

If the base measures 20 cm and the height measures 15 cm,
then what is the area of the triangle? Explain your thinking.

30. **Generalize** Write the rule for this sequence. Then write the next
(1, 14) four numbers.

$$\frac{1}{16}, \frac{1}{8}, \frac{3}{16}, \frac{1}{4}, \frac{5}{16}, \frac{3}{8} \text{ ———, ———, ———, ———, } \dots$$

California Mathematics Content Standards
NS **1.0**, **1.4** Calculate given percentages of quantities
and solve problems involving discounts at sales,
interest earned, and tips.
MR **1.0, 1.3** Determine when and how to break a
problem into simpler parts.
MR **2.0, 2.3** Estimate unknown quantities graphically
and solve for them by using logical reasoning and
arithmetic and algebraic techniques.

• Writing Fractions and Decimals as Percents, Part 2

Power Up

facts   Power Up L

mental math

    a. Number Sense: $30 \cdot 50$

    b. Percent: 50% of 24

    c. Decimals: 100×1.25

    d. Money: Shelby purchased a DVD for $14.75. If she paid with a $20 bill, how much change should she receive?

    e. Probability: Kody flipped a quarter and it landed heads up. If he flips the quarter again, what is the probability it will land tails up?

    f. Algebra: If $b = 600$, what is $\frac{b}{30}$?

    g. Algebra: If $n = 8$, what does $5n$ equal?

    h. Calculation: $\sqrt{36} + 4, \times 3, + 2, \div 4, + 1, \sqrt{}^{1}$

problem solving

Choose an appropriate problem-solving strategy to solve this problem. The basketball team's points-per-game average is 88 after its first four games. How many points does the team need to score during its fifth game to have a points-per-game average of 90?

New Concept

Since Lesson 73 we have practiced changing a fraction or decimal to a percent by writing an equivalent fraction with a denominator of 100.

$$\frac{3}{5} = \frac{60}{100} = 60\%$$

$$0.4 = 0.40 = \frac{40}{100} = 40\%$$

---

[1] Read $\sqrt{}$ as "find the square root."

In this lesson we will practice another method of changing a fraction to a percent. Since 100% equals 1, we can multiply a fraction by 100% to form an equivalent number. Here we multiply $\frac{3}{5}$ by 100%:

$$\frac{3}{5} \times \frac{100\%}{1} = \frac{300\%}{5}$$

Then we simplify and find that $\frac{3}{5}$ equals 60%.

$$\frac{300\%}{5} = 60\%$$

We can use the same procedure to change decimals to percents. Here we multiply 0.375 by 100%.

$$0.375 \times 100\% = 37.5\%$$

<table><tr><td>To change a number to a percent, multiply the number by 100%.</td></tr></table>

Thinking Skills

Justify

How can you use mental math to change a decimal to a percent?

Example 1

Change $\frac{1}{3}$ to a percent.

We multiply $\frac{1}{3}$ by 100%.

$$\frac{1}{3} \times \frac{100\%}{1} = \frac{100\%}{3}$$

To simplify, we divide 100% by 3 and write the quotient as a mixed number.

$$\begin{array}{r} 33\frac{1}{3}\% \\ 3\overline{)100\%} \\ \underline{9} \\ 10 \\ \underline{9} \\ 1 \end{array}$$

Example 2

Write 1.2 as a percent.

We multiply 1.2 by 100%.

$$1.2 \times 100\% = \textbf{120\%}$$

In some applications a percent may be greater than 100%. If the number we are changing to a percent is greater than 1, then the percent is greater than 100%.

Example 3

Write $2\frac{1}{4}$ as a percent.

We show two methods below.

Method 1: We split the whole number and fraction. The mixed number $2\frac{1}{4}$ means "$2 + \frac{1}{4}$." We change each part to a percent and then add.

$$2 + \frac{1}{4}$$

$$200\% + 25\% = \mathbf{225\%}$$

Method 2: We change the mixed number to an improper fraction. The mixed number $2\frac{1}{4}$ equals the improper fraction $\frac{9}{4}$. We then change $\frac{9}{4}$ to a percent.

$$\frac{9}{\overset{}{\underset{1}{4}}} \times \frac{\overset{25}{100\%}}{1} = \mathbf{225\%}$$

Example 4

Write $2\frac{1}{6}$ as a percent.

Method 1 shown in Example 3 is quick, if we can recall the percent equivalent of a fraction. Method 2 is easier if the percent equivalent does not readily come to mind. We will use method 2 in this example. We write the mixed number $2\frac{1}{6}$ as the improper fraction $\frac{13}{6}$ and multiply by 100%.

$$\frac{13}{6} \times \frac{100\%}{1} = \frac{1300\%}{6}$$

Now we divide 1300% by 6 and write the quotient as a mixed number.

$$\frac{1300\%}{6} = \mathbf{216\frac{2}{3}\%}$$

Example 5

Twenty of the thirty students on the bus were girls. What percent of the students on the bus were girls?

We first find the fraction of the students that were girls. Then we convert the fraction to a percent.

Girls were $\frac{20}{30}$ $\left(\text{or } \frac{2}{3}\right)$ of the students on the bus. Now we multiply the fraction by 100%, which is the percent name for 1. We can use either $\frac{20}{30}$ or $\frac{2}{3}$, as we show below.

$$\frac{20}{\underset{3}{30}} \times \overset{10}{100}\% = \frac{200\%}{3} = 66\frac{2}{3}\%$$

or

$$\frac{2}{3} \times 100\% = \frac{200\%}{3} = 66\frac{2}{3}\%$$

We find that $\mathbf{66\frac{2}{3}\%}$ of the students on the bus were girls.

Lesson Practice   Change each decimal number to a percent by multiplying by 100%:

    a. 0.5   **b.** 0.06   **c.** 0.125

    d. 0.45   **e.** 1.3   **f.** 0.025

    g. 0.09   **h.** 1.25   **i.** 0.625

Change each fraction or mixed number to a percent by multiplying by 100%:

j. $\frac{2}{3}$

k. $\frac{1}{6}$

l. $\frac{1}{8}$

m. $1\frac{1}{4}$

n. $2\frac{4}{5}$

o. $1\frac{1}{3}$

p. What percent of this rectangle is shaded?

q. (Connect) What percent of a yard is a foot?

* 1. (Analyze) Ten of the thirty students on the bus were boys. What percent of the
(92) students on the bus were boys?

2. (Connect) On the Celsius scale water freezes at 0°C and boils at 100°C. What
(10) temperature is halfway between the freezing and boiling temperatures
of water?

3. (Connect) If the length of segment *AB* is $\frac{1}{3}$ the length of segment *AC*, and if
(RF9) segment *AC* is 12 cm long, then how long is segment *BC*?

4. What percent of this group is shaded?
(92)

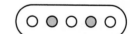

* 5. Change $1\frac{2}{3}$ to a percent by multiplying
(92) $1\frac{2}{3}$ by 100%.

* 6. Change 1.5 to a percent by multiplying 1.5 by 100%.
(92)

7. $6.4 - 6\frac{1}{4}$ (Begin by writing $6\frac{1}{4}$ as a decimal number.)
(69)

8. $\frac{2}{5}x = \frac{2}{3}$
(91)

9. How much is $\frac{3}{4}$ of 360?
(65)

Tommy placed a cylindrical can of spaghetti sauce on the counter.
He measured the diameter of the can and found that it was about
8 cm. Use this information to answer problems **10** and **11**.

10. The label wraps around the circumference of the can.
(42) How long does the label need to be?

* 11. (Analyze) How many square centimeters of countertop does the can
(84) occupy?

12. $3\frac{1}{2} + 1\frac{3}{4} + 4\frac{5}{8}$
(56)

13. $\frac{9}{10} \cdot \frac{5}{6} \cdot \frac{8}{9}$
(67)

14. Multiply 36 inches by 1 foot per 12 inches.
(89)

$$\frac{36 \text{ in.}}{1} \times \frac{1 \text{ ft}}{12 \text{ in.}} =$$

15. $8.47 + 95¢ + $12
(RF1)

16. $37.5 \div 100$
(41)

17. $\dfrac{3}{7} = \dfrac{21}{x}$
(83)

18. $33\dfrac{1}{3} \div 100$
(64)

19. If ninety percent of the answers were correct, then what percent were incorrect?
(27)

20. Write the decimal number one hundred twenty and three hundredths.
(RF19)

21. Arrange these numbers in order from least to greatest:
(74)

$$-2.5, \dfrac{2}{5}, \dfrac{5}{2}, -5.2$$

22. **Conclude** A pyramid with a square base has how many edges?
(RF25)

23. What is the area of this parallelogram?
(66)

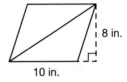

8 in.

10 in.

* 24. **Multiple Choice** The parallelogram in problem **23** is divided into two congruent triangles. Both triangles may be described as which of the following?
(90)

A acute   **B** right   **C** obtuse

25. During the year, the temperature ranged from $-37°F$ in winter to $103°F$ in summer. How many degrees was the range of temperature for the year?
(9)

* 26. **Model** The coordinates of the three vertices of a triangle are (0, 0), (0, −4), and (−4, 0). Graph the triangle and find its area.
(RF27, 77)

27. Latondra's first nine test scores are shown below.
(Inv. 4)

$$21, 25, 22, 19, 22, 24, 20, 22, 24$$

a. What is the mode of these scores?

b. What is the median of these scores?

28. Evaluate $a(b + c)$ if $a = 2$, $b = 3$, and $c = 4$.
(88)

29. Sandra filled the aquarium with 16 quarts of water. How many gallons of water did Sandra pour into the aquarium?
(Inv. 5, 81)

a. Create a function table showing the number of quarts in 1, 2, and 3 gallons. Graph the ordered pairs in the function table and estimate the answer.

b. Determine a rule for the function table and use it to write and solve an algebraic equation to answer the question.

30. A bag contains lettered tiles, two for each letter of the alphabet. What is the probability of drawing a tile with the letter A? Express the probability ratio as a fraction and as a decimal rounded to the nearest hundredth.
(54, 69)

California Mathematics Content Standards

AF 2.0, 2.1 Convert one unit of measurement to
another (e.g., from feet to miles, from centimeters
to inches).
AF 2.0, 2.3 Solve problems involving rates, average
speed, distance, and time.

• Formulas and Substitution

Power Up

facts	Power Up K
mental math	**a. Number Sense:** $234 - 50$
	b. Percent: 25% of 24
	c. Money: Jerome bought paint for $7.99 and a paintbrush for $2.47. What was the total cost?
	d. Decimals: $1.2 \div 100$
	e. Estimation: Estimate the area of this rectangle.
	f. Number Sense: 30×25
	g. Algebra: If $x = 5$, what does $5x$ equal?
	h. Calculation: $8 \times 9, + 3, \div 3, \sqrt{}, \times 6, + 3, \div 3, - 10$

4.8 cm
7.1 cm

problem solving	Choose an appropriate problem-solving strategy to solve this problem. Denzel put 2 purple marbles, 7 red marbles, and 1 brown marble in a bag and shook the bag. If he reaches in and chooses a marble without looking, what is the probability that he chooses a red marble? A purple or brown marble? What is the probability of *not* choosing a red marble? If Denzel does choose a red marble, but gives it away, what is the probability he will choose another red marble?

New Concept

A formula is a literal equation that describes a relationship between two or more variables. Formulas are used in mathematics, science, economics, the construction industry, food preparation—anywhere that measurement is used.

To use a formula, we replace the letters in the formula with measures that are known. Then we solve the equation for the measure we wish to find.

Example 1

Thinking Skills

Analyze

What measures do you know? What unknown measure do you need to find?

Use the formula $d = rt$ to find t when d is 36 and r is 9.

This formula describes the relationship between distance (d), rate (r), and time (t). We replace d with 36 and r with 9. Then we solve the equation for t.

$d = rt$	formula
$36 = 9t$	substituted
$t = 4$	divided by 9

Example 2

Use the formula $d = rt$ to find the distance Anthony traveled after driving for $2\frac{1}{2}$ hours at 60 miles per hour. Explain why your answer is reasonable.

We are given the time and the rate. We replace t with $2\frac{1}{2}$ hr and r with $60\frac{mi}{hr}$.

$d = rt$	Formula
$d = \left(60\frac{mi}{hr}\right)\left(2\frac{1}{2}\,hr\right)$	Substituted
$d = \left(\frac{60\,mi}{1\,hr}\right)\left(\frac{5\,hr}{2}\right)$	Rewrote numbers as improper fractions
$d = \left(\frac{\overset{30}{\cancel{60}}\,mi}{1\,\cancel{hr}}\right)\left(\frac{5\,\cancel{hr}}{\underset{1}{\cancel{2}}}\right)$	Cancelled
$d = 150\,mi$	Multiplied

Anthony traveled **150 mi.**

This answer is reasonable because for each hour that Anthony drove, he traveled 60 miles. In two hours, he drove 120 miles, and in a half hour he drove another 30 miles. Together this is 150 miles.

Example 3

Alexandria biked 64 miles in 4 hours. What was her average speed?

During her bike ride, Alexandria may not have traveled at a constant speed. Some stretches may have been fast and others slow. Her average speed is the rate at which she would have completed the trip if she had ridden at a constant rate. We can find her average speed with the distance formula.

We are told the distance and time. We replace d with 64 miles and t with 4 hours.

$d = rt$	Formula
$64\,mi = r \times 4\,hr$	Substituted
$\frac{64\,mi}{4\,hr} = r$	Divided both sides by 4 hr
$r = 16\frac{mi}{hr}$	Simplified

Alexandria's average speed was **16 miles per hour.** The rate at which she rode may have been faster at some points and slower at others, but averaging 16 miles per hour for 4 hours, she completed the 64-mile trip.

Example 4

Use the formula $F = 1.8C + 32$ to find F when C is 37.

This formula is used to convert measurements of temperature from degrees Celsius to degrees Fahrenheit. We replace C with 37 and simplify.

$F = 1.8C + 32$	formula
$F = 1.8(37) + 32$	substituted
$F = 66.6 + 32$	multiplied
$F = \mathbf{98.6}$	added

Thus, 37 degrees Celsius equals 98.6 degrees Fahrenheit.

Discuss How could we use the formula to convert from degrees Fahrenheit to degrees Celsius?

Lesson Practice

a. Use the formula $A = bh$ to find b when A is 20 and h is 4.

b. Use the formula $A = \frac{1}{2}bh$ to find b when A is 20 and h is 4.

c. Use the formula $d = rt$ to find t when d is 100 and r is 60.

d. Use the formula $F = 1.8C + 32$ to find F when C is -40.

e. **Connect** The formula for converting from Fahrenheit to Celsius is often given as $F = \frac{9}{5}C + 32$. How are $\frac{9}{5}$ and 1.8 related?

Written Practice

Distributed and Integrated

1. What is the total price of a $45.79 item when 7% sales tax is added to
(35) the price?

* 2. Jeff is 1.67 meters tall. How many centimeters tall is Jeff ? (Multiply
(89) 1.67 meters by 100 centimeters per meter.)

* 3. **Analyze** If $\frac{5}{8}$ of the 40 seeds sprouted, how many seeds did not sprout?
(75)

4. The ratio of wigglers to sliders was 7 to 3. If there were 21 sliders, how many
(78) wigglers were there?

* 5. Change $\frac{1}{6}$ to its percent equivalent by multiplying $\frac{1}{6}$ by 100%.
(92)

* 6. **Analyze** What is the percent equivalent of 2.5?
(92)

7. How much money is 30% of $12.00?
(35)

8. There are 12 markers in a box: 4 blue, 4 red, and 4 yellow. Without looking, Lee
(80) gives one marker to Keondra and takes one for himself. What is the probability that both markers will be blue?

9. **Evaluate** The circumference of the front tire on Elizabeth's bike is
(RF16, 8) 6 feet. How many complete turns does the front wheel make as Elizabeth rides down her 30-foot driveway?

10. **Multiple Choice** The expression $2(3 + 4)$ equals which of the following?
(88)

A $(2 \cdot 3) + 4$ **B** $(2 \cdot 3) + (2 \cdot 4)$

C $2 + 7$ **D** $23 + 24$

11. $\dfrac{3}{4} + \dfrac{3}{5}$ **12.** $18\dfrac{1}{8} - 12\dfrac{1}{2}$
(55)   (58)

13. $3\dfrac{3}{4} \times 2\dfrac{2}{3} \times 1\dfrac{1}{10}$ **14.** $\dfrac{2^5}{2^3}$
(67)   (72)

15. How many fourths are in $2\dfrac{1}{2}$?
(64)

16. $12 + 8.75 + 6.8$ **17.** $(1.5)^2$
(41)   (32, 41)

18. $6\dfrac{2}{5} \div 0.8$ (decimal answer)
(69)

19. **Estimate** Find the sum of $6\dfrac{1}{4}$, 4.95, and 8.21 by rounding each number to
(46) the nearest whole number before adding. Explain how you arrived at your answer.

* 20. **Analyze** The diameter of a round tabletop is 60 inches.
(84)

a. What is the radius of the tabletop?

b. What is the area of the tabletop? (Use 3.14 for π.)

21. Arrange these numbers in order from least to greatest:
(73)

$$\dfrac{1}{4}, 4\%, 0.4$$

Find each unknown number.

22. $y + 3.4 = 5$ **23.** $\dfrac{4}{8} = \dfrac{x}{12}$
(39)   (83)

24. $\dfrac{1}{2}x = \dfrac{1}{3}$
(81)

* 25. **Connect** $\overline{AB}$ is 24 mm long. $\overline{AC}$ is 42 mm long. How long is $\overline{BC}$?
(RF9)

A   B   C

26. The formula $c = 2.54n$ is used to convert inches (n) to cm (c).
(93) Find c when n is 12.

27. What is the ratio of a pint of water to a quart of water?
(81)

* 28. The formula for the area of a parallelogram is $A = bh$. If the base of a
(87) parallelogram is 1.2 m and the height is 0.9 m, what is the area of the parallelogram? How can estimation help you check your answer?

* 29. Multiply 2.5 liters by 1000 milliliters per liter.
(89)

$$\frac{2.5 \text{ liters}}{1} \times \frac{1000 \text{ milliliters}}{1 \text{ liter}}$$

30. If this spinner is spun once, what is the probability that the arrow will
(54, 69) end up pointing to an even number? Express the probability ratio as a
fraction and as a decimal.

*Real-World
Connection*

The local university football stadium seats 60,000 fans, and
average attendance at home games is 48,500. It has been determined
that an average fan consumes 2.25 beverages per game.

a. If each beverage is served in a cup, about how many cups are used during an
average game? Express your answer in scientific notation.

b. Next week is the homecoming game, which is always sold out. A box of cups
contains 1×10^3 cups. How many boxes of cups will be needed for
the game?

✎ *California Mathematics Content Standards*

MG 2.0, 2.1 Identify angles as vertical, adjacent, complementary, or supplementary and provide descriptions of these terms.

MR 2.0, 2.4 Use a variety of methods, such as words, numbers, symbols, charts, graphs, tables, diagrams, and models, to explain mathematical reasoning.

• Transversals

Power Up

facts	Power Up M
mental math	**a. Number Sense:** $50 \cdot 70$
	b. Number Sense: $572 + 150$
	c. Percent: 50% of 80
	d. Decimals: 100×0.02
	e. Statistics: Zach plays soccer. He scored 5 goals in 10 games. Find the mean number of goals Zach scored per game.
	f. Number Sense: $\frac{640}{20}$
	g. Algebra: If $r = 8\frac{1}{2}$, what is $r + r$?
	h. Calculation: $4 \times 5, + 1, \div 3, \times 8, - 1, \div 5, \times 4, - 2, \div 2$
problem solving	Choose an appropriate problem-solving strategy to solve this problem. What are the next four numbers in this sequence: $\frac{1}{12}, \frac{1}{6}, \frac{1}{4}, \frac{1}{3}, \cdots$

New Concept

A line that intersects two or more other lines is a **transversal.** In this drawing, line r is a transversal of lines s and t.

Math Language

Parallel lines are lines in the same plane that do not intersect and are always the same distance apart.

In the drawing, lines s and t are not parallel. However, in this lesson we will focus on the effects of a transversal intersecting parallel lines.

Below we show parallel lines *m* and *n* intersected by transversal *p*. Notice that eight angles are formed. In this figure there are four obtuse angles (numbered 1, 3, 5, and 7) and four acute angles (numbered 2, 4, 6, and 8).

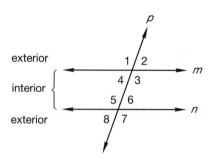

Thinking Skills

Verify

Why does every pair of supplementary angles in the diagram contain one obtuse and one acute angle?

Notice that obtuse angle 1, and acute angle 2, together form a straight line. These angles are **supplementary,** which means their measures total 180°. So if ∠1 measures 110°, then ∠2 measures 70°. Also notice that ∠2 and ∠3 are supplementary. If ∠2 measures 70°, then ∠3 measures 110°. Likewise, ∠3 and ∠4 are supplementary, so ∠4 would measure 70°.

There are names to describe some of the angle pairs. For example, we say that ∠1 and ∠5 are **corresponding angles** because they are in the same relative positions. Notice that ∠1 is the "upper left angle" from line *m*, while ∠5 is the "upper left angle" from line *n*.

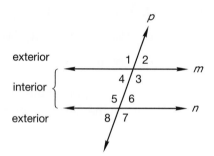

Which angle corresponds to ∠2?

Which angle corresponds to ∠7?

Since lines *m* and *n* are parallel, line *p* intersects line *m* at the same angle as it intersects line *n*. So the corresponding angles are congruent. Thus, if we know that ∠1 measures 110°, we can conclude that ∠5 also measures 110°.

The angles between the parallel lines (numbered 3, 4, 5, and 6 in the figure on previous page) are **interior angles.** Angle 3 and ∠5 are on opposite sides of the transversal and are called **alternate interior angles.**

Name another pair of alternate interior angles.

Alternate interior angles are congruent if the lines intersected by the transversal are parallel. So if ∠5 measures 110°, then ∠3 also measures 110°.

Angles not between the parallel lines are **exterior angles.** Angle 1 and ∠7, which are on opposite sides of the transversal, are **alternate exterior angles.**

Name another pair of alternate exterior angles.

Alternate exterior angles formed by a transversal intersecting parallel lines are congruent. So if the measure of ∠1 is 110°, then the measure of ∠7 is also 110°.

While we practice the terms for describing angle pairs, it is useful to remember the following.

> When a transversal intersects parallel lines, all acute angles formed are equal in measure, and all obtuse angles formed are equal in measure.

Thus any acute angle formed will be supplementary to any obtuse angle formed.

Example 1

Transversal *w* intersects parallel lines *x* and *y*.

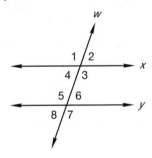

a. Name the pairs of corresponding angles.

b. Name the pairs of alternate interior angles.

c. Name the pairs of alternate exterior angles.

d. If the measure of ∠1 is 115°, then what are the measures of ∠5 and ∠6?

a. ∠1 and ∠5, ∠2 and ∠6, ∠3 and ∠7, ∠4 and ∠8

b. ∠4 and ∠6, ∠3 and ∠5

c. ∠1 and ∠7, ∠2 and ∠8

d. If ∠1 measures **115°**, then ∠5 also measures **115°** and ∠6 measures **65°**.

Example 2

Refer to the diagram from Example 1 to answer the following. Select all correct answers.

a. ∠1 and ∠2 are

    complementary angles   supplementary angles

    adjacent angles   vertical angles

b. ∠1 and ∠3 are

    complementary angles   supplementary angles

    adjacent angles   vertical angles

c. What angle is supplementary to ∠1 but not adjacent to ∠1?

    ∠2   ∠3   ∠4   ∠6

a. Angles *1* and *2* are **supplementary** (together measure 180°) and **adjacent** (share a common side).

b. Angles *1* and *3* are **vertical angles.**

c. Angle *1* is obtuse and ∠6 is acute. Since lines *x* and *y* are parallel, ∠*1* and ∠*6* are supplementary. Also, ∠*6* is not adjacent to ∠*1* since they do not share a common side. From this list, only **∠6** fits the criteria. Angles *2* and *4* are adjacent to ∠*1*. Angle *3* is not supplementary to ∠*1* since it is also obtuse.

Example 3

Draw a pair of parallel lines intersected by a transversal so that all angles formed are right angles. Which angles are equal in measure? Which angles are supplementary?

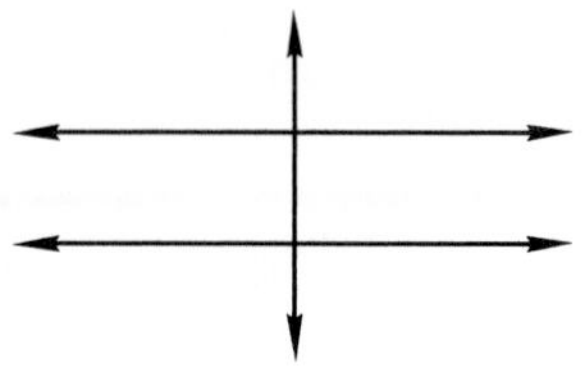

All of the right angles formed are equal in measure, and any pair of right angles is a pair of supplementary angles.

Lesson Practice

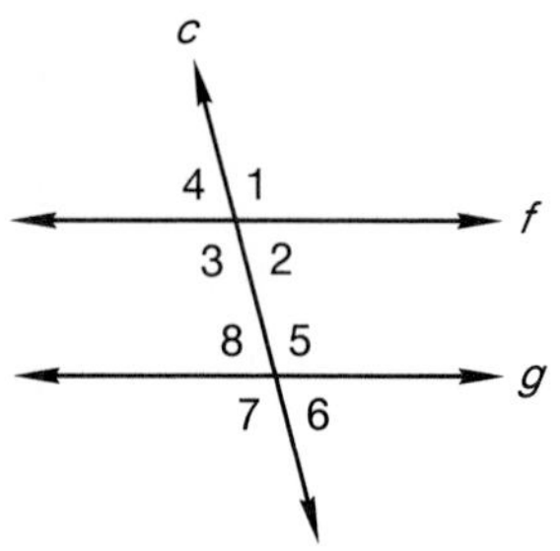

a. Which line in the figure at right is a transversal?

b. Which angle is an alternate interior angle to ∠3?

c. Which angle corresponds to ∠8?

d. Which angle is an alternate exterior angle to ∠7?

e. (**Conclude**) If the measure of ∠1 is 105°, what is the measure of each of the other angles in the figure?

Written Practice

Distributed and Integrated

* 1. When the sum of 2.0 and 2.0 is subtracted from the product of 2.0 and 2.0, what
(4, 41) is the difference?

2. A 4.2-kilogram object weighs the same as how many objects that each weigh
(44) 0.42 kilogram?

* 3. If the mean of 8 numbers is 12, what is the sum of the 8 numbers?
(10)

4. (**Conclude**) What is the name of a quadrilateral that has one pair of sides
(59) that are parallel and one pair of sides that are not parallel?

* 5. a. Write 0.15 as a percent.
(92)

    b. Write 1.5 as a percent.

* 6. Write $\frac{5}{6}$ as a percent.
(92)

* 7. **Multiple Choice** Three of the numbers below are equivalent. Which one is not
(92) equivalent to the others?

    A 1   **B** 100%   **C** 0.1   **D** $\frac{100}{100}$

8. 11^3 **9.** How much is $\frac{5}{6}$ of 360?
(71)   (65)

* 10. Evaluate $x(y + z)$ when $x = 0.3$, $y = 0.4$, and $z = 0.5$.
(88)

11. $x - 7 = -2$
(91)

12. The formula below may be used to convert temperature measurements from
(93) degrees Celsius (C) to degrees Fahrenheit (F). Find F to the nearest degree
when C is 17.

$$F = 1.8C + 32$$

13. Factor and reduce: $\dfrac{(45)(54)}{81}$
(63)

14. $\dfrac{30}{0.08}$ **15.** $16\frac{2}{3} \div 100$
(44)   (64)

16. $2\frac{1}{2} + 3\frac{1}{3} + 4\frac{1}{6}$ **17.** $6 \times 5\frac{1}{3} \times \frac{3}{8}$
(56)   (67)

18. $\frac{2}{5}$ of \$12.00   **19.** $0.12 \times \$6.50$
(15)   (33)

20. $5.3 - 3\frac{3}{4}$ (decimal answer)
(69)

* 21. What is the ratio of the number of cents in a dime to the number of cents
(16) in a quarter?

Find each unknown number:

* 22. $4n = 6 \cdot 14$ *** 23.** $0.3n = 12$
(85)   (85)

24. **Model** Draw a segment $1\frac{3}{4}$ inches long. Label the endpoints R and T. Then find
(RF9) and mark the midpoint of $\overline{RT}$. Label the midpoint S. What are the lengths of $\overline{RS}$
and $\overline{ST}$?

* 25. Solve this proportion: $\dfrac{6}{9} = \dfrac{36}{w}$
(83)

* 26. Multiply 4 hours by 6 dollars per hour:
(89)

$$\frac{4 \text{ hours}}{1} \times \frac{6 \text{ dollars}}{1 \text{ hours}}$$

* 27. **Connect** The coordinates of the vertices of a parallelogram are (0, 0),
(RF27, 66) (6, 0), (4, 4), and (–2, 4). What is the area of the parallelogram?

28. **Estimate** The saying "A pint's a pound the world around" refers to the fact
(81) that a pint of water weighs about one pound. About how many pounds does
a gallon of water weigh?

29. *l* ∥ *m*. Use the figure to answer the following:
(94)

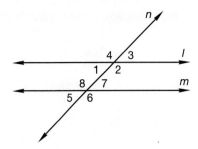

a. Which line is a transversal?

b. ∠1 is adjacent to which two angles?

30. What is the probability of rolling a prime number with one roll of a number cube?
(54. 69) Express the ratio as a fraction and as a decimal.

✎ *California Mathematics Content Standards*

MG 2.0, 2.1 Identify angles as vertical, adjacent, complementary, or supplementary and provide descriptions of these terms.

MG 2.0, 2.2 Use the properties of complementary and supplementary angles and the sum of the angles of a triangle to solve problems involving an unknown angle.

MG 2.0, 2.3 Draw quadrilaterals and triangles from given information about them (e.g., a quadrilateral having equal sides but no right angles, a right isosceles triangle).

• Sum of the Angle Measures of Triangles and Quadrilaterals

Power Up

facts　　Power Up L

mental math

   a. Number Sense: $60 \cdot 80$

   b. Percent: 25% of 80

   c. Decimals: $17.5 \div 100$

   d. Geometry: What is the perimeter of a regular pentagon whose sides are each 1.2 cm long?

   e. Measurement: Valerie jogged 1.5 kilometers and then walked 300 meters. Altogether, how many meters did she jog and walk?

   f. Number Sense: 30×55

   g. Algebra: If $w = 10$, what does $7w$ equal?

   h. Calculation: $6 \times 8, + 1, \sqrt{}, \times 5, + 1, \sqrt{}, \times 3, \div 2, \sqrt{}$

problem solving

Choose an appropriate problem-solving strategy to solve this problem. Carlos reads 5 pages in 4 minutes, and Malik reads 4 pages in 5 minutes. If they both begin reading 200-page books at the same time and do not stop until they are done, how many minutes before Malik finishes will Carlos finish?

New Concept

If we extend a side of a polygon, we form an **exterior angle.** In this figure ∠1 is an exterior angle, and ∠2 is an **interior angle.** Notice that these angles are supplementary. That is, the sum of their measures is 180°.

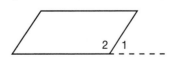

Thinking Skills

Verify

Act out the turns Elizabeth made to verify the number of degrees.

Recall that a full turn measures 360°. So if Elizabeth makes three turns to get around a park, she has turned a total of 360°. Likewise, if she makes four turns to get around a park, she has also turned 360°.

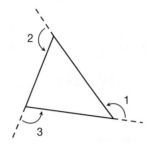

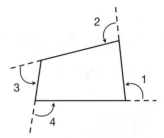

The sum of the measures of angles 1, 2, and 3 is 360°.

The sum of the measures of angles 1, 2, 3, and 4 is 360°.

If Elizabeth makes three turns to get around the park, then each turn averages 120°.

$$\frac{360°}{3 \text{ turns}} = 120° \text{ per turn}$$

If she makes four turns to get around the park, then each turn averages 90°.

$$\frac{360°}{4 \text{ turns}} = 90° \text{ per turn}$$

Recall that these turns correspond to exterior angles of the polygons and that the exterior and interior angles at a turn are supplementary. Since the exterior angles of a triangle average 120°, the interior angles must average 60°. A triangle has three interior angles, so the sum of the interior angles is 180° (3 × 60° = 180°).

> The sum of the interior angles of a triangle is 180°.

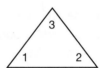

The sum of angles 1, 2, and 3 is 180°.

Since the exterior angles of a quadrilateral average 90°, the interior angles must average 90°. So the sum of the four interior angles of a quadrilateral is 360° (4 × 90° = 360°).

> The sum of the interior angles of a quadrilateral is 360°.

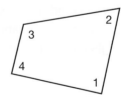

The sum of angles 1, 2, 3, and 4 is 360°.

Example 1

What is the unknown angle measure x in $\triangle ABC$?

The letter x is used to represent the unknown angle measure.

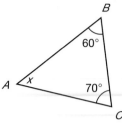

The measures of the interior angles of a triangle total 180°.

$$x + 60° + 70° = 180°$$

Since the measures of $\angle B$ and $\angle C$ total 130°, $m\angle x =$ is **50°**.

Example 2

Find the unknown angle measures a and b.

The angle which measures a is supplementary to the angle which measures 145°.

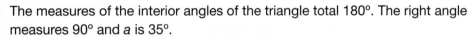

$$a + 145° = 180°$$

$$a = 35°$$

The measures of the interior angles of the triangle total 180°. The right angle measures 90° and a is 35°.

$$90° + 35° + b = 180°$$

$$125° + b = 180°$$

$$b = 55°$$

Example 3

Find x and y.

Since $\angle A$ is a right angle, the angle which measures x is complementary to the angle that measures 30°.

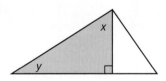

$$x + 30° = 90°$$

$$x = 60°$$

The unknown angles are part of a right triangle. The three angle measures total 180°. We found that x is 60°, and the right angle measures 90°.

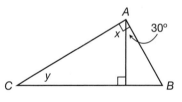

$$60° + 90° + y = 180°$$

$$150° + y = 180°$$

$$y = 30°$$

Example 4

What is m∠T in quadrilateral QRST?

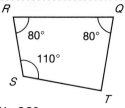

The measures of the interior angles of a quadrilateral total 360°.

$$m\angle T + 80° + 80° + 110° = 360°$$

The measures of ∠Q, ∠R, and ∠S total 270°. So m∠T is **90°**.

Lesson Practice

Quadrilateral *ABCD* is divided into two triangles by segment *AC*. Use for problems **a–c.**

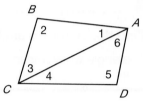

a. What is the sum of m∠1, m∠2, and m∠3?

b. What is the sum of m∠4, m∠5, and m∠6?

c. (**Generalize**) Draw a quadrilateral. What is the sum of the measures of the four interior angles of the quadrilateral?

d. Draw a scalene right triangle. If one of the acute angles measures 30°, what are the measures of the other two angles.

e. Draw a regular quadrilateral. What is the measure of each interior angle of a regular quadrilateral?

f. (**Model**) Nashawn made five left turns as she ran around the park. Draw a sketch that shows the turns in her run around the park. Then find the mean number of degrees in each turn.

Written Practice *Distributed and Integrated*

1. How many quarter-pound hamburgers can be made from 100 pounds of ground beef ?
(44)

* 2. (**Connect**) On the Fahrenheit scale water freezes at 32°F and boils at 212°F. What temperature is halfway between the freezing and boiling temperatures of water?
(10)

3. Estimate the value of πd when $\pi \approx 3.14159$ and *d* is 9.847 meters.
(88)

4. Compare: $\frac{5}{8} \bigcirc 0.675$
(74)

* 5. Write $2\frac{1}{4}$ as a percent.
(92)

* 6. Write $1\frac{2}{5}$ as a percent.
(92)

* 7. Write 0.7 as a percent.
(92)

* 8. Write $\frac{7}{8}$ as a percent.
(92)

9. Use division by primes to find the prime factors of 320. Then write the prime
(71) factorization of 320 using exponents.

10. $x - \frac{2}{3} = \frac{5}{6}$
(91)

11. **a.** Solve for h: $A = \frac{1}{2}bh$
(93)

   b. Use the solution from part **a** to find h when $A = 16$ and $b = 8$.

12. $6\frac{3}{4} + 5\frac{7}{8}$
(55)

13. $6\frac{1}{3} - 2\frac{1}{2}$
(58)

14. $2\frac{1}{2} \div 100$
(64)

15. $6.93 + 8.429 + 12$
(RF21)

16. $(1 - 0.1)(1 \div 0.1)$
(41)

17. $4.2 + \frac{7}{8}$ (decimal answer)
(69)

18. Jovita bought $3\frac{1}{3}$ cubic yards of mulch for the garden. She will need
(68) 2.5 cubic yards for the flowerbeds. How much mulch is left for Jovita to
use for her vegetable garden? Write your answer as a fraction.

19. **Analyze** If 80% of the 30 students passed the test, how many students
(35) did not pass?

* 20. Compare: $\frac{1}{2} \div \frac{1}{3} \bigcirc \frac{1}{3} \div \frac{1}{2}$
(45)

* 21. **Predict** What is the next number in this sequence?
(1)
$$\ldots, 1000, 100, 10, 1, \ldots$$

Find each unknown number:

22. $a + 60 + 70 = 180$
(5)

23. $\frac{7}{4} = \frac{w}{44}$
(83)

* 24. The perimeter of this square is 48 in. What is the area of one of the triangles?
(77)

Refer to the table below to answer problems **25–27.**

Mark's Personal Running Records

Distance	Time (minutes:seconds)
$\frac{1}{4}$ mile	0:58
$\frac{1}{2}$ mile	2:12
1 mile	5:00

25. If Mark set his 1-mile record by keeping a steady pace, then what was his $\frac{1}{2}$-mile
(RF11) time during the 1-mile run?

26. **Multiple Choice** What is a reasonable expectation for the time it would take
(RF11) Mark to run 2 miles?

   A 9:30   **B** 11:00   **C** 15:00

* 27. 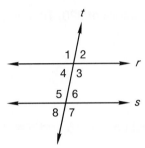 **Formulate** Write a question that relates to this table and answer the
(RF11)   question.

* 28. Transversal *t* intersects parallel lines *r* and *s*. Angle 2 measures 78°.
(94)

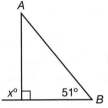

 a. **Analyze** Which angle corresponds to ∠2?

 b. Find the measures of ∠5 and ∠8.

29. **a.** Find m∠A.
(95)
 b. Find *x*.

30. What is the probability of rolling a composite
(54, 61)   number with one roll of a number cube?

✎ *California Mathematics Content Standards*
NS **1.0**, **1.4** Calculate given percentages of quantities and solve problems involving discounts at sales, interest earned, and tips.
MR **3.0, 3.3** Develop generalizations of the results obtained and the strategies used and apply them in new problem situations.

• Fraction-Decimal-Percent Equivalents

facts	Power Up M
mental math	**a. Number Sense:** $364 + 250$
	b. Money: A bag of sunflower seeds costs $0.89. Keiko paid for a bag with a $5 bill. How much change should she receive?
	c. Decimals: 100×0.015
	d. Percent: 50% of one yard is how many feet?
	e. Measurement: How many pints are in 2 quarts?
	f. Estimation: Estimate 48×61 by rounding each factor and then multiplying.
	g. Algebra: If $h - 95 = 5$, what is h?
	h. Calculation: $6 \times 6, -1, \div 5, \times 8, -1, \div 11, \times 8, \times 2, +1, \sqrt{}$

problem solving

Choose an appropriate problem-solving strategy to solve this problem. Copy this factor tree and fill in the missing numbers:

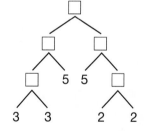

New Concept

Fractions, decimals, and percents are three ways to express parts of a whole. An important skill is being able to change from one form to another. This lesson asks you to complete tables that show equivalent fractions, decimals, and percents.

Complete the table.

Fraction	Decimal	Percent
$\frac{1}{2}$	a.	b.
c.	0.3	d.
e.	f.	40%

The numbers in each row should be equivalent. For $\frac{1}{2}$ we write a decimal and a percent. For 0.3 we write a fraction and a percent. For 40% we write a fraction and a decimal.

a. $\frac{1}{2} = 2\overline{)1.0}^{\,0.5}$

b. $\frac{1}{2} \times \frac{100\%}{1} = \mathbf{50\%}$

c. $0.3 = \frac{3}{10}$

d. $0.3 \times 100\% = \mathbf{30\%}$

e. $40\% = \frac{40}{100} = \frac{2}{5}$

f. $40\% = 0.40 = \mathbf{0.4}$

Lesson Practice

Connect Complete the table.

Fraction	Decimal	Percent
$\frac{3}{5}$	a.	b.
c.	0.8	d.
e.	f.	20%
$\frac{3}{4}$	g.	h.
i.	0.12	j.
k.	l.	5%

Written Practice   *Distributed and Integrated*

1. When the sum of $\frac{1}{2}$ and $\frac{1}{4}$ is divided by the product of $\frac{1}{2}$ and $\frac{1}{4}$, what is the
   (4, 50)   quotient?

* 2. **Analyze** Tzara is $5\frac{1}{2}$ feet tall. She is how many inches tall?
   (89)

3. If $\frac{4}{5}$ of the 200 runners finished the race, how many runners did not finish the
   (75)   race?

* 4. Lines *p* and *q* are parallel.
(94)

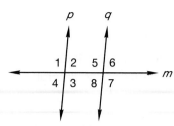

a. Which angle is an alternate interior angle to ∠2?

b. If ∠2 measures 85°, what are the measures of ∠6 and ∠7?

5. $x - 3\frac{1}{3} = 2\frac{1}{2}$
(91)

* 6. (**Estimate**) Use a ruler to measure the diameter of a quarter to
(14, 42) the nearest sixteenth of an inch. How can you use that information
to find the radius and the circumference of the quarter?

7. **Multiple Choice** (**Connect**) Which of these bicycle wheel parts is
(42) the best model of the circumference of the wheel?

    A spoke   **B** axle   **C** tire

* 8. (**Predict**) As this sequence continues, each term equals the sum
(1) of the two previous terms. What is the next term in this sequence?

$$1, 1, 2, 3, 5, 8, 13, \ldots$$

9. If there is a 20% chance of rain, what is the probability that it will not rain?
(54)

* 10. Write $1\frac{1}{3}$ as a percent.
(92)

* 11. (**Analyze**) $0.08w = \$0.60$
(85)

12. $\dfrac{1 - 0.001}{0.03}$
(44)

13. $\dfrac{3\frac{1}{3}}{100}$
(64)

14. The following formula can be used to find the area, *A*, of a trapezoid.
(93) The lengths of the parallel sides are *a* and *b*, and the height, *h*, is the
perpendicular distance between the parallel sides.

$$A = \frac{1}{2}(a + b)h$$

Use this formula to find the area of the trapezoid.

15. $6\frac{1}{2} + 4.95$ (decimal)
(69)

16. $2\frac{1}{6} - 1.5$ (fraction)
(68)

17. (**Explain**) If a shirt costs $19.79 and the sales-tax rate is 6%, what is the total
(35) price including tax? Explain how you can check your answer using estimation.

* 18. What fraction of a foot is 3 inches?
(89)

19. What percent of a meter is 3 centimeters?
(73)

20. The ratio of children to adults in the theater was 5 to 3. If there were 45
(86) children, how many adults were there?

21. Arrange these numbers in order from least to greatest:
(9, 14)

$$1, -1, 0, \frac{1}{2}, -\frac{1}{2}$$

22. (**Classify**) These two triangles together form a quadrilateral
(59) with only one pair of parallel sides. What type of quadrilateral
is formed?

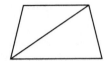

* 23. (**Conclude**) Do the triangles in this quadrilateral appear to be
(RF24) congruent or not congruent?

* 24. a. (**Analyze**) What is the measure of ∠A in △ABC?
(95)

b. (**Analyze**) What is the measure of the exterior angle
marked x?

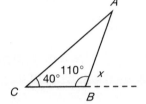

25. Write 40% as a
(27, 69)

a. simplified fraction.

b. simplified decimal number.

26. The diameter of this circle is 20 mm. What is the area of the
(84) circle? (Use 3.14 for π.)

27. Complete the table:
(96)

Fraction	Decimal	Percent
a.	0.4	b.
$\frac{1}{4}$	c.	d.
e.	f.	10%

* 28. Multiply 120 inches by 1 foot per 12 inches.
(89)

$$\frac{120 \text{ in.}}{1} \times \frac{1 \text{ ft}}{12 \text{ in.}}$$

29. A bag contains 20 red marbles and 15 blue marbles.
(16, 54)

a. What is the ratio of red marbles to blue marbles?

b. If one marble is drawn from the bag, what is
the probability that the marble will be blue?

* 30. (**Conclude**) An architect drew a set of plans for a house. In the plans, the roof is
(90) supported by a triangular framework. When the house is
built, two sides of the framework will be 19 feet long and
the base will be 33 feet long. Classified by side length,
what type of triangle will be formed?

LESSON 97

California Mathematics Content Standards
NS **2.0, 2.3** Solve addition, subtraction, multiplication, and division problems, including those arising in concrete situations, that use positive and negative integers and combinations of these operations.
MR **2.0, 2.4** Use a variety of methods, such as words, numbers, symbols, charts, graphs, tables, diagrams, and models, to explain mathematical reasoning.

• Algebraic Addition Activity

facts Power Up K

mental math

   a. Number Sense: 517 − 250

   b. Percent: How many seconds is 50% of one minute?

   c. Money: $7.99 + $7.58

   d. Geometry: Angles *A* and *B* each measure 64°. What is m∠*C*?

   e. Decimals: 0.1 ÷ 100

   f. Measurement: The top of the desk is 2 feet 5 inches above the floor. How many inches is that?

   g. Algebra: If *y* = 3.7, what is *y* + *y*?

   h. Calculation: 5 × 9, − 1, ÷ 2, − 1, ÷ 3, × 10, + 2, ÷ 9, − 2, ÷ 2

problem solving

Choose an appropriate problem-solving strategy to solve this problem. Cezar and his friends played three games that are scored from 1–100. His lowest score was 70 and his highest score is 100. What is Cezar's lowest possible three-game average? What is his highest possible three-game average?

Thinking Skills

Discuss

What addition equation shows that a positive charge and a negative charge neutralize each other?

One model for the addition of signed numbers is the electrical-charge model, which is used in the Sign Game. In this model, signed numbers are represented by positive and negative charges that can neutralize each other when they are added. The game is played with sketches, as shown here. The first two levels may be played with two color counters.

Activity

Sign Game

In the Sign Game pairs of positive and negative charges become neutral. After determining the neutral pairs we count the signs that remain and then write our answer. There are four skill levels to the game. Be sure you are successful at one level before moving to the next level.

level 1

Positive and negative signs are placed randomly on a "screen." When the game begins positive and negative pairs are neutralized so we cross out the signs as shown. (Appropriate sound effects strengthen the experience!)

Use counters to act out the game. Choose one color to represent positive charges and another to represent negative charges. Remove all pairs of counters that have different colors. These are neutralized charges.

Before   After

+   +   +
  −   +
      +   +
+   −
+   −   +

Two positives remain.

After marking positive-negative pairs we count the remaining positives or negatives. In the example shown above, two positives remain. With counters, two counters of one color remain. See whether you can determine what will remain on the three practice screens below:

− +   +   + −   −   −
  −   −   −
    +   −   −
+ − +   +   +   −   −
              −   −
− +

level 2

Positives and negatives are displayed in counted clusters or stacked counters. The suggested strategy is to combine the same signs first. So +3 combines with +1 to form +4, and −5 combines with −2 to form −7. Then determine how many of which charge (sign) remain.

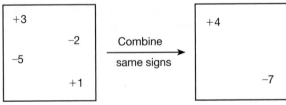

+3   +4
        −2   Combine
−5   ───────→
        same signs
        +1   −7

Three negatives, or −3, remain.

There were three more negatives than positives, so −3 remain. With counters, stacks of equal height and different colors are removed until only one color (or no counters) remains. See whether you can determine how many of which charge will remain for the three practice screens below:

+5	+3	+1
−4 −6	+4	−3
+3	+1	+4
+4	+2	−2

level 3

Positive and negative clusters can be displayed with two signs, one sign, or no sign. The first step is to change a double sign to one sign. A cluster with no sign, with "− −," or with "+ +" is a positive cluster. A cluster with "+ −" or with "− +" is a negative cluster. If a cluster has parentheses, look through the parentheses to see the sign. With counters, for "− −" invert a negative to a positive, and for "− +" invert a positive to a negative.

Reading Math

A negative sign indicates the opposite of a number. −3 means "the opposite of 3." Likewise, −(−3) means "the opposite of −3," which is 3. −(+3) means "the opposite of +3," which is −3.

Examples of Positives   **Examples of Negatives**

$-(-3) = +3$ $-(+2) = -2$

$--2 = +2$ $+(-3) = -3$

$4 = +4$ $+-1 = -1$

$++1 = +1$ $-+4 = -4$

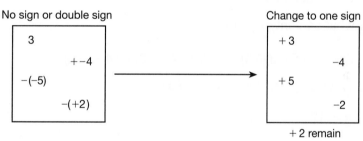

No sign or double sign   Change to one sign

3
+−4
−(−5)
−(+2)

+3
−4
+5
−2

+2 remain

See whether you can determine how many of which charge remain for the following practice screens:

− −3	+(−3)	−(+6)
+(−5)	−2	− −3
− +6		+4
+(+4)	−(+4)	−(+2)
		+−6

level 4

Extend Level 3 to a line of clusters without using a screen.

$$-3 + (-4) - (-5) - (+2) + (+6)$$

Use the following steps to find the answer:

Step 1: Change to single signs: $-3 - 4 + 5 - 2 + 6$

Step 2: Combine same signs: $-9 + 11$

Step 3: Find what remains: $+2$

Simplify:

   a. $-2 + -3 - -4 + -5$

   b. $-3 + (+2) - (+5) - (-6)$

   c. $+3 + -4 - +6 + +7 - -1$

   d. $2 + (-3) - (-9) - (+7) + (+1)$

   e. $3 - -5 + -4 - +2 + +8$

   f. $(-10) - (+20) - (-30) + (-40)$

Written Practice   *Distributed and Integrated*

1. (**Analyze**) A foot-long ribbon can be cut into how many $1\frac{1}{2}$-inch lengths?
(64)

2. Twenty paintings were sold. If the ratio of sold to unsold was 4 to 9, how many
(78) were unsold?

* **3.** (**Analyze**) If $\frac{3}{8}$ of the group voted yes and $\frac{3}{8}$ voted no, then what fraction of the
(75) group did not vote?

* **4.** (**Connect**) Nine months is
(25, 73)
   a. what fraction of a year?

   b. what percent of a year?

5. **a.** Compare: $(0.3)(0.4) + (0.3)(0.5) \bigcirc 0.3(0.4 + 0.5)$
(88)
   b. What property is illustrated in **a**?

6. $x - \frac{2}{5} = 12$
(91)

7. If $\frac{1}{5}$ of the pie was eaten, what percent of the pie was left?
(73)

* **8.** Write the percent form of $\frac{1}{7}$.
(92)

9. $6\frac{3}{4} - 6.2$ (decimal answer)
(69)

10. $5 \cdot 4 \cdot 3 \cdot 2 \cdot 1 \cdot 0$
(2)

11. $\dfrac{4.5}{0.18}$ **12.** $-3 + -4 + 6 + 4$
(44)   (97)

13. $5 + -4 + -6 + 3$
(97)

* **14.** (**Analyze**) Solve this proportion: $\dfrac{15}{20} = \dfrac{24}{n}$
(83)

15. $12\frac{1}{2} \times 1\frac{3}{5} \times 5$ **16.** $(4.2 \times 0.05) \div 7$
(67)   (41)

17. If the sales-tax rate is 7%, what is the tax on a $111.11 purchase?
(35)

18. **Analyze** The table shows the percent of the population aged 25–64 with some
(Inv. 4) senior high school education. The figures are for the year 2001. Use the table to
answer **a–c**.

Country	Percent
Peru	44%
Iceland	57%
Poland	46%
Italy	43%
Greece	51%
Chile	46%
Luxembourg	53%

    a. Find the mode of the data.

    b. If the data were arranged from least to greatest, which country or countries would have the
middle score?

    c. **Explain** What is the term used for the answer to problem **b?** Will this quantity always
be the same as the mode in every set of data? Explain.

19. Write the prime factorization of 900 using exponents.
(71)

20. Think of two different prime numbers, and write them on your paper. Then write
(12) the greatest common factor (GCF) of the two prime numbers.

21. **Explain** The perimeter of a square is 2 meters. How many centimeters long
(76, 87) is each side? Explain your thinking.

* 22. **a.** What is the area of this triangle?
(77, 90)

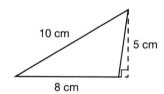

    b. **Classify** Is this an acute, right, or obtuse triangle?

* 23. **a.** What is the measure of ∠B in quadrilateral *ABCD?*
(95)
    b. What is the measure of the exterior angle at *D?*

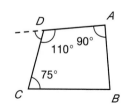

Complete the table to answer problems **24–26.**

	Fraction	Decimal	Percent
* **24.** (96)	a.	0.6	b.
* **25.** (96)	a.	b.	15%
* **26.** (96)	$\frac{3}{10}$	a.	b.

* **27.** **Model** Draw $\overline{AC}$ $1\frac{1}{4}$ inches long. Find and mark the midpoint of $\overline{AC}$, and label
(RF9,
14) the midpoint B. What are the lengths of $\overline{AB}$ and $\overline{BC}$?

28. There are 32 cards in a bag. Eight of the cards have letters written on them.
(54) What is the chance of drawing a card with a letter written on it?

29. Compare: 1 gallon ◯ 4 liters
(81)

30. **a.** Solve for d: $C = \pi d$
(93)
    b. Use the solution from part **a** to find d when $C = 62.8$ (Use 3.14 for π).

Early Finishers
Real-World Connection

Jesse displays trophies on 4 shelves in the family room. Two of the 6 trophies on each shelf are for soccer. How many trophies are NOT for soccer?

Write one equation and use it to solve the problem.

✎ *California Mathematics Content Standards*
NS **2.0, 2.3** Solve addition, subtraction, multiplication, and division problems, including those arising in concrete situations, that use positive and negative integers and combinations of these operations.
MR **2.0, 2.4** Use a variety of methods, such as words, numbers, symbols, charts, graphs, tables, diagrams, and models, to explain mathematical reasoning.

• Addition of Integers

 **Power Up**

facts	Power Up L
mental math	**a. Number Sense:** $40 \cdot 50$
	b. Percent: 50% of 48
	c. Money: Gwen had $20. She spent $18.72 on groceries. How much money did she have left over?
	d. Decimals: 12.5×100
	e. Number Sense: $\frac{360}{40}$
	f. Measurement: How many cups are in 2 pints?
	g. Algebra: If $\sqrt{g} = 6$, what is g?
	h. Calculation: $8 \times 8, -1, \div 9, \times 4, +2, \div 2, +1, \sqrt{}, \sqrt{}$
problem solving	Choose an appropriate problem-solving strategy to solve this problem. If two people shake hands, there is one handshake. If three people shake hands, there are three handshakes. If four people shake hands with one another, we can picture the number of handshakes by drawing four dots (for people) and connecting the dots with segments (for handshakes). Then we count the segments (six). Use this method to count the number of handshakes that will take place between Bill, Phil, Jill, Lil, and Wil.

New Concept

Math Language

Integers consist of the counting numbers (1, 2, 3, …), the negative counting numbers (−1, −2, −3, …), and 0.

In this lesson we will practice adding integers.

The dots on this number line mark the integers from negative five to positive five (−5 to +5).

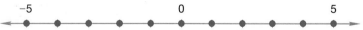

If we consider a rise in temperature of five degrees as a positive five (+5) and a fall in temperature of five degrees as a negative five (−5), we can use the scale on a thermometer to keep track of the addition.

Thinking Skills

Analyze

How is a thermometer like a number line? How is it different?

Imagine that the temperature is 0°F. If the temperature falls five degrees (−5) and then falls another five degrees (−5), the resulting temperature is ten degrees below zero (−10°F). When we add two negative numbers, the sum is negative.

$$-5 + -5 = -10$$

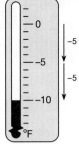

Imagine a different situation. We will again start with a temperature of 0°F. First the temperature falls five degrees (−5). Then the temperature rises five degrees (+5). This brings the temperature back to 0°F. The numbers −5 and +5 are opposites. When we add opposites, the sum is zero.

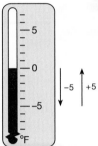

Starting from 0°F, if the temperature rises five degrees (+5) and then falls ten degrees (−10), the temperature will fall through zero to −5°F. The sum is less than zero because the temperature fell more than it rose.

$$+5 + -10 = -5$$

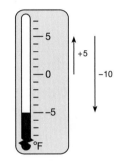

Example 1

Math Language

Opposites are numbers that can be written with the same digits but with opposite signs. They are the same distance, in opposite directions, from zero on the number line.

Add: $+8 + -5$

We will illustrate this addition on a number line. We begin at zero and move eight units in the positive direction (to the right). From +8 we move five units in the negative direction (to the left) to +3.

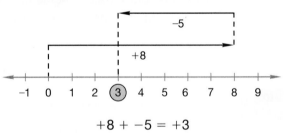

$$+8 + -5 = +3$$

The sum is **+3,** which we write as 3.

Example 2

Thinking Skills

Generalize

When two negative integers are added, is the sum negative or positive?

Add: $-5 + -3$

Again using a number line, we start at zero and move in the negative direction, or to the left, five units to -5. From -5 we continue moving left three units to -8.

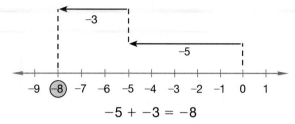

$$-5 + -3 = -8$$

The sum is **-8.**

Example 3

Add: $-6 + +6$

We start at zero and move six units to the left. Then we move six units to the right, returning to **zero.**

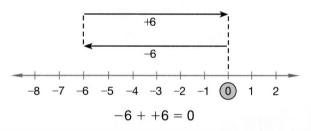

$$-6 + +6 = 0$$

Example 4

Add: $(+6) + (-6)$

Sometimes positive and negative numbers are written with parentheses. The parentheses help us see that the positive or negative sign is the sign of the number and not an addition or subtraction operation.

$$(+6) + (-6) = \mathbf{0}$$

Negative 6 and positive 6 are **opposites.** Opposites are numbers that can be written with the same digits but with opposite signs. The opposite of 3 is -3, and the opposite of -5 is 5 (which can be written as $+5$).

On a number line, we can see that any two opposites lie equal distances from zero. However, they lie on opposite sides of zero from each other.

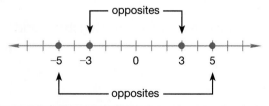

If opposites are added, the sum is zero.

$$-3 + +3 = 0 \qquad -5 + +5 = 0$$

Instead of subtracting 6 from 10, we can add the opposite of 6 to 10. The opposite of 6 is −6.

$$10 + -6$$

In both problems the answer is 4. Adding the opposite of a number to subtract is called **algebraic addition.** We change subtraction to addition by adding the opposite of the subtrahend.

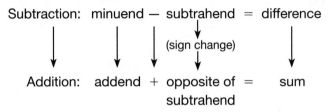

Subtraction: minuend − subtrahend = difference

(sign change)

Addition: addend + opposite of = sum
                       subtrahend

Lesson Practice

Model Find each sum. Draw a number line to show the addition for problems **a** and **b.** Solve problems **c–h** mentally.

a. −3 + +4   **b.** −3 + −4

c. −3 + +3   **d.** +4 + −3

e. (+3) + (−4)   **f.** (+10) + (−5)

g. (−10) + (−5)   **h.** (−10) + (+5)

Find the opposite of each number:

i. −8   **j.** 4   **k.** 0

Written Practice   *Distributed and Integrated*

1. If 0.6 is the divisor and 1.2 is the quotient, what is the dividend?
(5, 33)

2. If a number is twelve less than fifty, then it is how much more than twenty?
(4)

3. If the sum of four numbers is 14.8, what is the mean of the four numbers?
(10, 41)

* **4.** **Model** Illustrate this problem on a number line:
(98)
$$-3 + +5$$

* **5.** Find each sum mentally:
(98)
   a. −4 + +4   **b.** −2 + −3

   c. −5 + +3   **d.** +5 + −10

* **6.** Solve each subtraction problem using algebraic addition:
(98)
   a. −2 − −5   **b.** −3 − −3

   c. +2 − −3   **d.** −2 − +3

* **7.** **Analyze** What is the measure of each angle of an equilateral triangle?
(90, 95)

8. Quadrilateral *ABCD* is a parallelogram. If angle *A* measures 70°, what are the measures of angles *B, C,* and *D?*
(66)

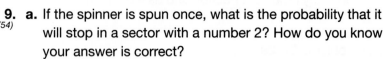

* 9. a. If the spinner is spun once, what is the probability that it will stop in a sector with a number 2? How do you know your answer is correct?
(54)

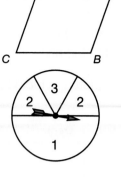

b. (Estimate) If the spinner is spun 30 times, about how many times would it be expected to stop in the sector with the number 3?

10. Find the volume of the rectangular prism at right.
(RF29)

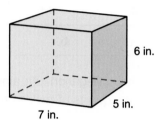

6 in.

7 in.   5 in.

11. Twelve of the 27 students in the class are boys. What is the ratio of girls to boys in the class?
(16)

12. Thirty number cubes are rolled simultaneously. Predict the number of 1s that will result.
(Inv. 8)

* 13. The fraction $\frac{2}{3}$ is equal to what percent?
(92)

14. If 20% of the students brought their lunch to school, then what fraction of the students did not bring their lunch to school?
(27)

15. $\frac{4}{3}x = 12$ **16.** $5\frac{7}{8} + 4\frac{3}{4}$ **17.** $1\frac{1}{2} \div 2\frac{1}{2}$
(91)   (55)   (64)

18. $5 - (3.2 + 0.4)$
(72)

19. (Estimate) If the diameter of a circular plastic swimming pool is 6 feet, then the area of the bottom of the pool is about how many square feet? Round to the nearest square foot. (Use 3.14 for π.)
(84)

* 20. Use the formula $b = \frac{2A}{h}$ to find *b* when *A* is 24 and *h* is 6.
(93)

21. Solve this proportion: $\frac{9}{12} = \frac{15}{x}$
(83)

Rectangle *ABCD* is 8 cm long and 6 cm wide. Segment *AC* is 10 cm long. Use this information to answer problems **22** and **23.**

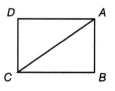

22. What is the area of triangle *ABC?*
(77)

23. What is the perimeter of triangle *ABC?*
(RF7)

24. Measure the diameter of a nickel to the nearest millimeter.
(RF16)

* 25. (Estimate) Calculate the circumference of a nickel. Round to the nearest millimeter. (Use 3.14 for π.)
(42)

26. A bag contains 12 marbles. Eight of the marbles are red and 4 are blue. If you
(54, 69) draw a marble from the bag without looking, what is the probability that the
marble will be blue? Express the probability ratio as a fraction and as a decimal
rounded to the nearest hundredth.

(**Connect**) Complete the table to answer problems **27–29.**

	Fraction	Decimal	Percent
* 27. (96)	$\frac{9}{10}$	a.	b.
* 28. (96)	a.	1.5	b.
* 29. (96)	a.	b.	4%

30. A full one-gallon container of milk was used to fill two one-pint containers. How
(81) many quarts of milk were left in the one-gallon container?

*Real-World
Connection*

These three prime factorizations represent numbers that are powers of
10. Simplify each prime factorization.

$$2^2 \times 5^2 \qquad\qquad 2^4 \times 5^4 \qquad\qquad 2^5 \times 5^5$$

Use exponents to write the prime factorization of another number that
is a power of 10.

California Mathematics Content Standards

NS **2.0**, **2.3** Solve addition, subtraction, multiplication, and division problems, including those arising in concrete situations, that use positive and negative integers and combinations of these operations.

MR 2.0, 2.4 Use a variety of methods, such as words, numbers, symbols, charts, graphs, tables, diagrams, and models, to explain mathematical reasoning.

• What Adding Two Negative Integers Means

We know that a negative integer plus a negative integer equals a negative integer. For example, $(-10) + (-20) = (-30)$. Why is this true? First we will look at a real world example.

If Sam owes his sister $10, Sam has -10 dollars. If he asks her for another $20 loan, he now owes her another $20.

$$(-\$10) + (-\$20) = (-\$30)$$

If Sam wants to owe his sister $0, he must pay back $10 and $20, or $30.

Now think about what a negative number is. For every number a, there is a number $-a$, such that $a + (-a) = 0$. For example, $10 + (-10) = 0$. You have seen that this is true on a number line in earlier lessons.

We will use this idea and the addition properties to see why it is true that $[(-10) + (-20)] = -[10+20]$. We are saying that $[(-10) + (-20)]$ is the opposite of $[10+20]$. If this is true, their sum should be zero.

We will begin by finding the sum of $[(-10)+(-20)]$ and $[10+20]$

$[(-10) + (-20)] + [10 + 20]$	
$(-10) + (-20) + 10 + 20$	Used Associative Property to ungroup the addends
$(-10) + 10 + (-20) + 20$	Used Commutative Property to reorder the addends
$[(-10) + 10] + [(-20) + 20]$	Used Associative Property to regroup the addends
$0 + 0$	The sum of a number and its opposite is zero.
0	Simplified

The sum of 0 implies that $[(-10) + (-20)]$ is the opposite of $[10 + 20]$. This means that $[10 + 20]$ is the opposite of $[(-10) + (-20)]$.

So it is true that $[(-10) + (-20)] = -[10 + 20]$

Predict Will the sum of any two negative integers always be a negative integer? Try finding the sums below to help you answer this question.

    a. $(-3) + (-6)$ **b.** $(-1) + (-4)$ **c.** $(-13) + (-2)$

California Mathematics Content Standards
NS **2.0**, **2.3** Solve addition, subtraction, multiplication, and division problems, including those arising in concrete situations, that use positive and negative integers and combinations of these operations.
MR 2.0, 2.4 Use a variety of methods, such as words, numbers, symbols, charts, graphs, tables, diagrams, and models, to explain mathematical reasoning.

• Subtraction of Integers

Power Up

facts   Power Up M

mental math

   a. Number Sense: 741 − 450

   b. Percent: 25% of 48

   c. Decimals: 37.5 ÷ 100

   d. Time: Goldie spoke on the phone for 8 minutes 40 seconds. How many seconds is that?

   e. Geometry: The perimeter of this equilateral triangle is 9.6 centimeters. What is the length of each side?

   f. Number Sense: 30 × 15

   g. Fractions: $3\frac{1}{3} + 3\frac{1}{3} + 3\frac{1}{3}$

   h. Calculation: $7 \times 7, + 1, \div 2, \sqrt{}, \times 4, - 2, \div 3, \times 5, + 3, \div 3$

problem solving

Choose an appropriate problem-solving strategy to solve this problem. If the last page of a section of large newspaper is page 36, what is the fewest number of sheets of paper that could be in that section?

New Concept

Recall that the graphs of −3 and 3 are the same distance from zero on the number line. The graphs are on the opposite sides of zero.

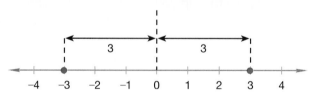

> **Math Language**
>
> A positive number and a negative number whose absolute values are equal are **opposites**.

This is why we say that 3 and −3 are the opposites of each other.

3 is the opposite of −3

−3 is the opposite of 3

We can read −3 as "the opposite of 3." Furthermore, −(−3) can be read as "the opposite of the opposite of 3." This means that −(−3) is another way to write 3.

There are two ways to simplify the expression 7 − 3. The first way is to let the minus sign signify subtraction. When we subtract 3 from 7, the answer is 4.

$$7 - 3 = 4$$

The second way is to use the thought process of **algebraic addition.** To use algebraic addition, we let the minus sign mean that −3 is a negative number and we treat the problem as an addition problem.

$$7 + (-3) = 4$$

Notice that we get the same answer both ways. The only difference is in the way we think about the problem.

We can also use algebraic addition to simplify this expression:

$$7 - (-3)$$

We use an addition thought and think that 7 is added to −(−3). This is what we think:

$$7 + [-(-3)]$$

Notice that we include −(−3) in brackets [ ], a symbol of inclusion like parentheses, in order to group the symbols and number.

The opposite of −3 is 3, so we can write

$$7 + [3] = 10$$

We will practice using the thought process of algebraic addition because algebraic addition can be used to simplify expressions that would be very difficult to simplify if we used the thought process of subtraction.

Example 1

Thinking Skills

Generalize

What is the sum of +7 and −7?

Simplify: −3 − (−2)

We think addition. We think we are to *add* −3 and −(−2). This is what we think:

$$(-3) + [-(-2)]$$

The opposite of −2 is 2 itself. So we have

$$(-3) + [2] = \mathbf{-1}$$

Example 2

Simplify: −(−2) − 5 − (+6)

We see three numbers. We think *addition,* so we have

$$[-(-2)] + (-5) + [-(+6)]$$

We simplify the first and third numbers and get
$$[+2] + (-5) + [-6] = -9$$
Note that this time we write 2 as +2. Either 2 or +2 may be used.

Example 3

Simplify: $-10 - -6$

This problem directs us to subtract a negative six from negative ten. Instead, we may add the opposite of negative six to negative ten.

$$-10 - \underbrace{-6}$$
$$\downarrow \quad \downarrow$$
$$-10 + +6 = -4$$

Example 4

Simplify: $(-3) - (+5)$

Instead of subtracting a positive five, we add a negative five.

$$(-3) - (+5)$$
$$\downarrow \quad \downarrow$$
$$(-3) + (-5) = -8$$

Lesson Practice

Generalize Use algebraic addition to simplify each expression.

a. $(-3) - (+2)$ **b.** $(-3) - (-2)$

c. $(+3) - (2)$ **d.** $(-3) - (+2) - (-4)$

e. $(-8) + (-3) - (+2)$ **f.** $(-8) - (+3) + (-2)$

g. Which is greater, $3 + (-6)$ or $3 - (-6)$? Explain.

Written Practice

Distributed and Integrated

* 1. For every three daisies there were a dozen dandelions. If there were 5 daisies,
(78)   how many dandelions were there?

2. **Connect** A shoe box is the shape of what geometric solid?
(RF25)

3. **Analyze** If the mean of six numbers is 12, what is the sum of the six
(10)   numbers?

4. If the diameter of a circle is $1\frac{1}{2}$ inches, what is the radius of the circle?
(RF16,
64)

* 5. What is the cost of 2.6 pounds of meat priced at $1.65 per pound?
(89)

6. Suppose $\overline{AC}$ is 12 cm long. If $\overline{AB}$ is $\frac{1}{4}$ the length of $\overline{AC}$, then how
(RF9) long is $\overline{BC}$?

* 7. Find each sum mentally:
(98)
   a. $-3 + -4$ **b.** $+5 + -5$

   c. $-6 + +3$ **d.** $+6 + -3$

* 8. Solve each subtraction problem using algebraic addition:
(98)
   a. $-3 - -4$ **b.** $+5 - -5$

   c. $-6 - +3$ **d.** $-6 - -6$

   e. (**Generalize**) Describe how to change a subtraction problem into an addition problem.

* 9. ✏️ (**Explain**) Two coins are tossed.
(Inv. 6)
   a. What is the probability that both coins will land heads up

   b. What is the probability that one of the coins will be heads and the other tails?

Complete the table to answer problems **10–12.**

	Fraction	Decimal	Percent
* 10. (96)	$\frac{3}{4}$	**a.**	**b.**
* 11. (96)	**a.**	1.6	**b.**
* 12. (96)	**a.**	**b.**	5%

13. $1\frac{1}{2} \times 4$ **14.** $6 \div 1\frac{1}{2}$ **15.** $x + \frac{1}{12} = \frac{1}{2}$
(62)   (64)   (91)

Find each unknown number:

16. $x + 2\frac{1}{2} = 5$ **17.** $\frac{8}{5} = \frac{40}{x}$
(39, 58)   (37)

18. $0.06n = \$0.15$ **19.** $6n = 21 \cdot 4$
(85)   (85)

20. (**Connect**) Nia's garage is 20 feet long, 20 feet wide, and 8 feet high.
(RF29)
   a. How many 1-by-1-by-1-foot boxes can she fit on the floor (bottom layer) of her garage?

   b. Altogether, how many boxes can Nia fit in her garage if she stacks the boxes 8 feet high?

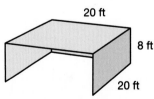

21. Multiple Choice (**Estimate**) If a roll of tape has a diameter of
(42) $2\frac{1}{2}$ inches, then removing one full turn of tape yields about how many inches? Choose the closest answer.

   A $2\frac{1}{2}$ in.   **B** 5 in.   **C** $7\frac{3}{4}$ in.   **D** $9\frac{1}{4}$ in.

22. Use the formula $h = \frac{2A}{b}$ to find h when A is 1.44 m² and b is 1.6 m.
(93)

Use the figure to answer problems **23** and **24**.

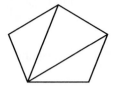

23. Together, these three triangles form what kind of polygon?
(RF24)

* 24. (**Generalize**) What is the sum of the measures of the angles of each
(95) triangle?

* 25. At 6 a.m. the temperature was −8°F. By noon the temperature was 15°F. The
(98) temperature had risen how many degrees?

26. (**Connect**) To what decimal number is the arrow pointing on the
(28) number line below?

27. What is the probability of rolling a perfect square with one roll of a number
(32, 54) cube?

* 28. (**Connect**) What is the area of a triangle with vertices located at (4, 0),
(RF27, 77) (0, −3), and (0, 0)?

* 29. (**Explain**) How can you convert 18 feet to yards?
(89)

30. If a gallon of milk costs $3.80, what is the cost per quart?
(81)

*Real-World
Connection*

The surface of the Dead Sea is approximately 408 meters below sea level. Its greatest depth is 330 meters. In contrast, Mt. Everest reaches a height of 8,850 meters. What is the difference in elevation between the summit of Mt. Everest and the bottom of the Dead Sea? Show your work.

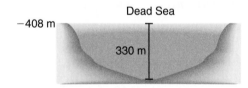

LESSON 100

◤ *California Mathematics Content Standards*

NS 1.0, 1.3 Use proportions to solve problems (e.g., determine the value of *N* if $\frac{4}{7} = \frac{N}{21}$, find the length of a side of a polygon similar to a known polygon). Use cross-multiplication as a method for solving such problems, understanding it as the multiplication of both sides of an equation by a multiplicative inverse.

MR 1.0, 1.3 Determine when and how to break a problem into simpler parts.

• Ratio Problems Involving Totals

facts   Power Up K

mental math

  a. Number Sense: 50 · 80

  b. Number Sense: 380 + 550

  c. Estimation: A rectangle has a length of 14.95 cm and a width of 5.02 cm. Estimate the perimeter.

  d. Functions: Tim earned $7.50 for each hour of work. He made this table to show the rate of pay. How much money would Tim earn for 4 hours of work?

Hours	1	2	3
Pay	$7.50	$15.00	$22.50

  e. Decimals: 0.8 × 100

  f. Algebra: If 3*t* = 18, what is *t*?

  g. Measurement: How many cups are in a quart?

  h. Calculation: 5 + 5, × 10, − 1, ÷ 9, + 1, ÷ 3, × 7, + 2, ÷ 2

problem solving   Choose an appropriate problem-solving strategy to solve this problem. How many different triangles of any size are in this figure?

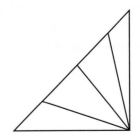

Math Language

A **proportion** is a statement that shows two ratios are equal.

In some ratio problems a total is used as part of the calculation. Consider this problem:

The ratio of boys to girls in a class was 5 to 4. If there were 27 students in the class, how many girls were there?

We begin by drawing a ratio box. In addition to the categories of boys and girls, we make a third row for the total number of students. We will use the letters *b* and *g* to represent the actual counts of boys and girls.

	Ratio	Actual Count
Boys	5	*b*
Girls	4	*g*
Total	9	27

In the ratio column we add the ratio numbers for boys and girls and get the ratio number 9 for the total. We were given 27 as the actual count of students. We will use two of the three rows from the ratio box to write a proportion. **We use the row we want to complete and the row that is already complete.** Since we are asked to find the actual number of girls, we will use the "girls" row. And since we know both "total" numbers, we will also use the "total" row. We solve the proportion below.

	Ratio	Actual Count
Boys	5	*b*
Girls	4	*g*
Total	9	27

$$\longrightarrow \frac{4}{9} = \frac{g}{27}$$

$$9g = 4 \cdot 27$$

$$g = 12$$

We find that there were 12 girls in the class. If we had wanted to find the number of boys, we would have used the "boys" row along with the "total" row to write a proportion.

Example 1

The ratio of football players to band members on the football field was 2 to 5. Altogether, there were 175 football players and band members on the football field. How many football players were on the field?

We use the information in the problem to make a table. We include a row for the total. The ratio number for the total is 7.

	Ratio	Actual Count
Football Players	2	f
Band Members	5	b
Total	7	175

Next we write a proportion using two rows of the table. We are asked to find the number of football players, so we use the "football players" row. We know both totals, so we also use the "total" row. Then we solve the proportion.

	Ratio	Actual Count
Football Players	2	f
Band Members	5	b
Total	7	175

$$\frac{2}{7} = \frac{f}{175}$$

$$7f = 2 \cdot 175$$

$$f = 50$$

We find that there were **50 football players** on the field.

Example 2

The ratio of basketball players to soccer players in the room was 5 to 7. If the basketball players and the soccer players in the room totaled 48, how many were basketball players?

We use the information in the problem to form a table. We include a row for the total number of players.

	Ratio	Actual Count
Basketball Players	5	b
Soccer Players	7	s
Total	12	48

$$\frac{5}{12} = \frac{b}{48}$$

$$12b = 5 \cdot 48$$

$$b = 20$$

To find the number of basketball players, we write a proportion from the "basketball players" row and the "total" row. We solve the proportion to find that there were **20 basketball players** in the room.

Represent Use ratio boxes to solve problems **a** and **b**.

a. Sparrows and crows perched on the wire in the ratio of 5 to 3. If the total number of sparrows and crows on the wire was 72, how many were crows?

b. Raisins and nuts were mixed by weight in a ratio of 2 to 3. If 60 ounces of mix were prepared, how many ounces of raisins were used?

c. **Model** Using 20 red and 20 yellow color tiles (or 20 shaded and unshaded circles) create a ratio of 3 to 2. How many of each color (or shading) do you have?

Written Practice *Distributed and Integrated*

On his first six tests, Cleavon had scores of 90%, 92%, 96%, 92%, 84%, and 92%. Use this information to answer problems **1** and **2**.

1. **a.** Which score occurred most frequently? That is, what is the mode of
(Inv. 4) the scores?

b. The difference between Cleavon's highest score and his lowest score is how many percentage points? That is, what is the range of the scores?

2. What was Cleavon's average score for the six tests? That is, what is the
(10) mean of the scores?

* 3. In basketball there are one-point baskets, two-point baskets, and three-point
(4, 85) baskets. If a team scored 96 points and made 18 one-point baskets and 6 three-point baskets, how many two-point baskets did the team make? Explain how you found your answer.

* 4. **Multiple Choice** **Analyze** Which ratio forms a proportion with $\frac{4}{7}$?
(82)
    A $\frac{7}{4}$ **B** $\frac{14}{17}$ **C** $\frac{12}{21}$ **D** $\frac{2}{3}$

* 5. Complete this proportion: Four is to five as what number is to twenty?
(83)

6. Arrange these numbers in order from least to greatest:
(28)

$$-1, 1, 0.1, -0.1, 0$$

7. **a.** $(-4) - (-6)$ **b.** $(-4) - (+6)$
(99)
    c. $(-6) - (-4)$ **d.** $(+6) - (-4)$

8. The area of the square in this figure is 100 mm².
(84)
    a. What is the radius of the circle?

    b. What is the diameter of the circle?

    c. What is the area of the circle? (Use 3.14 for π.)

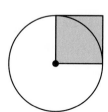

Connect Complete the table to answer problems **9–11.**

	Fraction	Decimal	Percent
* 9. (96)	$\frac{4}{25}$	a.	b.
* 10. (99)	a.	0.01	b.
* 11. (99)	a.	b.	90%

12. $1\frac{2}{3} + 3\frac{1}{2} + 4\frac{1}{6}$
(56)

13. $\frac{5}{6} \times \frac{3}{10} \times 4$
(67)

14. $6\frac{1}{4} \div 100$
(64)

15. $6.437 + 12.8 + 7$
(41)

16. **Estimate** Convert $\frac{1}{7}$ to a decimal number by dividing 1 by 7. Stop dividing
(46, 69) after three decimal places, and round your answer to two decimal places.

17. An octagon has how many more sides than a pentagon?
(RF24)

18. $\frac{1}{2}x = \frac{3}{7}$
(91)

* 19. **Analyze** Sector 2 on this spinner is a 90° sector. If the spinner is spun
(Inv. 6) twice, what is the probability that it will stop in sector 2 both times?

20. If the spinner is spun 100 times, about how many times would it be expected
(54) to stop in sector 1?

21. Draw a ratio box for this problem. Then solve the problem using a proportion.
(100)
> The ratio of boys to girls in the class was 3 to 2. If there were
> 30 students in the class, how many were girls?

22. The average of four numbers is 5. What is their sum?
(10)

* 23. **Connect** When Andy was born, he weighed 8 pounds 4 ounces.
(RF10) Three weeks later he weighed 10 pounds 1 ounce. How many pounds
and ounces had he gained in three weeks?

* 24. Lines s and t are parallel.
(94)
    a. Which angle is an alternate interior angle to $\angle 5$?

    b. If the measure of $\angle 5$ is 76°, what are the measures of $\angle 1$ and $\angle 2$?

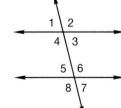

25. Which property is illustrated by the equation $x(y + z) = xy + xz$?
(88)

26. A spinner has 5 equal spaces marked one through 5. If it is spun 20 times,
(Inv. 8) how many times can we predict it will stop on 5?

27. **a.** What is the area of the parallelogram at right?
(66, 77)
    b. What is the area of the triangle?

    c. What is the combined area of the parallelogram and triangle?

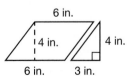

* 28. How many milligrams is half of a gram?
(76)

29. (**Model**) The coordinates of the endpoints of a line segment are (3, −1)
(RF27) and (3, 5). The midpoint of the segment is the point halfway between the
endpoints. What are the coordinates of the midpoint?

30. (**Estimate**) Tania took 10 steps to walk across the tetherball circle and
(42) 31 steps to walk around the tetherball circle. Use this information to find
the approximate number of diameters in the circumference of the tetherball
circle.

California Mathematics Content Standards

NS 10 1.3 Use proportions to solve problems (e.g., determine the value of N if 4/7 = N/ 21, find the length of a side of a polygon similar to a known polygon). Use cross-multiplication as a method for solving such problems, understanding it as the multiplication of both sides of an equation by a multiplicative inverse.

MG 2.0, 2.2 Use the properties of complementary and supplementary angles and the sum of the angles of a triangle to solve problems involving an unknown angle.

Focus on
Similar Figures

Triangles that are the same shape and size are **congruent.** The two triangles below are congruent. Each triangle has three angles and three sides. The angles and sides of triangle *ABC* **correspond** to the angles and sides of triangle *XYZ.*

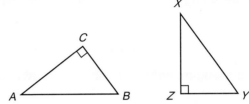

By rotating, translating, and reflecting triangle *ABC,* we could position it on top of triangle *XYZ.* Then their **corresponding parts** would be in the same place.

A corresponds to X.	$\overline{AB}$ corresponds to $\overline{XY}$.
B corresponds to Y.	$\overline{BC}$ corresponds to $\overline{YZ}$.
C corresponds to Z.	$\overline{AC}$ corresponds to $\overline{XZ}$.

If two figures are congruent, their corresponding parts are congruent. So the measures of the corresponding parts are equal.

Example 1

These triangles are congruent. What is the perimeter of each?

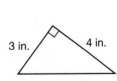

We will rotate the triangle on the left.

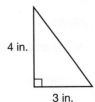

Now we can more easily see that the unmarked side on the left-hand triangle corresponds to the 5-inch side on the right-hand triangle. Since the triangles are congruent, the measures of the corresponding parts are equal. So each triangle has sides that measure 3 inches, 4 inches, and 5 inches. Adding, we find that the perimeter of each triangle is **12 inches.**

3 in. + 4 in. + 5 in. = 12 in.

Figures that have the same shape but are not necessarily the same size are **similar**. Three of these four triangles are similar:

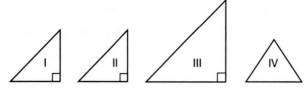

Triangles I and II are similar. They are also congruent. Remember, congruent figures have the same shape *and* size. Triangle III is similar to triangles I and II. It has the same shape but not the same size as triangles I and II. Notice that the corresponding angles of similar figures have the same measure. Triangle IV is not similar to the other triangles.

Analyze Can we reduce or enlarge Triangle IV to make it match the other triangles in the diagram? Explain.

Example 2

> **The two triangles below are similar. What is the measure of angle A?**
>
>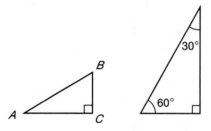
>
> We will rotate and reflect triangle *ABC* so that the corresponding angles are easier to see.
>
>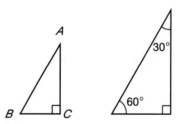
>
> We see that angle *A* in triangle *ABC* corresponds to the 30° angle in the similar triangle. Since corresponding angles of similar triangles have the same measure, the measure of angle *A* is **30°**.

Here are two important facts about similar polygons.

1. The corresponding angles of similar polygons are congruent.

2. The corresponding sides of similar polygons are proportional.

The first fact means that even though the sides of similar polygons might not have matching lengths, the corresponding angles do match. The second fact means that similar figures are related by a scale factor. The scale factor is a number. Multiplying the side length of a polygon by the scale factor gives the side length of the corresponding side of the similar polygon.

Example 3

The two rectangles below are similar. What is the ratio of corresponding sides? By what scale factor is rectangle *ABCD* larger than rectangle *EFGH*?

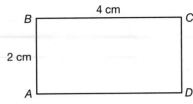

First, we find the ratios of corresponding sides.

Side *AB*: Side *EF* = $\frac{2}{1}$

Side *BC*: Side *FG* = $\frac{4}{2} = \frac{2}{1}$

In all similar polygons, such as these two rectangles, the ratios of corresponding sides are **equal.**

The sides of rectangle *ABCD* are 2 times larger than the sides of rectangle *EFGH*. So rectangle *ABCD* is larger than rectangle *EFGH* by a scale factor of **2.**

1. **Verify** "All squares are similar." True or false?

2. **Verify** "All similar triangles are congruent." True or false?

3. **Verify** "If two polygons are similar, then their corresponding angles are equal in measure." True or false?

4. These two triangles are congruent. Which side of triangle *PQR* is the same length as $\overline{AB}$?

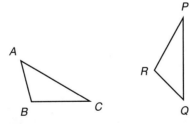

5. These two pentagons are similar. The scale factor for corresponding sides is 3. How long is segment *AE*? How long is segment *IJ*?

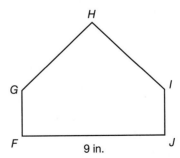

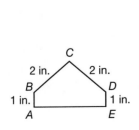

6. One way to draw a triangle similar to the one shown is to trace along two sides, and either extend or shorten the sides by the factor you choose. Here, a similar triangle was drawn by tracing $\overline{AB}$ and $\overline{AC}$ as guides for the sides of a larger triangle.

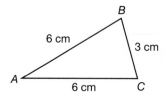

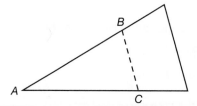

a. Draw a triangle similar to triangle *ABC* by tracing $\overline{AB}$ and $\overline{AC}$ as guides, then doubling the side lengths. Measure the three sides of the triangle. By what scale factor is the triangle you drew larger than triangle *ABC*?

b. Draw a triangle similar to triangle *ABC* by tracing $\overline{AB}$ and $\overline{AC}$ as guides but shortening the side lengths by one-half. Measure the sides of the triangle.
By what scale factor is the triangle you drew smaller than triangle *ABC*?

We have said that one property of similar figures is that the lengths of corresponding sides are proportional

This means that ratios formed by corresponding sides are equal, as we illustrate with the two triangles below.

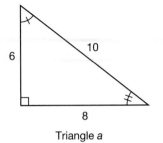

Triangle *a*   Triangle *b*

The lengths of the corresponding sides of triangles *a* and *b* are 6 and 3, 8 and 4, and 10 and 5. These pairs of lengths can be written as equal ratios.

$$\frac{\text{triangle } a}{\text{triangle } b} \quad \frac{6}{3} = \frac{8}{4} = \frac{10}{5}$$

Notice that each of these ratios equals 2. If we choose to put the lengths of the sides of triangle *b* on top, we get three ratios, each equal to $\frac{1}{2}$.

$$\frac{\text{triangle } b}{\text{triangle } a} \quad \frac{3}{6} = \frac{4}{8} = \frac{5}{10}$$

We can write proportions using equal ratios in order to find the lengths of unknown sides of similar triangles.

Example 4

> **Estimate the length *a*. Then use a proportion to find *a*.**
>
>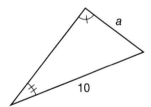
>
> The tick marks indicate two pairs of congruent angles in the triangles. Because the sum of the interior angles of every triangle is 180°, we know that the unmarked angles are also congruent. Thus the two triangles are similar, and the lengths of the corresponding sides are proportional.
>
> The side of length 6 in the smaller triangle corresponds to the side of length 10 in the larger triangle. Thus the side lengths of the larger triangle are not quite double the side lengths of the smaller triangle. Since the side of length *a* in the larger triangle corresponds to the side of length 3 in the smaller triangle, *a* should be a little less than 6. We estimate *a* to be 5.

We now use corresponding sides to write a proportion and solve for *a*. We decide to write the ratios so that the sides from the smaller triangle are on top.

$$\frac{6}{10} = \frac{3}{a} \qquad \text{equal ratios}$$

$$6a = 30 \qquad \text{cross multiplied}$$

$$a = 5 \qquad \text{solved}$$

7. The two triangles are similar. Use a proportion to find lengths *a* and *b*.

 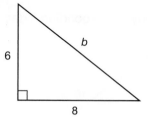

8. The two trapezoids are similar. Use a proportion to find lengths *a* and *b*.

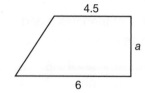

California Mathematics Content Standards
MG 2.0, 2.1 Identify angles as vertical, adjacent, complementary, or supplementary and provide descriptions of these terms.
MG 2.0, 2.2 Use the properties of complementary and supplementary angles and the sum of the angles of a triangle to solve problems involving an unknown angle.
MR 3.0, 3.2 Note the method of deriving the solution and demonstrate a conceptual understanding of the derivation by solving similar problems.

• Complex Shapes

facts   Power Up N

mental math

    a. Measurement: There are 16 ounces in one pound. How many ounces are in $1\frac{1}{2}$ pounds?

    b. Number Sense: 920 − 550

    c. Percent: 25% of 100

    d. Money: $18.99 + $5.30

    e. Decimals: 3.75 ÷ 100

    f. Geometry: The area of this rectangle is 35 in.². What is the length *l*?

    g. Measurement: Each bottle contains 0.5 liter of water. How many bottles are needed for 2 liters of water?

    h. Calculation: Find half of 100, − 1, $\sqrt{}$, × 5, + 1, $\sqrt{}$, × 3, + 2, ÷ 2

problem solving

Choose an appropriate problem-solving strategy to solve this problem. Michelle's grandfather taught her this math method for converting kilometers to miles: "Divide the kilometers by 8, and then multiply by 5." Michelle's grandmother told her to, "Just multiply the kilometers by 0.6." Use both methods to convert 80 km to miles and compare the results. If one kilometer is closer to 0.62 miles, whose method produces the more accurate answer?

In this lesson we will practice finding the perimeters, areas, and interior angle measures of complex shapes. First, we will consider perimeter. The figure below is an example of a complex shape. Notice that the lengths of two of the sides are not given. We will first find the lengths of these sides; then we will find the perimeter of the shape. (In this book, assume that corners that look square are square.)

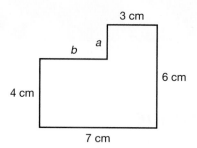

We see that the figure is 7 cm long. The sides marked b and 3 cm together equal 7 cm. So b must be 4 cm.

$$b + 3 \text{ cm} = 7 \text{ cm}$$
$$b = 4 \text{ cm}$$

The width of the figure is 6 cm. The sides marked 4 cm and a together equal 6 cm. So a must equal 2 cm.

$$4 \text{ cm} + a = 6 \text{ cm}$$
$$a = 2 \text{ cm}$$

We have found that b is 4 cm and a is 2 cm.

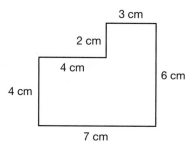

We add the lengths of all the sides and find that the perimeter is 26 cm.

$$6 \text{ cm} + 7 \text{ cm} + 4 \text{ cm} + 4 \text{ cm} + 2 \text{ cm} + 3 \text{ cm} = 26 \text{ cm}$$

Example 1

Find the perimeter of this figure.

To find the perimeter, we add the lengths of the six sides. The lengths of two sides are not given in the illustration. We will write two equations to find the lengths of these sides. The length of the figure is 10 inches. The sides parallel to the 10-inch side have lengths of 4 inches and m inches. Their combined length is 10 inches, so m must equal 6 inches.

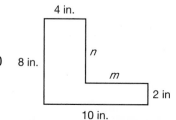

$$4 \text{ in.} + m = 10 \text{ in.}$$
$$m = 6 \text{ in.}$$

The width of the figure is 8 inches. The sides parallel to the 8-inch side have lengths of n inches and 2 inches. Their combined measures equal 8 inches, so n must equal 6 inches.

$$n + 2 \text{ in.} = 8 \text{ in.}$$
$$n = 6 \text{ in.}$$

We add the lengths of the six sides to find the perimeter of the complex shape.

$$10 \text{ in.} + 8 \text{ in.} + 4 \text{ in.} + 6 \text{ in.} + 6 \text{ in.} + 2 \text{ in.} = \textbf{36 in.}$$

Example 2

Which of these figures has a greater perimeter? First make a guess, then compute the perimeters. (Estimate to the nearest hundredth.)

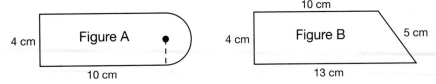

Figure A looks like a combination of a rectangle and a half-circle. The diameter of the circle is 4 cm, so its radius is 2 cm.

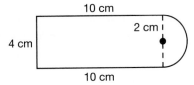

To find the length of the curved portion of the figure, we compute half of the circumference of the circle. We use the estimate $\pi \approx 3.14$.

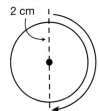

$$C = 2\pi r$$
$$C \approx 2(3.14)(2 \text{ cm})$$
$$C \approx 12.56 \text{ cm}$$

half of circumference
≈ 6.28 cm

Since the circumference is 12.56 cm, the length of the curved portion is 12.56 cm ÷ 2, which is 6.28 cm.

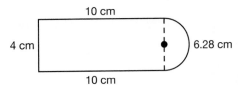

The perimeter is 10 cm + 4 cm + 10 cm + 6.28 cm = **30.28 cm.**

All the side lengths of figure B are given, so we add them to find the perimeter.

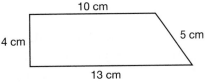

The perimeter is 10 cm + 4 cm + 13 cm + 5 cm = **32 cm.**

The perimeter of **figure B** is greater.

Now we will practice finding the area of complex shapes. One way to find the area of a complex shape is to divide the shape into two or more parts, find the area of each part, and then add the areas. Think of how the shape below could be divided into two rectangles.

Example 3

Thinking Skills

Classify

Based on the number of sides and their lengths, what is the geometric name of this figure?

Find the area of this figure.

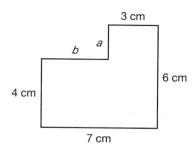

We will show two ways to divide this shape into two rectangles. We use the skills we learned in Example 1 to find that side a is 2 cm and side b is 4 cm. We extend side b with a dashed line segment to divide the figure into two rectangles.

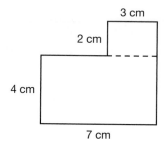

The length and width of the smaller rectangle are 3 cm and 2 cm, so its area is 6 cm². The larger rectangle is 7 cm by 4 cm, so its area is 28 cm². We find the combined area of the two rectangles by adding.

$$6 \text{ cm}^2 + 28 \text{ cm}^2 = \textbf{34 cm}^2$$

A second way to divide the figure into two rectangles is to extend side a.

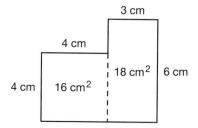

Thinking Skills

Define

Based on the number of sides and their lengths, what is the geometric name of this figure?

Extending side a forms a 4-cm by 4-cm rectangle and a 3-cm by 6-cm rectangle. Again we find the combined area of the two rectangles by adding.

$$16 \text{ cm}^2 + 18 \text{ cm}^2 = 34 \text{ cm}^2$$

Either way we divide the figure, we find that its area is 34 cm².

Example 4

Which of these figures has greater area? First make a guess, then compute the areas. (Estimate to the nearest hundredth.)

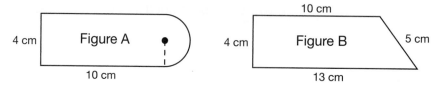

We divide figure A into a rectangle and a half circle. The radius is 2 cm.

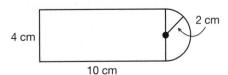

Area of Rectangle	Area of Circle
$A = bh$	$A = \pi r^2$
$A = (10 \text{ cm})(4 \text{ cm})$	$A \approx 3.14(2 \text{ cm})^2$
$A = 40 \text{ cm}^2$	$A \approx 12.56 \text{ cm}^2$

Since the area of the circle is about 12.56 cm², the area of the half-circle is about 12.56 cm² ÷ 2, which is 6.28 cm². The total area is 40 cm² + 6.28 cm² = **46.28 cm²**.

We divide **figure B** into a rectangle and a triangle.

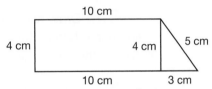

The side with length 13 cm is split into two parts, one which is a side of the rectangle (and has the same length as the opposite side) and one which is a side of the triangle (with the remaining 3 cm). The other side of the triangle has length 4 cm, equal to the length of the side of the rectangle opposite it.

Area of Rectangle	Area of Triangle
$A = bh$	$A = \frac{1}{2} bh$
$A = (10 \text{ cm})(4 \text{ cm})$	$A = \frac{1}{2}(3 \text{ cm})(4 \text{ cm})$
$A = 40 \text{ cm}^2$	$A = 6 \text{ cm}^2$

The total area is 40 cm² + 6 cm² = **46 cm²**.

The area of **figure A** is greater.

We can always use what we know about angles to find angle measures in complex shapes.

Example 5

Find angle measures *a* and *b*.

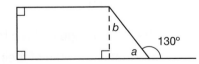

To find *a*, we note that these adjacent angles are supplementary.
$$a + 130° = 180°$$
$$\boldsymbol{a = 50°}$$

The measures of the angles of a triangle add to 180°. Since one angle is a right angle (with a measure of 90°), the other two angles are complementary.

$$a + b = 90°$$
$$50° + b = 90°$$
$$\textbf{b = 40°}$$

Lesson Practice

Find the perimeter of each complex shape:

a.

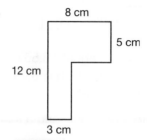

b.

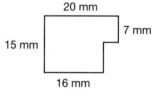

c. **Model** Draw two ways to divide this figure into two rectangles. Then find the area of the figure each way.

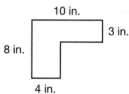

d. This trapezoid can be divided into a rectangle and a triangle. Find the area of the trapezoid.

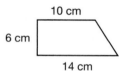

e. Find the perimeter and area of the figure. (Estimate to the nearest hundredth.

f. Find the angle measures.

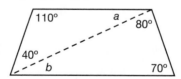

Written Practice *Distributed and Integrated*

1. When the sum of $\frac{1}{2}$ and $\frac{1}{3}$ is divided by the product of $\frac{1}{2}$ and $\frac{1}{3}$, what is
(4, 45) the quotient?

2. The average age of three men is 24 years.
(10)
a. What is the sum of their ages?

b. If two of the men are 22 years old, how old is the third?

3. A string one yard long is formed into the shape of a square.
(32)
a. How many inches long is each side of the square?

b. How many square inches is the area of the square?

4. Complete this proportion: Five is to three as thirty is to what number?
(83)

5. Mr. Cho has 30 books. Fourteen of the books are mysteries. What is the ratio of
(16) mysteries to non-mysteries?

* 6. (**Analyze**) In another class of 33 students, the ratio of boys to girls is 4 to 7. How
(100) many girls are in that class?

7. Evaluate $x(x + y)$ for $x = 0.5$ and $y = 0.6$
(88)

* 8. Roscoe complained that he had a "ton" of homework.
(54, 81)
    a. How many pounds is a ton?

    b. (**Conclude**) What is the probability that Roscoe would literally have a ton of
    homework?

(**Connect**) Complete the table to answer problems **9–11.**

	Fraction	Decimal	Percent
* 9. (96)	$\frac{1}{100}$	**a.**	**b.**
* 10. (96)	**a.**	0.4	**b.**
* 11. (96)	**a.**	**b.**	8%

12. $10\frac{1}{2} \div 3\frac{1}{2}$
(64)

13. $(6 + 2.4) \div 0.04$
(41)

Find each unknown number:

14. $7\frac{1}{2} + 6\frac{3}{4} + n = 15\frac{3}{8}$
(39, 55)

15. $x - 1\frac{3}{4} = 7\frac{1}{2}$
(39, 55)

16. (**Verify**) Instead of dividing $10\frac{1}{2}$ by $3\frac{1}{2}$, Guadalupe doubled both numbers before
(38) dividing. What was Guadalupe's division problem and its quotient?

17. (**Estimate**) Mariabella used a tape measure to find the circumference and the
(42) diameter of a plate. The circumference was about 35 inches, and the diameter
was about 11 inches. Find the approximate number of diameters in the
circumference. Round to the nearest tenth.

18. $\frac{3}{4}c = 12$
(91)

19. (**List**) Name the prime numbers between 40 and 50.
(11)

* 20. (**Analyze**) Calculate mentally:
(98)
    a. $-3 + -8$ **b.** $-3 - -8$

    c. $-8 + +3$ **d.** $-8 - +3$

* 21. (**Conclude**) In $\triangle ABC$ the measure of $\angle A$ is 40°.
(95) Angles B and C are congruent. What is the measure
of $\angle C$?

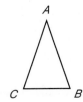

* 22. **a.** What is the perimeter of this triangle?
(RF7, 77)

b. What is the area of this triangle?

c. What is the ratio of the length of the 20 mm side to the length of the longest side? Express the ratio as a fraction and as a decimal.

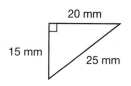

23. **Analyze** The Simpsons rented a trailer that was 8 feet long and 5 feet wide.
(RF29) If they load the trailer with 1-by-1-by-1-foot boxes to a height of 3 feet, how many boxes can be loaded onto the trailer?

24. What is the probability of drawing the queen of spades from a normal deck
(54) of 52 cards?

* 25. **a.** **Connect** What temperature is shown on this thermometer?
(81, 98)

b. If the temperature rises 12°F, what will the temperature be?

* 26. Find the perimeter of the figure below.
(101)

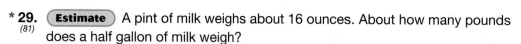

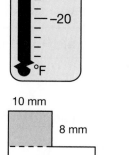

27. **a.** **Analyze** What is the area of the shaded rectangle?
(77)

b. What is the area of the unshaded rectangle?

c. What is the combined area of the two rectangles?

28. **Connect** What are the coordinates of the point halfway between
(RF27) (−3, −2) and (5, −2)?

* 29. **Estimate** A pint of milk weighs about 16 ounces. About how many pounds
(81) does a half gallon of milk weigh?

30. In a bag are 12 marbles: 3 red, 4 white, and 5 blue. One marble is drawn from a
(80) bag and not replaced. A second and third marble are drawn without replacing.

a. What is the probability of drawing a red, a white, and a blue marble in that order?

b. What is the probability of drawing a blue, a white, and a red marble in that order?

California Mathematics Content Standards

NS **1.0**, **1.3** Use proportions to solve problems (e.g., determine the value of N if $\frac{4}{7} = \frac{N}{21}$, find the length of a side of a polygon similar to a known polygon). Use cross-multiplication as a method for solving such problems, understanding it as the multiplication of both sides of an equation by a multiplicative inverse.

NS **1.0**, **1.4** Calculate given percentages of quantities and solve problems involving discounts at sales, interest earned, and tips.

• Using Proportions to Solve Percent Problems

facts	Power Up L
mental math	**a. Fractions:** $\frac{1}{4}$ of 16

b. Percent: 25% of 16

c. Money: $100.00 − $47.50

d. Number Sense: 50 × 15

e. Measurement: How many pounds is 80 ounces?

f. Time: A direct flight from San Francisco to San Diego takes about $1\frac{1}{2}$ hours. How many minutes is that?

g. Algebra: If $48 + t = 50.1$, what is t?

h. Calculation: 12 + 12, + 1, $\sqrt{}$, × 3, + 1, $\sqrt{}$, × 2, + 2, × 5

problem solving

Choose an appropriate problem-solving strategy to solve this problem. "Casting out nines" is a technique for checking long multiplication. To cast out nines, we sum the digits of each number from left to right and "cast out" (subtract) 9 from the resulting sums. For instance:

$$6{,}749 \quad 6 + 7 + 4 + 9 = 26 \qquad 26 - 9 = 17;\ 17 - 9 = \mathbf{8}$$
$$\underline{\times \quad 85} \quad 8 + 5 = 13 \qquad\qquad 13 - 9 = \mathbf{4}$$
$$573{,}665 \quad 5 + 7 + 3 + 6 + 6 + 5 = 32 \quad 32 - 9 = 23;\ 23 - 9 = 14;\ 14 - 9 = \mathbf{5}$$

To verify the product is correct, we multiply the 8 and the 4 (8 × 4 = 32), add the resulting digits (3 + 2 = 5), and compare the result to the product after casting out nines. The number 573,665 results in 5 after casting out nines, so the product is most likely correct. If the numbers had been different, we would know that our original product was incorrect. Matching results after casting out nines does not always guarantee that our product is correct, but the technique catches most random errors.

Check 1234 × 56 = 69,106 by casting out nines

Math Language

Percent means per hundred. 30% means 30 out of 100

We know that a percent can be expressed as a fraction with a denominator of 100.

$$30\% = \frac{30}{100}$$

A percent can also be regarded as a ratio in which 100 represents the total number in the group, as we show in the following example.

Example 1

Thirty percent of the cars in the parade are antique cars. If 12 vehicles are antique cars, how many vehicles are in the parade in all?

We construct a ratio box. The ratio numbers we are given are 30 and 100. We know from the word *percent* that 100 represents the ratio total. The actual count we are given is 12. Our categories are "Antiques" and "not Antiques."

	Percent	Actual Count
Antiques	30	12
Not Antiques		
Total	100	

Since the ratio total is 100, we calculate that the ratio number for "not Antiques" is 70. We use n to stand for "not Antiques" and t for "total" in the actual-count column. We use two rows from the table to write a proportion. Since we know both numbers in the "Antiques" row, we use the numbers in the "Antiques" row for the proportion. Since we want to find the total number of vehicles, we also use the numbers from the "total" row. We will then solve the proportion using cross products.

	Percent	Actual Count
Antiques	30	12
Not Antiques	70	n
Total	100	t

$$\frac{30}{100} = \frac{12}{t}$$

$$30t = 12 \cdot 100$$

$$t = \frac{\overset{4}{\cancel{12}} \cdot \overset{10}{\cancel{100}}}{\underset{\underset{1}{3}}{\cancel{30}}}$$

$$t = 40$$

We find that a total of **40 vehicles** were in the parade.

In the above problem we did not need to use the 70% who were "not Antiques." In the next example we will need to use the "not" percent in order to solve the problem.

Example 2

Only 40% of the team members played in the game. If 24 team members did not play, then how many did play?

We construct a ratio box. The categories are "played," "did not play," and "total." Since 40% played, we calculate that 60% did not play. We are asked for the actual count of those who played. So we use the "played" row and the "did not play" row (because we know both numbers in that row) to write the proportion.

	Percent	Actual Count
Played	40	p
Did Not Play	60	24
Total	100	t

$$\frac{40}{60} = \frac{p}{24}$$

$$60p = 40 \cdot 24$$

$$p = \frac{\overset{4}{\cancel{40}} \cdot \overset{4}{\cancel{24}}}{\underset{\underset{1}{6}}{\cancel{60}}}$$

$$p = 16$$

We find that **16 team members** played in the game.

Example 3

Buying the shoes on sale, Nathan paid $45.60, which was 60% of the full price. What was the full price of the shoes?

Nathan paid 60% instead of 100%, so he saved 40% of the full price. We are given what Nathan paid. We are asked for the full price, which is the 100% price.

	Percent	Actual Count
Paid	60%	$45.60
Saved	40%	s
Full Price	100%	f

$$\frac{60}{100} = \frac{45.60}{f}$$

$$60f = 4560$$

$$f = 76$$

Full price for the shoes was **$76.**

Lesson Practice

Model Solve these percent problems using proportions. Make a ratio box for each problem.

a. Forty percent of the cameras in a store are digital cameras. If 24 cameras are not digital, how many cameras are in the store in all?

b. Seventy percent of the team members played in the game. If 21 team members played, how many team members did not play?

c. **Model** Referring to problem **b,** what proportion would we use to find the number of members on the team?

d. Joan walked 0.6 miles in 10 minutes. How far can she walk in 25 minutes at that rate? Write and solve a proportion to find the answer.

e. (**Formulate**) Create and solve your own percent problem using the method shown in this lesson.

Written Practice
Distributed and Integrated

* 1. How far would a car travel in $2\frac{1}{2}$ hours at 50 miles per hour?
(89)

$$\frac{50 \text{ mi}}{1 \text{ hr}} \times \frac{2\frac{1}{2} \text{ hr}}{1}$$

2. (**Connect**) A map of California is drawn to a scale of 1 inch = 50 miles. San Jose and Bakersfield are about 5 inches apart on the map. According to this map, what is the distance between San Jose and Bakersfield?
(89)

3. The ratio of humpback whales to orcas was 2 to 1. If there were 900 humpback whales, how many orcas were there?
(86)

4. Name the property illustrated: $5(6 + 7) = 30 + 35$
(88)

* 5. Calculate mentally:
(97)
    a. $+10 + -10$ **b.** $-10 - -10$ **c.** $+6 + -5 - -4$

* 6. (**Estimate**) On Earth a 1-kilogram object weighs about 2.2 pounds. A rock weighs 50 kilograms. About how many pounds does the rock weigh?
(76)

* 7. Sonia has only dimes and nickels in her coin jar; they are in a ratio of 3 to 5. If she has 120 coins in the jar, how many are dimes?
(100)

* 8. (**Analyze**) The airline sold 25% of the seats on the plane at a discount. If 45 seats were sold at a discount, how many seats were on the plane? How do you know your answer is correct?
(102)

(**Connect**) Complete the table to answer protblems **9–11.**

	Fraction	Decimal	Percent
* 9. (96)	$\frac{3}{50}$	**a.**	**b.**
* 10. (96)	**a.**	0.04	**b.**
* 11. (96)	**a.**	**b.**	150%

12. $4\frac{1}{12} + 5\frac{1}{6} + 2\frac{1}{4}$
(56)

13. $\frac{4}{5} \times 3\frac{1}{3} \times 3$
(67)

14. 0.125×80
(33)

15. $(1 + 0.5) \div (1 - 0.5)$
(41)

16. Solve: $\dfrac{c}{12} = \dfrac{3}{4}$
(83)

17. What is the total cost of an $8.75 purchase after 8% sales tax is added?
(35)

18. Write the decimal number one hundred five and five hundredths.
(RF19)

* 19. (**Conclude**) The measure of $\angle A$ in quadrilateral $ABCD$ is 115°. What are the measures of $\angle B$ and $\angle C$?
(95)

20. Write the prime factorization of 500 using exponents.
(71)

21. (**Estimate**) A quart is a little less than a liter, so a gallon is a little less than how many liters?
(81)

22. Diane will spin the spinner twice. What is the probability that it will stop in sector 2 both times?
(Inv. 6)

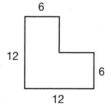

23. The perimeter of this isosceles triangle is 18 cm. What is the length of its longest side?
(RF7)

24. What is the area of the triangle in problem **23?**
(77)

25. The temperature was −5°F at 6:00 a.m. By noon the temperature had risen 12 degrees. What was the noontime temperature?
(98)

26. The weather report stated that the chance of rain is 30%. Use a decimal number to express the probability that it will not rain.
(54)

*27. Find the perimeter and area of this figure. Dimensions are in inches.
(101)

*28. $\dfrac{1}{4}x = 5$
(91)

29. A room is 15 feet long and 12 feet wide.
(23, 89)

    a. The room is how many **yards** long and wide?

    b. What is the area of the room in square yards?

*30. (**Explain**) Latrivius rolled a die and it turned up 6. If he rolls the die again, what is the probability that it will turn up 6?
(54)

✎ *California Mathematics Content Standards*

NS 2.0, 2.3 Solve addition, subtraction, multiplication, and division problems, including those arising in concrete situations, that use positive and negative integers and combinations of these operations.

MR 1.0, 1.2 Formulate and justify mathematical conjectures based on a general description of the mathematical question or problem posed.

• Multiplying and Dividing Integers

facts	Power Up O
mental math	**a. Fractions:** $\frac{3}{4}$ of 40
	b. Percent: 75% of 40
	c. Number Sense: 543 − 250
	d. Money: Notebooks are regularly priced at 75¢ each. If the sale is "buy 2 notebooks and get 1 free," how many notebooks can Sonia get with $1.50?
	e. Decimals: 87.5 ÷ 100
	f. Number Sense: $\frac{500}{20}$
	g. Measurement: How many milliliters are in 10 liters?
	h. Calculation: $6 \times 6, -1, \div 5, \times 6, -2, \div 5, \times 4, -2, \times 3$

problem solving

Choose an appropriate problem-solving strategy to solve this problem. On a balanced scale are a 25-gram mass, a 100-gram mass, and five identical blocks marked *x,* which are distributed as shown. What is the mass of each block marked *x*? Write an equation illustrated by this balanced scale.

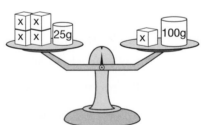

New Concept

We know that when we multiply two positive numbers the product is positive.

$$(+3)(+4) = +12$$

positive × positive = positive

Reading Math

$(+3)(+4)$ is the same as 3×4 or $3 \cdot 4$.

Notice that when we write $(+3)(+4)$ there is no $+$ or $-$ sign between the sets of parentheses.

When we multiply a positive number and a negative number, the product is negative. We show an example on this number line by multiplying 3 and -4.

$$3 \times -4 \text{ means } (-4) + (-4) + (-4)$$

We write the multiplication this way:

$$(+3)(-4) = -12$$

Positive three times *negative* four equals negative 12.

positive × negative = negative

When we multiply two negative numbers, the product is positive. Consider this sequence of equations:

1. Three times 4 is 12. $\qquad\qquad\qquad 3 \times 4 = 12$

2. "Three times the opposite of 4" is $\qquad 3 \times -4 = -12$
   "the opposite of 12."

3. The opposite of "3 times the opposite of 4" $\quad -3 \times -4 = +12$
   is the opposite of "the opposite of 12."

negative × negative = positive

Recall that we can rearrange the numbers of a multiplication fact to make two division facts.

Thinking Skills

Discuss

How is multiplying integers similar to multiplying whole numbers? How is it different?

Multiplication Facts	Division Facts	
$(+3)(+4) = +12$	$\dfrac{+12}{+3} = +4$	$\dfrac{+12}{+4} = +3$
$(+3)(-4) = -12$	$\dfrac{-12}{+3} = -4$	$\dfrac{-12}{-4} = +3$
$(-3)(-4) = +12$	$\dfrac{+12}{-3} = -4$	$\dfrac{+12}{-4} = -3$

Studying these nine facts, we can summarize the results in two rules:

1. If the two numbers in a multiplication or division problem have the **same sign,** the answer is positive.

2. If the two numbers in a multiplication or division problem have **different signs,** the answer is negative.

Calculate mentally:

a. (+8)(+4)   b. (+8) ÷ (+4)

c. (+8)(−4)   d. (+8) ÷ (−4)

e. (−8)(+4)   f. (−8) ÷ (+4)

g. (−8)(−4)   h. (−8) ÷ (−4)

a. +32   b. +2

c. −32   d. −2

e. −32   f. −2

g. +32   h. +2

Lesson Practice

Predict First predict which problems will have a positive answer and which will have a negative answer. Then simplify each problem.

a. (−5)(+4)   b. (−5)(−4)

c. (+5)(+4)   d. (+5)(−4)

e. $\dfrac{+12}{-2}$ f. $\dfrac{+12}{+2}$ g. $\dfrac{-12}{+2}$ h. $\dfrac{-12}{-2}$

Written Practice

Distributed and Integrated

1. The first three prime numbers are 2, 3, and 5. Their product is 30. What is the
(11)   product of the next three prime numbers?

2. On the map 2 cm equals 1 km. What is the actual length of a street that is
(89)   10 cm long on the map?

3. Between 8 p.m. and 9 p.m. the station broadcasts 8 minutes of commercials.
(16)   What was the ratio of commercial time to noncommercial time during that
       hour?

4. Multiple Choice   a. Which of the following triangles appears to have a right angle as
(90)   one of its angles?

b. In which triangle do all three sides appear to be the same length?

A   B    C

* 5. **Conclude** If the two acute angles of a right triangle are congruent, then what is
(20, 95)   the measure of each acute angle?

6. Ms. Hernandez is assigning each student in her class one of the fifty U.S.
(54) states on which to write a report. What is the probability that Manuela will be
assigned one of the 5 states that has coastline on the Pacific Ocean? Express the
probability ratio as a fraction and as a decimal.

Solve:

7. $\frac{3}{5}x = \frac{6}{7}$
(91)

* 8. $\frac{8}{n} = \frac{4}{2.5}$
(83, 102)

(**Connect**) Complete the table to answer problems **9–11.**

	Fraction	Decimal	Percent
9. (96)	$\frac{5}{8}$	**a.**	**b.**
10. (96)	**a.**	1.25	**b.**
11. (96)	**a.**	**b.**	70%

12. **a.** If the spinner is spun once, what is the
(54) probability that it will stop on a number
less than 4?

b. If the spinner is spun 100 times, how
many times would it be expected to stop
on a prime number?

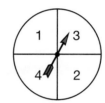

13. Convert 200 centimeters to meters by completing this multiplication:
(89)
$$\frac{200 \text{ cm}}{1} \cdot \frac{1 \text{ m}}{100 \text{ cm}}$$

14. $(6.2 + 9) - 2.79$
(41)

15. **a.** $\frac{-12}{+4}$
(103)

b. $\frac{+12}{-4}$

16. **a.** $(-4)(-2)$
(103)

b. $(+4)(-2)$

17. Write the fraction $\frac{2}{3}$ as a decimal number rounded to the hundredths place.
(46, 69)

18. The Zamoras rent a storage room that is 10 feet wide, 12 feet long, and 8 feet
(RF29) high. How many cube-shaped boxes 1 foot on each edge can the Zamoras
store in the room?

* 19. These two rectangles are similar.
(Inv. 10)
a. What is the scale factor from the
smaller rectangle to the larger rectangle?

b. What is the scale factor from the larger
rectangle to the smaller rectangle?

20. $0.12m = \$4.20$
(85)

* 21. Calculate mentally:
(98)
    a. $+7 + -8$ **b.** $-7 + +8$

    c. $-7 - +8$ **d.** $-7 - -8$

* 22. Triangles I and II are congruent. What is the area of each triangle?
(77,
Inv. 10)

23. Use the formula $d = r \cdot t$ to find the distance when the rate is 50 miles per hour
(89, 93) and the time is 1.5 hours.

* 24. This trapezoid has been divided into a rectangle and a
(101) triangle. What is the area of the trapezoid?

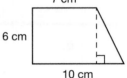

25. (**Estimate**) A soup label must be long enough to wrap
(42) around a can. If the diameter of the can is 7 cm, then the label must be at least how long? Round up to the nearest whole number.

* 26. (**Analyze**) The triangles below are similar. What is the measure of angle A?
(Inv. 10)

27. The ratio of almonds to cashews in the mix was 9 to 2. Horace counted 36
(86) cashews in all. How many almonds were there?

28. Write the length of the segment below in
(RF6,
76)   **a.** millimeters.

  b. centimeters.

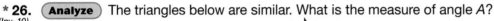

29. $6\frac{2}{3} \div 100$
(64)

30. **a.** $(-4) + (-5) - (-6)$
(99)
  b. $(-2) + (-3) - (-4) - (+5)$

California Mathematics Content Standards

NS **2.0**, **2.3** Solve addition, subtraction, multiplication, and division problems, including those arising in concrete situations, that use positive and negative integers and combinations of these operations.

MR **2.0, 2.4** Use a variety of methods, such as words, numbers, symbols, charts, graphs, tables, diagrams, and models, to explain mathematical reasoning.

• Two Ways to Think About Multiplying Negative Integers

A negative integer times a negative integer equals a positive integer. Why is this true? Here are two ways to think about the answer to this question.

1. Look for a Pattern

We can use some multiplication problems for which we know the answers and look for a pattern. Notice that as the second factor decreases by 1 the products increase by 2 each time.	$(-2) \times 3 = -6$ $(-2) \times 2 = -4$ $(-2) \times 1 = -2$ $(-2) \times 0 = 0$
Now we can continue this pattern with negative integers. Notice that the products now are positive integers.	$(-2) \times (-1) = 2$ $(-2) \times (-2) = 4$ $(-2) \times (-3) = 6$

The pattern indicates that the product of two negative numbers will be a positive number.

2. Use the Properties of Multiplication

First we must show that $(-1) \times (-1) = 1$.

Steps	Reasons
$[(-1) \times (-1)] - 1 = ?$	If $(-1) \times (-1) = 1$, then we will get 0 when we subtract 1 from $[(-1) \times (-1)]$.
$[(-1) \times (-1)] - 1 = [(-1) \times (-1)] + (-1)$	Adding (-1) is equal to subtracting 1.
$[(-1) \times (-1)] - 1 = [(-1) \times (-1)] + [(-1) \times 1]$	Identity Property: $(-1) = [(-1) \times 1]$
$[(-1) \times (-1)] - 1 = (-1) \times [(-1) + 1]$	Distributive Property: collect terms
$[(-1) \times (-1)] - 1 = (-1) \times 0$	Simplified: $[(-1) + 1] = 0$
$[(-1) \times (-1)] - 1 = 0$	Zero Property: $(-1) \times 0 = 0$

Now that we know $(-1) \times (-1) = 1$, we will look at $(-2) \times (-3)$.

Steps	Reasons
$(-2) \times (-3)$	The integers we want to multiply.
$(-2) \times (-3) = [(-1) \times 2] \times [(-1) \times 3]$	$(-2) = (-1) \times 2; (-3) = (-1) \times 3$
$(-2) \times (-3) = (-1) \times 2 \times (-1) \times 3$	Associative Property
$(-2) \times (-3) = (-1) \times (-1) \times 2 \times 3$	Commutative Property
$(-2) \times (-3) = 1 \times 2 \times 3$	$(-1) \times (-1) = 1$
$(-2) \times (-3) = 2 \times 3$	Simplified $1 \times 2 \times 3$

The properties show that the product of two negative numbers is a positive number.

Discuss Which way of thinking about multiplying two negative integers makes more sense to you?

Apply Determine the sign of each product or quotient. Recall that the division of integers follows the same rules for multiplication of integers.

a. $(-2) \times (-3)$

b. $(-5) \times (+6)$

c. $(-6) \times (+8)$

d. $(-4) \times (-5)$

e. $(-22) \times (-56) \times (-7)$

f. $(-22) \times (-12) \times (+8)$

g. $(-12) \div (-3)$

h. $(-16) \div (+3)$

i. $(+36) \div (+18)$

j. $(+56) \div (-8)$

k. $(-42) \div (-21)$

✎ *California Mathematics Content Standards*

SDAP 1.0, 1.1 Compute the range, mean, median, and mode of data sets.

MR 2.0, 2.6 Indicate the relative advantages of exact and approximate solutions to problems and give answers to a specified degree of accuracy.

MR 3.0, 2.7 Make precise calculations and check the validity of the results from the context of the problem.

MR 3.0, 3.1 Evaluate the reasonableness of the solution in the context of the original situation.

• Applications Using Division

Power Up

facts   Power Up P

mental math

   a. Fractions: $\frac{1}{4}$ of 80

   b. Fractions: $\frac{3}{4}$ of 80

   c. Percent: 50% of 200

   d. Percent: 25% of 200

   e. Money: Jason wants to save a total of $120 over the next four months. On average, how much money does he need to save each month?

   f. Estimation: Roberto read 22 pages in 30 minutes. About how many pages can he expect to read in 1 hour?

   g. Algebra: If $t = 0$, what is $t - 2$?

   h. Calculation: $9 \times 6, - 4, \div 5, \times 8, \div 2, \div 2, + 4, + 1, \times 4$

problem solving

Choose an appropriate problem-solving strategy to solve this problem. Sonya, Sid, and Sinead met at the gym on Monday. Sonya goes to the gym every two days. The next day she will be at the gym is Wednesday. Sid goes to the gym every three days. The next day Sid will be at the gym is Thursday. Sinead goes to the gym every four days. She will next be at the gym on Friday. What will be the next day that Sonya, Sidney, and Sinead are at the gym on the same day?

New Concept

When a division problem has a remainder, there are several ways to write the answer: with a remainder, as a mixed number, or as a decimal number.

$$\begin{array}{r} 3\ R\ 3 \\ 4\overline{)15} \end{array} \qquad \begin{array}{r} 3\frac{3}{4} \\ 4\overline{)15} \end{array} \qquad \begin{array}{r} 3.75 \\ 4\overline{)15.00} \end{array}$$

How a division answer should be written depends upon the question to be answered. In real-world applications we sometimes need to round an answer up, and we sometimes need to round an answer down. The quotient of 15 ÷ 4 rounds up to 4 and rounds down to 3.

Example 1

One hundred students are to be assigned to 3 classrooms. How many students should be in each class so that the numbers are as balanced as possible?

Dividing 100 by 3 gives us 33 R 1. Assigning 33 students per class totals 99 students. We add the remaining student to one of the classes, giving that class 34 students. We write the answer **33, 33,** and **34.**

Example 2

Matinee movie tickets cost $8. Jim has $30. How many tickets can he buy?

We divide 30 dollars by 8 dollars per ticket. The quotient is $3\frac{3}{4}$ tickets.

$$\frac{30 \text{ dollars}}{8 \text{ dollars per ticket}} = 3\frac{3}{4} \text{ tickets}$$

Jim cannot buy $\frac{3}{4}$ of a ticket, so we round down to the nearest whole number. Jim can buy **3 tickets.**

Example 3

Fifteen children need a ride to the fair. Each car can transport 4 children. How many cars are needed to transport 15 children?

We divide 15 children by 4 children per car. The quotient is $3\frac{3}{4}$ cars.

$$\frac{15 \text{ children}}{4 \text{ children per car}} = 3\frac{3}{4} \text{ cars}$$

Three cars are not enough. Four cars will be needed. One of the cars will be $\frac{3}{4}$ full. We round $3\frac{3}{4}$ cars up to **4 cars.**

Example 4

Dale cut a 10-foot board into four equal lengths. How long was each of the four boards?

We divide 10 feet by 4.

$$\frac{10 \text{ ft}}{4} = 2\frac{2}{4} \text{ ft} = 2\frac{1}{2} \text{ ft}$$

Each board was **$2\frac{1}{2}$ ft** long.

Example 5

Kimberly is on the school swim team. At practice she swam the 50 m freestyle three times. Her times were 37.53 seconds, 36.90 seconds, and 36.63 seconds. What was the mean of her three times?

To find the mean we add the three times and divide by 3.

$$
\begin{array}{r}
37.53 \\
36.90 \\
\underline{36.63} \\
111.06
\end{array}
\qquad
\begin{array}{r}
37.02 \\
3\overline{)111.06}
\end{array}
$$

Kimberly's mean time was **37.02 seconds.**

Lesson Practice

a. Ninety students were assigned to four classrooms as equally as possible. How many students were in each of the four classrooms?

b. Movie tickets cost $9.50. Aluna has $30.00. How many movie tickets can she buy?

c. Twenty-eight children need a ride to the fair. Each van can carry six children. How many vans are needed?

d. Corinne folded an $8\frac{1}{2}$ in. by 11 in. piece of paper in half. Then she folded the paper in half again as shown. After the two folds, what are the dimensions of the rectangle that is formed? How can you check your answer?

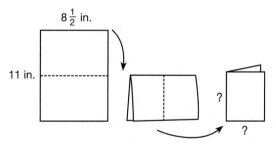

e. Kevin ordered four books at the book fair for summer reading. The books cost $6.95, $7.95, $6.45, and $8.85. Find the average (mean) price of the books.

Written Practice

Distributed and Integrated

1. When the greatest four-digit number is divided by the greatest two-digit number,
(RF2) what is the quotient?

2. The ratio of the length to the width of the Alamo is about 5 to 3. If the width of the
(86) Alamo is approximately 63 ft, about how long is the Alamo?

3. A box of crackers in the shape of a square prism had a length, width,
(RF29) and height of 4 inches, 4 inches, and 10 inches respectively. How many
cubic inches was the volume of the box?

* 4. **(Analyze)** A full turn is 360°. How many degrees is $\frac{1}{6}$ of a turn?
(RF17)

5. Three friends purchase a 3-pack of toys. The total bill comes to $10.81.
(104) How much money should each friend pay so that their contributions are
as equal as possible?

Refer to these triangles to answer problems **6–7:**

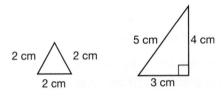

* 6. Sketch a triangle similar to the equilateral triangle. Make the scale factor from
(90, the equilateral triangle to your sketch 3. What is the perimeter of the triangle
Inv. 10) you sketched?

7. What is the area of the right triangle?
(77, 90)

* 8. **(Model)** Draw a ratio box for this problem. Then solve the problem
(102) using a proportion.

*Ms. Mendez is sorting her photographs. She notes that 12 of the
photos are black and white and that 40% of the photos are color.
How many photos does she have?*

(Connect) Complete the table to answer problems **9–11.**

	Fraction	Decimal	Percent
9. *(96)*	$2\frac{3}{4}$	**a.**	**b.**
10. *(96)*	**a.**	1.1	**b.**
11. *(96)*	**a.**	**b.**	64%

12. $24\frac{1}{6} + 23\frac{1}{3} + 22\frac{1}{2}$
(56)

13. $\left(1\frac{1}{5} \div 2\right) \div 1\frac{2}{3}$
(64)

14. $9 - (6.2 + 2.79)$
(72)

15. $0.36m = \$63.00$
(85)

16. Find 6.5% of $24.89 by multiplying 0.065 by $24.89. Round the product to the
(35, 46) nearest cent.

17. (Estimate) Round the quotient to the nearest thousandth:
(46)
$$0.065 \div 4$$

18. Write the prime factorization of 1000 using exponents.
(71)

* 19. (Verify) "All squares are similar." True or false?
(Inv. 10)

20. **a.** $\frac{-15}{-3}$
(103)

    b. $(+4)(-3)$

* 21. What is the perimeter of this polygon?
(101)

* 22. What is the area of this polygon?
(101)

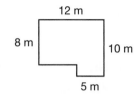

23. **a.** $(-6) - (-4) + (+12)$
(99)

    b. $(-5) + (-2) - (-7) - (+9)$

24. Triangles I and II are congruent. The perimeter of each triangle is 24 cm. What is
(RF7, the length of the shortest side of each triangle?
Inv. 10)

25. (Estimate) The first Ferris wheel was built in 1893 for the world's fair in Chicago.
(42) The diameter of the Ferris wheel was 250 ft. Find the circumference of the original
Ferris wheel to the nearest hundred feet.

26. Use a ruler to draw $\overline{AB}$ $1\frac{3}{4}$ inches long. Then draw a dot at the midpoint of $\overline{AB}$,
(RF9) and label the point M. How long is $\overline{AM}$?

27. (Model) Use a compass to draw a circle on a coordinate plane. Make the center
(RF16, of the circle the origin, and make the radius five units. At which two points does
RF27) the circle cross the x-axis?

28. What is the area of the circle in problem **27**? (Use 3.14 for π.)
(84)

29. $-3 + -4 - -5 - +7$
(97)

* 30. If Freddy tosses a coin four times, what is the probability that the coin will
(Inv. 6) turn up heads, tails, heads, tails in that order?

LESSON

105

📐 California Mathematics Content Standards

NS 1.0, 1.2 Interpret and use ratios in different contexts (e.g., batting averages, miles per hour) to show the relative sizes of two quantities, using appropriate notations ($\frac{a}{b}$, *a* to *b*, *a:b*).

AF 2.0, 2.1 Convert one unit of measurement to another (e.g., from feet to miles, from centimeters to inches).

• Unit Multipliers

facts	Power Up O
mental math	**a. Number Sense:** $200 \cdot 40$
	b. Number Sense: $567 - 150$
	c. Fractions: $\frac{1}{4}$ of 120
	d. Fractions: $\frac{1}{8}$ of 120
	e. Decimals: $7.5 \div 100$
	f. Estimation: If 704 pennies will fit into a half-liter bottle, about how many pennies will fit into a 2-liter bottle?
	g. Measurement: How many quarts are in 2 gallons?
	h. Calculation: $6 \times 8, +1, \sqrt{}, \times 3, -1, \div 2, \times 10, -1, \div 9$
problem solving	Choose an appropriate problem-solving strategy to solve this problem. Copy this problem and fill in the missing digits:

$$\frac{\square}{4} + \frac{\square}{6} = \frac{11}{12}$$

A **unit multiplier** is a fraction that equals 1 and that is written with two different units of measure. Recall that when the numerator and denominator of a fraction are equal (and are not zero), the fraction equals 1. Since 1 foot equals 12 inches, we can form two unit multipliers with the measures 1 foot and 12 inches.

$$\frac{1 \text{ ft}}{12 \text{ in.}} \qquad \frac{12 \text{ in.}}{1 \text{ ft}}$$

Each of these fractions equals 1 because the numerator and denominator of each fraction are equal.

We can use unit multipliers to help us convert from one unit of measure to another. If we want to convert 60 inches to feet, we can multiply 60 inches by the unit multiplier $\frac{1 \text{ ft}}{12 \text{ in.}}$.

$$\frac{\overset{5}{\cancel{60 \text{ in.}}}}{1} \times \frac{1 \text{ ft}}{\underset{1}{\cancel{12 \text{ in.}}}} = 5 \text{ ft}$$

Example 1

a. **Write two unit multipliers using these equivalent measures:**

3 ft = 1 yd

b. **Which unit multiplier would you use to convert 30 yards to feet?**

a. We use the equivalent measures to write two fractions equal to 1.

$$\frac{3 \text{ ft}}{1 \text{ yd}} \qquad \frac{1 \text{ yd}}{3 \text{ ft}}$$

b. We want the units we are changing **from** to appear in the denominator and the units we are changing **to** to appear in the numerator. To convert 30 yards to feet, we use the unit multiplier that has yards in the denominator and feet in the numerator.

$$\frac{30 \text{ yd}}{1} \times \frac{3 \text{ ft}}{1 \text{ yd}}$$

Here we show the work. Notice that the yards "cancel," and the product is expressed in feet.

$$\frac{30 \text{ }\cancel{\text{yd}}}{1} \times \frac{3 \text{ ft}}{1 \text{ }\cancel{\text{yd}}} = 90 \text{ ft}$$

Example 2

Convert 30 feet to yards using a unit multiplier.

We can form two unit multipliers.

$$\frac{1 \text{ yd}}{3 \text{ ft}} \text{ and } \frac{3 \text{ ft}}{1 \text{ yd}}$$

We are asked to convert from feet to yards, so we use the unit multiplier that has feet in the denominator and yards in the numerator.

$$\frac{\overset{10}{\cancel{30 \text{ ft}}}}{1} \times \frac{1 \text{ yd}}{\cancel{3 \text{ ft}}} = 10 \text{ yd}$$

Thirty feet converts to **10 yards.**

a. Write two unit multipliers for these equivalent measures:

$$1 \text{ gal} = 4 \text{ qt}$$

b. Which unit multiplier from problem **a** would you use to convert 12 gallons to quarts?

c. Write two unit multipliers for these equivalent measures:

$$1 \text{ m} = 100 \text{ cm}$$

d. Which unit multiplier from problem **c** would you use to convert 200 centimeters to meters?

e. Use a unit multiplier to convert 12 quarts to gallons.

f. Use a unit multiplier to convert 200 meters to centimeters.

g. Use a unit multiplier to convert 60 feet to yards (1 yd = 3 ft).

h. Use a unit multiplier to convert 7 feet to inches.

Written Practice

Distributed and Integrated

1. Eighty students will be assigned to three classrooms. How many students
(104) should be in each class so that the numbers are as balanced as possible?
(Write the numbers.)

* 2. Four friends went out to lunch. Their bill was $45. If the friends divide the bill
(104) equally, how much will each friend pay?

3. Shauna bought a sheet of 39¢ stamps at the post office for $15.60. How many
(44) stamps were in the sheet?

4. Eight cubes were used to build this 2-by-2-by-2 cube. How many cubes are
(RF29) needed to build a cube that has three cubes along each edge?

5. Convert 72 hours to days using a unit multiplier.
(105)

6. (**Estimate**) Use a centimeter ruler and an inch ruler to answer this question.
(46, 76) Twelve inches is closest to how many centimeters? Round the answer to the nearest centimeter.

* 7. The ratio of raisins to peanuts in the trail mix was 5 to 4. If there are 40
(78) raisins, how many peanuts are there?

* 8. (**Conclude**) If two angles of a triangle measure 70° and 80°, then what is the
(95) measure of the third angle?

Connect Complete the table to answer problems **9–11.**

	Fraction	Decimal	Percent
9. (96)	$\frac{11}{20}$	**a.**	**b.**
10. (96)	**a.**	1.5	**b.**
11. (96)	**a.**	**b.**	1%

* 12. (98) **Analyze** Calculate mentally:

    a. $-6 + -12$ **b.** $-6 - -12$

    c. $-12 + +6$ **d.** $-12 - +6$

13. (64) $6\frac{1}{4} \div 100$ **14.** (85) $0.3m = \$4.41$

* 15. (102) **Analyze** Kim scored 15 points, which was 30% of the team's total. How many points did the team score in all?

* 16. (104) Andrea received the following scores in a gymnastic event.

6.7	7.6	6.6	6.7	6.5	6.7	6.8

    The highest score and the lowest score are not counted.
    What is the mean of the remaining scores?

17. (7) **Formulate** Refer to problem **16** to write a comparison question about the scores Andrea received from the judges. Then answer the question.

* 18. (101) What is the area of the quadrilateral below?

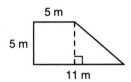

19. (RF25) What is the ratio of vertices to edges on a pyramid with a square base?

* 20. (RF26) **Conclude** Line *r* is called a line of symmetry because it divides the equilateral triangle into two mirror images. Which other line is also a line of symmetry?

* 21. (103) **a.** $(-3)(+7)$

    b. $\frac{(+24)}{(-8)}$

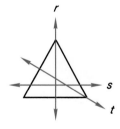

22. (71) Write the prime factorization of 600 using exponents.

23. (RF25) **Conclude** You need to make a three-dimensional model of a soup can using paper and tape. What 3 two-dimensional shapes do you need to cut to make the model?

24. The price of an item is 89¢. The sales-tax rate is 7%. What is the total for the
(35) item, including tax?

25. The probability of winning a prize in the drawing is one in a million. What is the
(54) probability of not winning a prize in the drawing?

26. (**Conclude**) Triangles *ABC* and *CDA* are congruent.
(Inv. 10) Which angle in triangle *ABC* corresponds to ∠*D* in
triangle *CDA*?

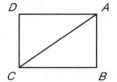

27. a. $(-5) + (-6) - (+7)$
(99)

 b. $(-15) - (-24) - (+8)$

28. Malik used a compass to draw a circle with a radius of 5 centimeters. What was
(42) the circumference of the circle? (Use 3.14 for π.)

29. Solve this proportion: $\dfrac{10}{16} = \dfrac{25}{y}$
(83)

30. (**Evaluate**) The formula $d = rt$ shows that the distance traveled (d) equals the rate
(89, 93) (r) times the time (t) spent traveling at that rate. (Here, *rate* means "speed.") This
function table shows the relationship between distance and time when the rate is 50.

t	1	2	3	4
d	50	100	150	200

Find the value of d in $d = rt$ when r is $\dfrac{50 \text{ mi}}{1 \text{ hr}}$ and t is 5 hr.

✎ *California Mathematics Content Standards*
NS 2.0, 2.4 Determine the least common multiple and
the greatest common divisor of whole numbers; use
them to solve problems with fractions (e.g., to find
a common denominator to add two fractions or to
find the reduced form for a fraction).
MR 3.0, 3.2 Note the method of deriving the solution
and demonstrate a conceptual understanding of
the derivation by solving similar problems.

• Writing Percents as Fractions, Part 2

Power Up

facts   Power Up P

mental math
    a. Number Sense: $429 + 350$

    b. Money: $\$60.00 - \$59.45$

    c. Percent: 10% of 200

    d. Decimals: 1.2×100

    e. Measurement: Cassidy had 1 pint of water. She poured out half the water. How much water was left?

    f. Estimation: Estimate 45×55 by rounding one number up and the other number down and then multiplying.

    g. Algebra: If $q = 5$, what is $q - 6$?

    h. Calculation: $5^2, -1, \div 4, \times 5, +2, \div 4, \times 3, +1, \sqrt{}$

problem solving
Choose an appropriate problem-solving strategy to solve this problem. Every whole number can be expressed as the sum of, *at most,* four square numbers. In the diagram, we see that 12 is made up of one 3×3 square and three 1×1 squares. The number sentence that represents the diagram is $12 = 9 + 1 + 1 + 1$.

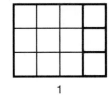
1

Diagram how 15, 18, and 20 are composed of four smaller squares, at most, and then write an equation for each diagram. (*Hint:* Diagrams do not have to be perfect rectangles.)

New Concept

Recall that a percent is a fraction with a denominator of 100. We can write a percent in fraction form by removing the percent sign and writing the denominator 100.

$$50\% = \frac{50}{100}$$

We then simplify the fraction to lowest terms. If the percent includes a fraction, we actually divide by 100 to simplify the fraction.

$$33\frac{1}{3}\% = \frac{33\frac{1}{3}}{100}$$

In this case we divide $33\frac{1}{3}$ by 100. We have performed division problems similar to this in the problem sets.

$$33\frac{1}{3} \div 100 = \frac{\overset{1}{\cancel{100}}}{3} \times \frac{1}{\underset{1}{\cancel{100}}} = \frac{1}{3}$$

We see that $33\frac{1}{3}\%$ equals $\frac{1}{3}$.

Example

Convert $3\frac{1}{3}\%$ to a fraction.

We remove the percent sign and write the denominator 100.

$$3\frac{1}{3}\% = \frac{3\frac{1}{3}}{100}$$

We perform the division.

$$3\frac{1}{3} \div 100 = \frac{\overset{1}{\cancel{10}}}{3} \times \frac{1}{\underset{10}{\cancel{100}}} = \frac{1}{30}$$

We find that $3\frac{1}{3}\%$ equals $\frac{1}{30}$.

Lesson Practice

a. Convert $66\frac{2}{3}\%$ to a fraction.

b. Convert $6\frac{2}{3}\%$ to a fraction.

c. Convert $12\frac{1}{2}\%$ to a fraction.

d. Write $14\frac{2}{7}\%$ as a fraction.

e. Write $83\frac{1}{3}\%$ as a fraction.

f. Convert 62.5% to a fraction.

g. Convert $16\frac{2}{3}\%$ to a fraction.

h. Convert 87.5% to a fraction.

Written Practice   *Distributed and Integrated*

* 1. Two hundred students are traveling by bus on a field trip. The maximum number of
(104)   students allowed on each bus is 84. How many buses are needed for the trip?

2. Montrelyn was walked ten times in 42 at-bats. Predict how many times Montrelyn
(Inv. 8) will be walked in the next 63 at-bats.

* 3. Calculate mentally:
(103)

    a. $(-2)(-6)$ **b.** $\dfrac{+6}{-2}$

    c. $\dfrac{-6}{-6}$ **d.** $(-2)(+6)$

* 4. Calculate mentally:
(98)

    a. $-2 + -6$ **b.** $-2 - -6$

    c. $+2 + -6$ **d.** $+2 - -6$

* 5. (**Analyze**) The chef chopped 27 carrots, which was 90% of the carrots in the
(102) bag. How many carrots remained?

6. Write $37\frac{1}{2}\%$ as a fraction.
(106)

7. Find 8% of $3.65 and round the product to the nearest cent.
(35)

* 8. Use a unit multiplier to convert 176 ounces to pounds
(105)

(**Connect**) Complete the table to answer problems **9–11.**

	Fraction	Decimal	Percent
9. (96)	$1\frac{4}{5}$	**a.**	**b.**
10. (96)	**a.**	0.6	**b.**
11. (96)	**a.**	**b.**	2%

Solve:

12. $5\frac{1}{2} - m = 2\frac{5}{6}$ *** 13.** $\dfrac{6}{10} = \dfrac{0.9}{n}$
(39, 58)   (83, 102)

14. a. $(-7) + (+5) + (-9)$ **15.** $0.05w = 8$
(99)   (85)

     b. $(16) + (-24) - (-18)$

16. All eight books in the stack are the same weight.
(8) Three books weigh a total of six pounds.

    a. How much does each book weigh?

    b. How much do all eight books weigh?

} 6 lb

* 17. Find the volume of a rectangular prism using the formula $V = lwh$ when the length
(RF29, 93) is 8 cm, the width is 5 cm, and the height is 2 cm.

18. How many millimeters is 1.2 meters (1 m =
(105) 1000 mm)?

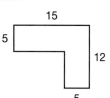

* 19. What is the perimeter of the polygon
(101) at right? (Dimensions are in millimeters.)

* 20. What is the area of the polygon in problem 19?
(101)

* 21. **Verify** If the pattern shown below were cut out and folded on the dotted lines, would it form a cube, a pyramid, or a cylinder? Explain how you know.
(RF25)

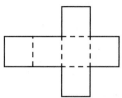

* 22. **Multiple Choice** Which one of these numbers is not a composite number?
(61)

     A 34   **B** 35   **C** 36   **D** 37

23. **Estimate** Debbie wants to decorate a cylindrical wastebasket by wrapping it with wallpaper. The diameter of the wastebasket is 12 inches. The length of the wallpaper should be at least how many inches? Round up to the next inch.
(42)

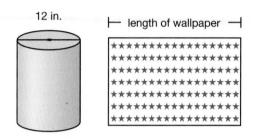

* 24. Use the formula $d = r \cdot t$ to find the distance traveled at a rate of 10 miles per hour for 1.5 hours.
(89, 93)

25. **Connect** Which arrow is pointing to $-\frac{1}{2}$?
(14)

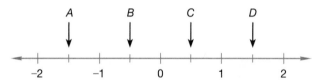

26. The ratio of nonfiction to fiction books on Shawna's bookshelf is 2 to 3. If the total number of books on her shelf is 30, how many nonfiction books are there?
(100)

* 27. **Connect** What are the coordinates of the point that is halfway between (−2, −3) and (6, −3)?
(RF27)

* 28. A set of 40 number cards contains one each of the counting numbers from 1 through 40. A multiple of 10 is drawn from the set and is not replaced. A second card is drawn from the remaining 39 cards. What is the probability that the second card will be a multiple of 10?
(54)

* 29. Combine the areas of the two triangles to find the area of this trapezoid.
(101)

30. $\frac{2}{3} + x = \frac{5}{6}$
(91)

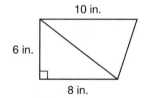

California Mathematics Content Standards
NS 2.0, 2.3 Solve addition, subtraction, multiplication, and division problems, including those arising in concrete situations, that use positive and negative integers and combinations of these operations.
AF 1.0, 1.3 Apply algebraic order of operations and the commutative, associative, and distributive properties to evaluate expressions; and justify each step in the process.
AF 1.0, 1.4 Solve problems manually by using the correct order of operations or by using a scientific calculator.

• Order of Operations with Positive and Negative Numbers

facts   Power Up Q

mental math
  a. **Number Sense:** $100 \cdot 100$

  b. **Percent:** 75% of 200

  c. **Money:** $12.89 + $9.99

  d. **Decimals:** $6 \div 100$

  e. **Probability:** There are 2 red marbles and 4 blue marbles in a bag. What is the probability of drawing a blue marble with one draw?

  f. **Geometry:** These angles are supplementary. What is m∠A?

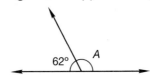

  g. **Measurement:** How many pints are in a gallon?

  h. **Calculation:** $10 \times 6, + 4, \sqrt{}, \times 3, + 1, \sqrt{}, \times 7, + 1, \sqrt{}$

problem solving
Choose an appropriate problem-solving strategy to solve this problem. Laura has nickels, dimes, and quarters in her pocket. She has half as many dimes as nickels and half as many quarters as dimes. If Laura has four dimes, then how much money does she have in her pocket?

To simplify expressions that involve several operations, we perform the operations in a prescribed order. We have practiced simplifying expressions with whole numbers. In this lesson we will begin simplifying expressions that contain negative integers as well.

Example 1

Thinking Skills

Generalize

Use the sentence _Please_ _excuse my dear_ _Aunt Sally_ to remember the correct order of operations.

Simplify: $(-2) + (-2)(-2) - \dfrac{(-2)}{(+2)}$

First we multiply and divide in order from left to right.

$$(-2) + \underbrace{(-2)(-2)}_{(+4)} - \underbrace{\dfrac{(-2)}{(+2)}}_{(-1)}$$

Then we add and subtract in order from left to right.

$$\underbrace{(-2) + (+4)}_{(+2)} \underbrace{- (-1)}_{- (-1)}$$

$$\mathbf{+3}$$

Expressions are made up of **terms** that are joined by addition and subtraction.

Mentally separating an expression into its terms can make an expression easier to simplify. Here is the same expression. This time we will use slashes to separate the terms.

$$(-2)\Big/ + (-2)(-2)\Big/ - \dfrac{(-2)}{(+2)}$$

First we simplify each term; then we combine the terms.

$$(-2)\Big/ + (-2)(-2)\Big/ - \dfrac{(-2)}{(+2)}$$

$$-2 \Big/ \quad +4 \quad \Big/ \quad + 1$$

$$\mathbf{+3}$$

Example 2

Simplify each term. Then combine the terms.

$$-3(2 - 4) - 4(-2)(-3) + \dfrac{(-3)(-4)}{2}$$

We separate the individual terms with slashes. The slashes precede plus and minus signs that are not enclosed by parentheses or other symbols of inclusion.

$$-3(2 - 4)\Big/ - 4(-2)(-3)\Big/ + \dfrac{(-3)(-4)}{2}$$

Next we simplify each term.

$$-3(2 - 4)\Big/ - 4(-2)(-3)\Big/ + \dfrac{(-3)(-4)}{2}$$

$$-3(-2) \quad \Big/ \quad + 8(-3) \quad \Big/ \quad + \dfrac{12}{2}$$

$$+6 \quad \Big/ \quad - 24 \quad \Big/ \quad + 6$$

Now we combine the simplified terms.

$$+6 - 24 + 6$$
$$-18 + 6$$
$$\mathbf{-12}$$

Example 3

Simplify: $(-2) - [(-3) - (-4)(-5)]$

There are only two terms, -2 and the bracketed quantity. Recall that brackets are a symbol of inclusion like parentheses. By the order of operations, we simplify within brackets first, multiplying and dividing before adding and subtracting.

$(-2) \Big/ - [(-3) - (-4)(-5)]$	Original expression
$(-2) \Big/ - [(-3) - (+20)]$	Multiplied
$(-2) \Big/ \qquad - (-23)$	Subtracted
$(-2) \Big/ \qquad + 23$	Simplified
$\mathbf{+21}$	Added

Signed numbers are often written without parentheses. To simplify such expressions we simply add algebraically from left to right.

$$-3 + 4 - 5 = -4$$

Another way to simplify this expression is to use the commutative and associative properties to rearrange and regroup the terms by their signs.

$-3 + 4 - 5$	Given
$+4 - 3 - 5$	Commutative Property
$+4 + (-3 - 5)$	Associative Property
$+4 - 8$	$-3 - 5 = -8$
-4	$+4 - 8 = -4$

Example 4

Simplify: $-2 + 3(-2) - 2(+4)$

To emphasize the separate terms, we first draw a slash before each plus or minus sign that is not enclosed.

$$-2 \Big/ +3(-2) \Big/ -2(+4)$$

Next we simplify each term.

$$-2 \Big/ +3(-2) \Big/ -2(+4)$$
$$-2 \qquad -6 \qquad -8$$

Then we algebraically add the terms.

$$-2 - 6 - 8 = \mathbf{-16}$$

Lesson Practice

Justify Simplify. Show steps and properties.

a. $(-3) + (-3)(-3) - \dfrac{(-3)}{(+3)}$ **b.** $(-3) - [(-4) - (-5)(-6)]$

c. $(-2)[(-3) - (-4)(-5)]$ **d.** $(-5) - (-5)(-5) + (-5)$

e. $-3 + 4 - 5 - 2$ **f.** $-2 + 3(-4) - 5(-2)$

g. $-3(-2) - 5(2) + 3(-4)$ **h.** $-4(-3)(-2) - 6(-4)$

Written Practice

Distributed and Integrated

* 1. (104) **Analyze** For cleaning the yard, four teenagers were paid a total of $75.00. If they divide the money equally, how much money will each teenager receive?

* 2. (76) **Multiple Choice** Which of the following is the best estimate of the length of a bicycle?

    A 0.5 m   **B** 2 m   **C** 6 m   **D** 36 m

3. (54) If the chance of rain is 80%, what is the probability that it will not rain? Express the answer as a decimal.

* 4. (100) The ratio of students who walk to school to students who ride a bus to school is 5 to 3. If there are 120 students, how many students walk to school?

* 5. (107) **a.** $(-3) - \left[(-2) - (+2) - \dfrac{(-2)}{(-2)} \right]$

    b. $(-3) - \left[(-2) - (+4)(-5) \right]$

* 6. (103) Calculate mentally:

    a. $(-12)(+3)$ **b.** $(-12)(-3)$

    c. $\dfrac{-12}{+3}$ **d.** $\dfrac{-12}{-3}$

* 7. (98) Calculate mentally:

    a. $-12 + -3$ **b.** $-12 - -3$

    c. $+3 + -12$ **d.** $+3 - -12$

* 8. (74) **Explain** Describe a method for arranging these fractions from least to greatest:

$$\frac{3}{4}, \frac{3}{5}, \frac{4}{5}$$

Connect Complete the table to answer problems **9–11.**

	Fraction	Decimal	Percent
9. (96)	$\frac{1}{50}$	**a.**	**b.**
10. (96)	**a.**	1.75	**b.**
11. (96)	**a.**	**b.**	25%

12. $12\frac{1}{4}$ in. $- 3\frac{5}{8}$ in.
(58)

13. $3\frac{1}{3}$ ft $\times$ $2\frac{1}{4}$ ft
(62)

14. (3 cm)(3 cm)(3 cm)
(79)

15. 0.6 m $\times$ 0.5 m
(79)

16. Convert $13\frac{1}{3}\%$ to a fraction.
(106)

* 17. ✎ **Justify** Find the area of this trapezoid. Show and
(101) explain your work to justify your answer.

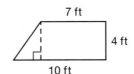
7 ft

4 ft

10 ft

18. Use a unit multiplier to convert 2.5 hours to minutes.
(105)

19. a. $(-8) + (-7) - (-15)$
(99)

  b. $(-15) + (11) - (24)$

20. A number cube is rolled three times. The results were 3, 5, and 3. For
(Inv. 8) predicting future rolls, should the experimental probability be used, or should the theoretical probability be used?

21. Cordell worked for three days and earned $240. At that rate, how much would
(86) he earn in ten days?

* 22. **Analyze** Seventy is the product of which three prime numbers?
(61)

23. Saturn is about 900 million miles from the Sun. Write that distance in standard
(RF12) notation.

* 24. Use the rhombus at the right for problems **a–c.**
(66)
  a. What is the perimeter of this rhombus?

  b. What is the area of this rhombus?

  c. **Analyze** If an acute angle of this rhombus measures 61°, then what is the measure of each obtuse angle?

8 in.

7 in.

8 in.

25. The ratio of quarters to dimes in Keiko's savings jar is 5 to 8. If there were 120
(86) quarters, how many dimes were there?

26. **Connect** The coordinates of the three vertices of a triangle are (0, 0), (0, 4), and
(RF27, 77) (4, 4). What is the area of the triangle?

The following list shows the ages of the children attending a luncheon. Use this information to answer problems **27** and **28**.

8, 9, 8, 8, 7, 9, 12, 12, 11, 16

27. What was the median age of the children attending the luncheon?
(Inv. 4)

28. What was the mean age of the children at the luncheon?
(Inv. 4)

29. The diameter of a playground ball is 10 inches. What is the
(42) circumference of the ball? (Use 3.14 for π.) How can estimation help you determine if your answer is reasonable?

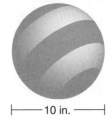

30. Find the value of A in $A = s^2$ when s is 10 m.
(87)

├───── 10 in. ─────┤

✎ *California Mathematics Content Standards*

NS **2.0**, **2.3** Solve addition, subtraction, multiplication, and division problems, including those arising in concrete situations, that use positive and negative integers and combinations of these operations.

AF 1.0, 1.2 Write and evaluate an algebraic expression for a given situation, using up to three variables.

• Evaluations with Positive and Negative Numbers

facts	Power Up O

mental math	**a. Number Sense:** $296 - 150$
	b. Percent: 25% of $20
	c. Money: $8.23 + $8.99
	d. Decimals: $75 \div 100$
	e. Number Sense: $\frac{800}{40}$
	f. Measurement: One quart is about 946 milliliters. If Janie has 1 quart of water, about how many more milliliters does she need to make 1 liter?
	g. Algebra: If $12t = 48$, what is t?
	h. Calculation: $8 \times 8, -4, \div 2, +5, \div 5, \times 8, -1, \div 5, \times 2, -1, \div 3$

problem solving

Choose an appropriate problem-solving strategy to solve this problem. If two people shake hands, there is one handshake. If three people shake hands, there are three handshakes. From this table can you predict the number of handshakes with 6 people? Draw a diagram or act it out to confirm your prediction.

Number in Group	Number of Handshakes
2	1
3	3
4	6
5	10
6	

We have practiced evaluating expressions such as

$$x - xy - y$$

with positive numbers in place of x and y. In this lesson we will practice evaluating such expressions with negative numbers as well. When evaluating expressions with signed numbers, it is helpful to first replace each variable with parentheses. This will help prevent making mistakes in signs.

Example 1

Evaluate: $x - xy - y$ **if** $x = -2$ **and** $y = -3$

We write parentheses for each variable.

$$( ) - ( )( ) - ( ) \qquad \text{Parentheses}$$

Now we write the proper numbers within the parentheses.

$$(-2) - (-2)(-3) - (-3) \qquad \text{Insert numbers}$$

By the order of operations, we multiply before adding.

$$(-2) - (+6) - (-3) \qquad \text{Multiplied}$$

Then we add algebraically from left to right.

$$(-8) - (-3) \qquad \text{Added } -2 \text{ and } -6$$

$$-5 \qquad \text{Added } -8 \text{ and } +3$$

Discuss What are some values for x and y that will result in a positive solution?

Example 2

Evaluate the expression and justify each step.

$mn - np + p$

if $m = -5, n = 2,$ **and** $p = -1$

We copy the expression, then write parentheses for each variable. We replace each variable with the given number.

$$mn - np + p \qquad \text{Original expression}$$

$$(-5)(2) - (2)(-1) + (-1) \qquad \text{Substituted the numbers}$$

$$(-10) - (-2) + (-1) \qquad \text{Multiplied}$$

$$-8 + (-1) \qquad \text{Added } -10 \text{ and } 2$$

$$-9 \qquad \text{Added } -8 \text{ and } -1$$

Lesson Practice

Generalize Evaluate each expression. Write parentheses as the first step.

a. $x + xy - y$ if $x = 3$ and $y = -2$

b. $-m + n - mn$ if $m = -2$ and $n = -5$

c. $a + bc - c$ if $a = -1, b = 2,$ and $c = -3$

d. Find $-x + y - xy$ if $x = -4$ and $y = 1$.

e. Find $m - n - mn$ if $m = -6$ and $n = -2$.

f. Find $ab + c - b$ if $a = -5, b = -3$ and $c = 1$.

g. Find $y + yx - x$ if $x = -7$ and $y = 3$.

h. Find $mv - m + v$ if $m = -10$ and $v = -9$.

* 1. Tickets to the matinee are $6 each. How many tickets can Maela buy with
(104) $20?

2. (Analyze) Maria ran four laps of the track at an even pace. If it took 6 minutes
(86) to run the first three laps, how long did it take to run all four laps?

3. Fifteen of the 25 members played in the game. What fraction of the members
(75) did not play?

4. Two fifths of the 160 acres were planted with alfalfa. How many acres were not
(75) planted with alfalfa?

5. Which digit in 94,763,581 is in the ten-thousands place?
(RF12)

* 6. **a.** Write two unit multipliers for these equivalent measures:
(105)

$$1 \text{ gallon} = 4 \text{ quarts}$$

b. Which of the two unit multipliers from part **a** would you use to convert
8 gallons to quarts? Why?

* 7. (Estimate) What is the sum of $36.43, $41.92, and $26.70 to the nearest
(46) dollar.

8. Evaluate: $m(m + n)$ if $m = -2$ and $n = -3$ **9.** $3\frac{1}{4}$ in. $+ 2\frac{1}{2}$ in. $+ 4\frac{5}{8}$ in.
(108)   (56)

(Connect) Complete the table to answer problems **10–12.**

	Fraction	Decimal	Percent
10. (96)	$\frac{1}{8}$	**a.**	**b.**
11. (96)	**a.**	0.9	**b.**
12. (96)	**a.**	**b.**	60%

13. $3.25 \div \frac{2}{3}$ (fraction answer)
(68)

14. Simplify:
(107)

a. $(-4)(-5) - (-4)(+3)$

b. $(-2)[(-3) - (-4)(+5)]$

* 15. $\frac{3}{2} = \frac{1.8}{m}$
(83, 102)

* 16. Calculate mentally:
(103)

a. $(-5)(-20)$ **b.** $(-5)(+20)$

c. $\frac{-20}{+5}$ **d.** $\frac{-20}{-5}$

* 17. The distance between San Francisco and Los Angeles is about 387 miles. If
(89)   Takara leaves San Francisco and travels 6 hours at an average speed of 55 miles
per hour, will she reach Los Angeles? How far will she travel?

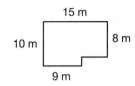

15 m

10 m

8 m

9 m

* 18. (**Analyze**) What is the area of this polygon?
(101)

* 19. What is the perimeter of this polygon?
(101)

20. Calculate mentally:
(97)

   a. $-5 + -20$ **b.** $-20 - -5$

   c. $-5 - -5$ **d.** $+5 - -20$

* 21. (**Conclude**) Transversal t intersects parallel lines q and r. Angle 1 is half the
(94)   measure of a right angle.

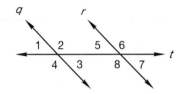

   a. Which angle corresponds to $\angle 1$?

   b. What is the measure of each obtuse angle?

* 22. (**Explain**) Fifty people responded to the survey, a number that represented
(102)   5% of the surveys mailed. How many surveys were mailed? Explain how you
found your answer.

* 23. Think of two different prime numbers, and write them on your paper. Then find
(11, 26)   the least common multiple (LCM) of the two prime numbers.

24. Convert $62\frac{1}{2}\%$ to a fraction
(106)

25. A classroom that is 30 feet long, 30 feet wide, and 10 feet high has a volume of
(RF29)   how many cubic feet?

* 26. Convert 8 quarts to gallons using a unit multiplier.
(105)

* 27. (**Analyze**) A circle was drawn on a coordinate plane. The coordinates of the
(RF27,   center of the circle were (1, 1). One point on the circle was (1, 3).
84)
   a. What was the radius of the circle?

   b. What was the area of the circle? (Use 3.14 for π.)

28. During one season, the highest number of points scored in one game by the local
(Inv. 4)   college basketball team was 95 points. During that same season, the range of the
team's scores was 35 points. What was the team's lowest score?

29. Use the formula $d = r \cdot t$ to find d when the rate is 300 miles per hour and the
(89, 93)   time is 5 hours.

30. $\frac{3}{5}x = 6$
(91)

California Mathematics Content Standards

NS **2.0**, **2.3** Solve addition, subtraction, multiplication, and division problems, including those arising in concrete situations, that use positive and negative integers and combinations of these operations.

AF **1.0, 1.3** Apply algebraic order of operations and the commutative, associative, and distributive properties to evaluate expressions; and justify each step in the process.

• Translating and Evaluating Expressions and Equations, Part 2

Power Up

facts Power Up Q

mental math

    a. Number Sense: $400 \cdot 30$

    b. Fractions: $\frac{1}{10}$ of $20

    c. Percent: 10% of $20

    d. Decimals: 0.5×100

    e. Money: The total cost of the sandwich and drink is $6.87. If Ramon pays with a $10 bill, how much change should he receive?

    f. Geometry: Angles *A* and *C* of this parallelogram measure 130° each. What is the measure of angles *B* and *D*?

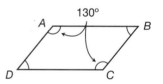

    g. Measurement: How many inches is $\frac{1}{4}$ of a foot?

    h. Calculation: 7^2, $+ 1$, $\div 2$, $\times 3$, $- 3$, $\div 8$, $\sqrt{}$

problem solving Choose an appropriate problem-solving strategy to solve this problem. One state uses a license plate that contains two letters followed by four digits. How many different license plates are possible if all of the letters and numbers are used?

New Concept

An essential skill in mathematics is the ability to translate language, situations, and relationships into mathematical form. Since the earliest lessons of this book we have practiced translating word problems into equations that we then solved. In this lesson we will practice translating other common patterns of language into algebraic form. We will also use our knowledge of geometric relationships to write equations to solve geometry problems.

The table on the next page shows examples of mathematical language translated into algebraic form. Notice that the word *number* is represented by a letter in italic form. We can use any letter to represent an unknown number.

Examples of Translations

English	Symbols
Twice a number	$2n$
Five more than a number	$x + 5$
Three less than a number	$a - 3$
Half of a number	$\frac{1}{2}h$ or $\frac{h}{2}$
The product of a number and seven	$7b$
Twelve is three times a number	$12 = 3n$
Seventeen is five more than twice a number.	$17 = 2n + 5$

The last translation in the examples above resulted in an equation that can be solved to find the unstated number. We will solve a similar equation in the next example.

Example 1

In one year, Darius read fifteen books, which is seven more than his friend read. How many books did his friend read? Write an equation and solve it.

We write an equation for the sentence,

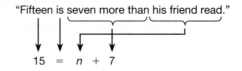

"Fifteen is seven more than his friend read."

$$15 = n + 7$$

Then we solve the equation.

$15 = n + 7$	Equation
$15 - 7 = n + 7 - 7$	Subtracted 7 from both sides
$8 = n$	Simplified

Darius's friend read **8 books.**

In Example 2, we will translate a situation involving variables into an expression. Then we will evaluate the expression for a given scenario.

Example 2

Eduardo plays a game in which he earns 50 points for each coin he collects and 200 points for each diamond he finds, but he loses 100 points for each trap he falls into.

a. Write an expression for the total number of points Eduardo earns in a game. Use *c* for the number of coins he collects, *d* for the number diamonds he finds, and *t* for the number traps he falls into.

b. Evaluate the expression to find the number of points Eduardo earned in a game when he found 70 coins and 5 diamonds, but fell into 10 traps.

a. The number of points Eduardo earns from collecting coins is 50 times the number he finds. $50c$

The number of points from diamonds is 200 times the number he finds. $200d$

The number of points he loses from traps is 100 times the number he falls into. $100t$

We add the number of points he earns from coins and diamonds and subtract the points he loses from traps.

$$50c + 200d - 100t$$

b. We evaluate the expression where c is 70, d is 5, and t is 10.

$50c + 200d - 100t$	Expression
$50(70) + 200(5) - 100(10)$	Substituted
$3500 + 1000 - 1000$	Multiplied
$4500 - 1000$	Added 3500 and 1000
3500	Subtracted 1000 from 4500

Eduardo earned **3500 points.**

We can also translate geometric relationships into algebraic expressions.

Example 3

This triangle has unknown height x.

a. Write an equation for the area of this triangle.

b. Suppose the area of this triangle is 30. Find x.

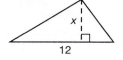

a. We use the formula for the area of a triangle.

$$A = \frac{1}{2}bh$$

The base is 12 and the height is x.

$$A = \frac{1}{2}(12)x$$

$$A = 6x$$

b. Since $A = 6x$ and the area is 30, we can write an equation.

$$30 = 6x$$

Solve for x by dividing both sides by 6.

$$\frac{30}{6} = \frac{6x}{6}$$

$$\frac{\overset{5}{\cancel{30}}}{\underset{1}{\cancel{6}}} = \frac{\overset{1}{\cancel{6}}x}{\underset{1}{\cancel{6}}}$$

$$5 = x$$

We find that **$x = 5$.**

A caterer charges $20 for each vegetable platter, $25 for each fruit platter, and $10 for each bowl of lemonade.

a. Write an algebraic expression that describes the total amount the caterer charges. Use *v* for the number of vegetable platters, *f* for the number of fruit platters, and *l* for the number of bowls of lemonade.

b. Evaluate the expression to find the caterer's charge when the caterer provided 4 vegetable platters, 4 fruit platters, and 8 bowls of lemonade.

For problems **c** and **d,** write and solve an equation.

c. Thirty-nine is three times a number. What is the number?

d. Jaime is 10, which is two years younger than Teresa. How old is Teresa?

e. Write an equation for the relationship of these angles and solve for *x*.

f. Amber's age is twice her sister's, who is 6. How old is Amber?

g. Ismael read a book containing 400 pages. This is 150 pages more than the book that Javier read. How many pages did Javier read?

h. On a certain day, there were 13.5 hours of sunlight, plus some hours without sunlight. How many hours were there without sunlight?

Written Practice   *Distributed and Integrated*

1. What is the total cost of a $12.60 item plus 7% sales tax?
(35)

* 2. Convert $16\frac{2}{3}\%$ to a fraction.
(106)

* 3. (**Model**) Draw a ratio box for this problem. Then solve the problem using a proportion.
(102)

Ines missed three questions on the test but answered 90% of the questions correctly. How many questions were on the test?

4. Sound travels about 331 meters per second in air. How far will it travel in 60 seconds?
(89)

$$\frac{331 \text{ m}}{1 \text{ s}} \cdot \frac{60 \text{ s}}{1}$$

5. Evaluate: $b^2 - 4ac$ if $a = -1$, $b = -2$, and $c = 3$.
(108)

* 6. If the radius of a circle is seventy-five hundredths of a meter, what is the diameter?
(RF16, 33)

7. **Estimate** Round the product of $3\frac{2}{3}$ and $2\frac{2}{3}$ to the nearest whole number.
(62)

* 8. In a bag are three red marbles and three white marbles. If two marbles are
(80) taken from the bag at the same time, what is the probability that both marbles
will be red?

Connect Complete the table to answer problems **9** and **10**.

	Fraction	Decimal	Percent
9. (96)	$2\frac{2}{5}$	a.	b.
10. (96)	a.	0.85	b.

Solve:

* 11. In the forest the ratio of deciduous trees to evergreen trees was 2 to 7. If there
(78) were 28 deciduous trees, how many evergreens were there?

* 12. $\dfrac{x}{7} = \dfrac{35}{5}$
(83)

* 13. Calculate mentally:
(103)
a. $(-3)(-15)$ 　　　　　　　　　　　**b.** $\dfrac{-15}{+3}$

c. $\dfrac{-15}{-3}$ 　　　　　　　　　　　**d.** $(+3)(-15)$

* 14. $-6 + -7 + +5 - -8$ 　　　　**15.** $0.12 \div (12 \div 0.4)$
(97) 　　　　　　　　　　　　　　　　(44)

* 16. Write $\frac{22}{7}$ as a decimal rounded to the hundredths place.
(46, 69)

17. **a.** $(-5) - (-2)\left[(-3) - (-4)\right]$
(107)

　　b. $\dfrac{(-3) + (-3)(+4)}{(+3) + (-4)}$

* 18. What is the area of this hexagon?
(101)

19. What is the perimeter of this hexagon?
(101)

20. What is the volume of this cube?
(RF29)

21. What is the mode of the number of
(Inv. 4) days in the twelve months of
the year?

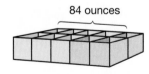

22. **Analyze** If seven of the containers can
(86) hold a total of 84 ounces, then how many
ounces can 10 containers hold?

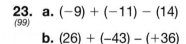

23. **a.** $(-9) + (-11) - (14)$
(99)

　　b. $(26) + (-43) - (+36)$

24. Round 58,697,284 to the nearest million.
(21)

25. **Connect** Which arrow is pointing to 0.4?
(28)

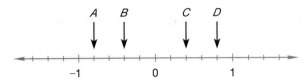

26. Use the formula $d = r \cdot t$ to find the distance traveled by a runner who ran
(89, 93) 10 miles per hour for $\frac{3}{4}$ of an hour.

* 27. **Connect** The coordinates of the vertices of a parallelogram are (0, 0), (5, 0),
(RF27, 66) (6, 3), and (1, 3). What is the area of the parallelogram?

* 28. **Multiple Choice** Which is the greatest weight?
(40)
   A 6.24 lb   **B** 6.4 lb   **C** 6.345 lb

* 29. $\frac{4}{5} x = 16$
(91)

30. **Analyze** Gilbert started the trip with a full tank of gas. He drove
(79) 323.4 miles and then refilled the tank with 14.2 gallons of gas. How can Gilbert
calculate the average number of miles he traveled on each gallon of gas?

Real-World Connection

A few students from the local high school decide to survey the types
of vehicles in the parking lot. Their results are as follows: 210 cars,
125 trucks, and 14 motorcycles.

a. Find the simplified ratios for cars to trucks, cars to motorcycles, and trucks
to motorcycles.

b. Find the fraction, decimal, and percent for the ratio of cars to the total number of
vehicles. Round to nearest thousandth and tenth of a percent.

LESSON

110

California Mathematics Content Standards

NS 2.0, **2.1** Solve problems involving addition, subtraction, multiplication, and division of positive fractions and explain why a particular operation was used for a given situation.

MR 2.0, 2.4 Use a variety of methods, such as words, numbers, symbols, charts, graphs, tables, diagrams, and models, to explain mathematical reasoning.

• Finding a Whole When a Fraction Is Known

Power Up

facts   Power Up P

mental math

    a. Number Sense: $90 \cdot 90$

    b. Fractions: $\frac{1}{3}$ of 300

    c. Fractions: $\frac{1}{6}$ of 300

    d. Percent: 50% of $90

    e. Decimals: $8 \div 100$

    f. Geometry: Adam measured the radius of the clock face as 15.2 cm. What is the diameter of the clock face?

    g. Algebra: If $c = 50$, what is $\frac{350}{c}$?

    h. Calculation: $10^2, -1, \div 9, \times 3, -1, \div 4, \times 7, +4, \div 3$

problem solving

Choose an appropriate problem-solving strategy to solve this problem. Copy the problem and fill in the missing digits.

Use only zeros or ones in the spaces.

$$
\begin{array}{r}
9\_ \\
\_\_\overline{)\_\_\_\_} \\
99 \\
\overline{\_\_} \\
\overline{\overline{\_\_}} \\
\_
\end{array}
$$

New Concept

Thinking Skills

Discuss

What information in the problem suggests that we divide the rectangle into 5 parts?

Consider the following fractional-parts problem:

Two fifths of the students in the class are boys. If there are ten boys in the class, how many students are in the class?

A diagram can help us understand and solve this problem. We have drawn a rectangle to represent the whole class. The problem states that two fifths are boys, so we divide the rectangle into five parts. Two of the parts are boys, so the remaining three parts must be girls.

__ students in the class

$\frac{3}{5}$ girls

$\frac{2}{5}$ boys

We are also told that there are ten boys in the class. In our diagram ten boys make up two of the parts. Since ten divided by two is five, there are five students in each part. All five parts together represent the total number of students, so there are 25 students in all. We complete the diagram.

25 students in the class

$\frac{3}{5}$ girls	5
	5
	5
$\frac{2}{5}$ boys	5
	5

Example 1

Three eighths of the townspeople voted. If 120 of the townspeople voted, how many people live in the town?

We are told that $\frac{3}{8}$ of the town voted, so we divide the whole into eight parts and mark off three of the parts. We are told that these three parts total 120 people. Since the three parts total 120, each part must be 40 (120 ÷ 3 = 40). Each part is 40, so all eight parts must be 8 times 40, which is **320 people.**

___ people live in the town.

$\frac{3}{8}$ voted.	40
	40
	40
$\frac{5}{8}$ did not vote.	40
	40
	40
	40
	40

Example 2

Six is $\frac{2}{3}$ of what number?

A larger number has been divided into three parts. Six is the total of two of the three parts. So each part equals three, and all three parts together equal **9.**

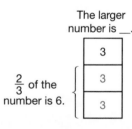

The larger number is __.

$\frac{2}{3}$ of the number is 6.

3
3
3

Lesson Practice

Model Solve. Draw a diagram for problems **a–c.**

a. Eight is $\frac{1}{5}$ of what number?

b. Eight is $\frac{2}{5}$ of what number?

c. Nine is $\frac{3}{4}$ of what number?

d. Sixty is $\frac{3}{8}$ of what number?

e. Three fifths of the students in the class were girls. If there were 18 girls in the class, how many students were in the class altogether?

f. Six sevenths of the team wore wristbands. If 18 wore wristbands, how many people were on the team?

g. Forty is two-thirds of what number?

h. Ninety is five-sixths of what number?

1. Jimarcus drew a right triangle with sides 6 inches, 8 inches, and 10 inches long. What was the area of the triangle?
(77, 90)

2. If 5 feet of ribbon costs $1.20, then 10 feet of ribbon would cost how much?
(86)

3. **a.** Six is what fraction of 15?
(15, 73)
b. Six is what percent of 15?

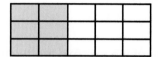

* 4. The multiple-choice question has four choices. Raidon knows that one of the choices must be correct, but he has no idea which one. If Raidon simply guesses, what is the chance that he will guess the correct answer?
(54)

5. If $\frac{2}{5}$ of the 30 students in the class buy lunch in the school cafeteria, what is the ratio of students who buy lunch in the school cafeteria to students who bring their lunch from home?
(16, 75)

6. 9 is $\frac{3}{5}$ of what number?
(110)

* 7. **Evaluate** The cost (c) of apples is related to its price per pound (p) and its weight (w) by this formula:
(89, 93)

$$c = pw$$

Find the cost when p is $\frac{\$1.25}{1 \text{ pound}}$ and w is 5 pounds.

8. Arrange these numbers in order from least to greatest:
(40)

$$9.9, \ 9.95, \ 9.925, \ 9.09$$

Connect Complete the table to answer problems **9** and **10**.

	Fraction	Decimal	Percent
* 9. (96)	$3\frac{3}{8}$	**a.**	**b.**
* 10. (96)	**a.**	**b.**	15%

Solve and check:

* 11. $9x + 17 = 80$
(85)

* 12. $\frac{x}{3} = \frac{1.6}{1.2}$
(83, 102)

* 13. $-6 + -4 - +3 - -8$
(97)

14. $6 + 3\frac{3}{4} + 4.6$ (decimal answer)
(69)

* 15. The Gateway Arch in St. Louis, Missouri, is approximately 210 yards tall. How tall
(105) is it in feet?

16. Use division by primes to find the prime factors of 648. Then write the prime
(71) factorization of 648 using exponents.

17. If a 32-ounce box of cereal costs $3.84, what is the cost per ounce?
(8)

* 18. Find the area of this trapezoid by combining the area of the rectangle
(101) and the area of the triangle.

19. The radius of a circle is 10 cm. Use 3.14
(42, 84) for π to calculate the

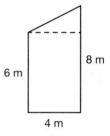

8 m

6 m

4 m

    a. circumference of the circle.

    b. area of the circle.

20. **Conclude** The volume of the pyramid is $\frac{1}{3}$ the volume of the cube.
(RF29) What is the volume of the pyramid?

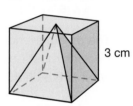

3 cm

21. Solve: $0.6y = 54$
(85)

* 22. Calculate mentally:
(103)

    a. $(-8)(-2)$ **b.** $(+8)(-2)$

    c. $\dfrac{+8}{-2}$ **d.** $\dfrac{-8}{-2}$

* 23. **Analyze** The 306 students were assigned to ten rooms so that there were 30 or
(104) 31 students in each room. How many rooms had exactly 30 students?

24. Two angles of a triangle measure 40° and 110°.
(95)

    a. What is the measure of the third angle?

    b. **Represent** Make a rough sketch of the triangle.

25. **Multiple Choice** **Estimate** If Anthony spins the spinner
(Inv. 7) 60 times, about how many times should he expect the
arrow to stop in sector 3?

    A 60 times   **B** 40 times

    C 20 times   **D** 10 times

* 26. **Analyze** An equilateral triangle and a square share a common side.
(RF7, 32) If the area of the square is 100 mm², then what is the perimeter of the
equilateral triangle?

* 27. Write $11\frac{1}{9}\%$ as a reduced fraction.
(106)

28. The heights of the five starters on the basketball team are listed below. Find the
_(Inv. 4) mean, median, and range of these measures.

$$181 \text{ cm}, \ 177 \text{ cm}, \ 189 \text{ cm}, \ 158 \text{ cm}, \ 195 \text{ cm}$$

29. a. $\dfrac{(-8) - (-6) - (4)}{-3}$
₍₁₀₇₎
    b. $-5(-4) -3(-2) - 1$

30. If $y = 4x -3$ and $x = -2$, then y equals what number?
₍₁₀₈₎

Real-World Connection

Gerard plays basketball on his high school team. This season he scored 372 of his team's 1488 points.

    a. What percent of the points were scored by Gerard's teammates?

    b. Gerard's team played 12 games. How many points did Gerard average per game?

California Mathematics Content Standards

NS **1.0**, **1.3** Use proportions to solve problems (e.g., determine the value of N if 4/7 = N/ 21, find the length of a side of a polygon similar to a known polygon). Use cross-multiplication as a method for solving such problems, understanding it as the multiplication of both sides of an equation by a multiplicative inverse.

AF 2.0, **2.2** Demonstrate an understanding that rate is a measure of one quantity per unit value of another quantity.

Focus on
Scale Factor: Scale Drawings and Models

The **scale** is the ratio that shows the relationship between a scale drawing or model and the actual object. The dimensions of similar figures are related by a **scale factor**, as shown below.

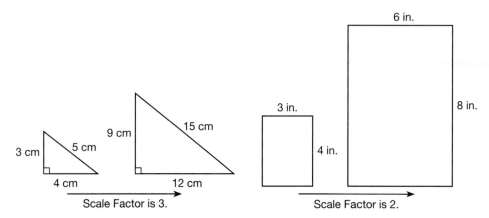

Scale Factor is 3.   Scale Factor is 2.

Similar figures are often used by manufacturers to design products and by architects to design buildings. Architects create **scale drawings** to guide the construction of a building. Sometimes, a **scale model** of the building is also constructed to show the appearance of the finished project.

Scale drawings, such as architectural plans, are two-dimensional representations of larger objects. Scale models, such as model cars and action figures, are three-dimensional representations of larger objects. In some cases, however, a scale drawing or model represents an object smaller than the model itself. For example, we might want to construct a large model of a bee in order to more easily portray its anatomy.

In scale drawings and models the **legend** gives the relationship between a unit of length in the drawing and the actual measurement that the unit represents. The drawing below shows the floorplan of Angela's studio apartment. The legend for this scale drawing is $\frac{1}{2}$ inch = 5 feet.

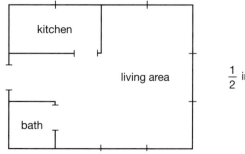

$\frac{1}{2}$ inch = 5 feet

If we measure the scale drawing above, we find that it is 2 inches long and $1\frac{1}{2}$ inches wide. Using these measurements, we can determine the actual dimensions of Angela's apartment. On the next page we show some relationships that are based on the scale drawing's legend.

$$\frac{1}{2} \text{ inch} = 5 \text{ feet (given)}$$
$$1 \text{ inch} = 10 \text{ feet}$$
$$1\frac{1}{2} \text{ inches} = 15 \text{ feet}$$
$$2 \text{ inches} = 20 \text{ feet}$$

Since the scale drawing is 2 inches long by $1\frac{1}{2}$ inches wide, we find that Angela's apartment is 20 feet long by 15 feet wide.

1. **Connect** What are the actual length and width of Angela's kitchen?

2. In the scale drawing each doorway measures $\frac{1}{4}$ inch wide. Since $\frac{1}{4}$ inch is half of $\frac{1}{2}$ inch, what is the actual width of each doorway in Angela's apartment?

3. **Connect** A dollhouse was built as a scale model of an actual house using 1 inch to represent 1.5 feet. What are the dimensions of a room in the actual house if the corresponding dollhouse room measures 8 in. by 10 in.?

4. A scale model of an airplane is built using 1 inch to represent 2 feet. The wingspan of the model airplane is 24 inches. What is the wingspan of the actual airplane in feet?

The lengths of corresponding parts of scale drawings or models and the objects they represent are proportional. Since the relationships are proportional, we can use a ratio box to organize the numbers and a proportion to find the unknown.

Connect To answer problems **5–8,** use a ratio box and write a proportion. Then solve for the unknown measurement either by using cross products or by writing an equivalent ratio. Make one column of the ratio box for the model and the other column for the actual object.

5. A scale model of a sports car is 7 inches long. The car itself is 14 feet long. If the model is 3 inches wide, how wide is the actual car?

6. **Explain** For the sports car in problem **5,** suppose the actual height is 4 feet. What is the height of the model? How do you know your answer is correct?

7. **Analyze** The femur is the large bone that runs from the knee to the hip. In a scale drawing of a human skeleton the length of the femur measures 3 cm, and the full skeleton measures 12 cm. If the drawing represents a 6-ft-tall person, what is the actual length of the person's femur?

8. The humerus is the bone that runs from the elbow to the shoulder. Suppose the humerus of a 6-ft-tall person is 1 ft long. How long should the humerus be on the scale drawing of the skeleton in problem **7?**

9. A scale drawing of a room addition that measures 28 ft by 16 ft is shown below. The scale drawing measures 7 cm by 4 cm.

    a. **(Connect)** Complete this legend for the scale drawing:

$$1 \text{ cm} = \underline{\hspace{1cm}} \text{ ft}$$

    b. **(Estimate)** What is the actual length and width of the bathroom, rounded to the nearest foot.

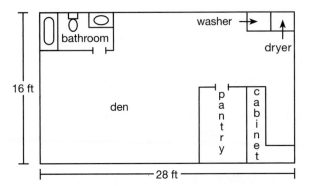

10. Noe will drive from San Francisco to Los Angeles.

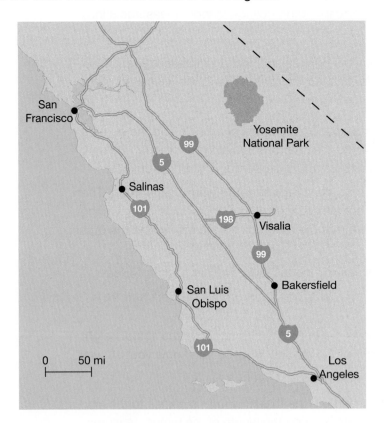

    a. On this map, each inch represents how many miles?

    b. Estimate the driving distance from San Francisco to Los Angeles. What highway should Noe take?

    c. Noe will stop in Visalia on his way to Los Angeles, taking California highway 198 to Visalia and back to Interstate 5. What distance does this add to his trip?

11. A natural history museum contains a 44-inch-long scale model of a *Stegosaurus* dinosaur. The actual length of the *Stegosaurus* was 22 feet. What should be the legend for the scale model of the dinosaur?

$$1 \text{ inch} = \underline{\hspace{1cm}} \text{ feet}$$

Maps, blueprints, and models are called *renderings.* If a rendering is smaller than the actual object it represents, then the dimensions of the rendering are a fraction of the dimensions of the actual object. This fraction is called the **scale** of the rendering.

To determine the scale of a rendering, we form a fraction using corresponding dimensions and the same units for both. Then we reduce.

$$\text{scale} = \frac{\text{dimension of rendering}}{\text{dimension of object}}$$

In the case of the *Stegosaurus* in problem **11,** the corresponding lengths are 44 inches and 22 feet. Before reducing the fraction, we will convert 22 feet to 264 inches. Then we write a fraction, using the length of the model as the numerator and the dinosaur's actual length as the denominator.

$$\text{scale} = \frac{44 \text{ inches}}{22 \text{ feet}} = \frac{44 \text{ inches}}{264 \text{ inches}} = \frac{1}{6}$$

So the model is a $\frac{1}{6}$ scale model. The reciprocal of the scale is the **scale factor.** So the scale factor from the model to the actual *Stegosaurus* is 6. This means we can multiply any dimension of the model by 6 to determine the corresponding dimension of the actual object.

12. What is the scale of the model car in problem **5?** What is the scale factor?

13. A scale may be written as a ratio that uses a colon. For example, we can write the scale of the *Stegosaurus* model as 1:6. Suppose that a toy company makes action figures of sports stars using a scale of 1:10. How many inches tall will a figure of a 6-ft-8-in. basketball player be?

14. In a scale drawing of a wall mural, the scale factor is 6. If the scale drawing is 3 feet long by 1.5 feet wide, what are the dimensions of the actual mural?

Investigate Further

a. (**Model**) Make a scale drawing of your bedroom's floor plan, where 1 in. = 3 ft. Include in your drawing the locations of doors and windows as well as major pieces of furniture. What is the scale factor you used?

b. (**Model**) Cut out and assemble the pieces from Activity 21 and 22 to make a scale model of the *Freedom 7* spacecraft. This spacecraft was piloted by Alan B. Shepard, the first American to go into space. With *Freedom 7* sitting atop a rocket, Shepard blasted off from Cape Canaveral, Florida, on May 5, 1961. Because he did not orbit (circle) the earth, the trip lasted only 15 minutes before splashdown in the Atlantic Ocean. Shepard was one of six astronauts to fly a Mercury spacecraft like *Freedom 7.* Each Mercury spacecraft could carry only one astronaut, because the rockets available in the early 1960s were not powerful enough to lift heavier loads.

Your completed *Freedom 7* model will have a scale of 1:24. After you have constructed the model, measure its length. Use this information to determine the length of an actual Mercury spacecraft.

c. (**Represent**) On a coordinate plane draw a square with vertices at (2, 2), (4, 2), (4, 4), and (2, 4). Then apply a scale factor of 2 to the square so that the dimensions of the square double but the point (3, 3) remains the center of the larger square. Draw the larger square on the same coordinate plane. What are the coordinates of its vertices?

d. (**Represent**) Using plastic straws, scissors, and string, make a scale model of a triangle whose sides are 3 ft, 4 ft, and 5 ft. For the model, let $\frac{3}{4}$ in. = 1 foot. What is the scale factor? How long are the sides of the model?

e. (**Model**) Draw these shapes on grid paper and measure the angles. Write the measure by the angle and then mark each angle as acute, obtuse, or right.

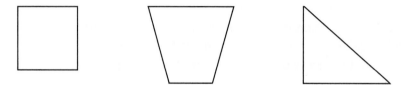

Now use the grid paper to draw each figure increased by a scale factor of 2. Then develop a mathematical argument proving or disproving that the angle classification of the images are the same as the given figures.

California Mathematics Content Standards
NS **1.0, 1.4** Calculate given percentages of quantities and solve problems involving discounts at sales, interest earned, and tips.
MR **2.0, 2.4** Use a variety of methods, such as words, numbers, symbols, charts, graphs, tables, diagrams, and models, to explain mathematical reasoning.

• Finding a Whole When a Percent Is Known

facts	Power Up Q
mental math	**a. Fractions:** $\frac{2}{3}$ of 24
	b. Number Sense: 7×35
	c. Percent: 50% of $52
	d. Money: Mizuka had $10.00. Then she spent $8.59 on dinner. How much money did she have left?
	e. Estimation: Estimate the area of a square with side lengths of 8.9 cm.
	f. Measurement: Convert 125 centimeters into meters.
	g. Powers/Roots: $2^3 + 4^2$
	h. Calculation: $8 \times 8, -1, \div 9, \times 4, +2, \div 2, \div 3, \times 5$
problem solving	Choose an appropriate problem-solving strategy to solve this problem. Twenty students in a homeroom class are signing up for fine arts electives. So far, 5 students have signed up for band, 6 have signed up for drama, and 12 signed up for art. There are 3 students who signed up for both drama and art, 2 students who signed up for both band and art, and 1 student who signed up for all three. How many students have not yet registered for an elective?

New Concept

We have solved problems like the following using a ratio box. In this lesson we will practice writing equations to help us solve these problems.

Thirty percent of the football fans in the stadium are waving team banners. There are 150 football fans waving banners. How many football fans are in the stadium in all?

The statement above tells us that 30% of the fans are waving banners and that the number is 150. We will write an equation using t to stand for the total number of fans.

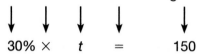

30% of the fans are waving banners.

$$30\% \times t = 150$$

Now we change 30% to a fraction or to a decimal. For this problem we choose to write 30% as a decimal.

$$0.3t = 150$$

Now we find t by dividing 150 by three tenths.

$$\overset{\displaystyle 500.}{03.)\overline{1500.}}$$

We find that there were 500 fans in all. We can use a model to represent the problem.

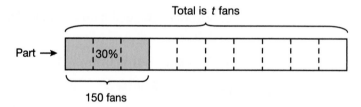

Total is t fans

Part → | 30% |

150 fans

Example 1

Thirty percent of what number is 120? Find the answer by writing and solving an equation. Model the problem with a sketch.

To translate the question into an equation, we translate the word *of* into a multiplication sign and the word *is* into an equal sign. For the words *what number* we write the letter n.

Thirty percent of what number is 120?

$$30\% \times n = 120$$

We may choose to change 30% to a fraction or a decimal number. We choose the decimal form.

$$0.3n = 120$$

Now we find n by dividing 120 by 0.3.

$$\overset{\displaystyle 400.}{03.)\overline{1200.}}$$

Thirty percent of **400** is 120.

Thinking Skills

Explain

How can we check the answer?

We were given a part (30% is 120) and were asked for the whole. Since 30% is $\frac{3}{10}$ we divide 100% into ten divisions instead of 100.

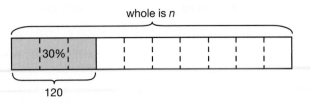

Example 2

Sixteen is 25% of what number? Solve with an equation and model with a sketch.

We translate the question into an equation, using an equal sign for *is*, a multiplication sign for *of,* and a letter for *what number*.

Sixteen is 25% of what number?

$$16 \quad = 25\% \times \quad n$$

Because of the way the question was asked, the numbers are on opposite sides of the equal sign as compared to example 1. We can solve the equation in this form, or we can rearrange the equation. Either form of the equation may be used.

$$16 = 25\% \times n$$

$$25\% \times n = 16$$

We will use the first form of the equation and change 25% to the fraction $\frac{1}{4}$.

$$16 = 25\% \times n$$

$$16 = \frac{1}{4}n$$

We find *n* by dividing 16 by $\frac{1}{4}$.

$$16 \div \frac{1}{4} = \frac{16}{1} \times \frac{4}{1} = 64$$

Sixteen is 25% of **64.**

We are given a part and asked for the whole. Since 25% is $\frac{1}{4}$ we model 100% with four sections of 25%.

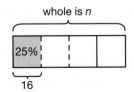

Thinking Skills

Discuss

Why did we use a fraction to solve this problem rather than a decimal?

Example 3

The county fair raised the price of admission by $4, which was a 20% increase. What was the original price? [entire problem bold face]

We can state this question in other words,

"$4 is 20% of what price?"

$$\downarrow\downarrow\ \downarrow\ \ \ \downarrow\qquad\qquad\downarrow$$

$$\$4 = 20\% \times \qquad p$$

We can write 20% as a fraction.

$$\$4 = \frac{1}{5}p$$

To find p, divide 4 by $\frac{1}{5}$.

$$4 \div \frac{1}{5}$$

$$4 \times \frac{5}{1}$$

$$20$$

The original price of admission was **$20.**

Lesson Practice

Formulate Translate each question into an equation and solve:

a. Twenty percent of what number is 120?

b. Fifty percent of what number is 30?

c. Twenty-five percent of what number is 12?

d. Twenty is 10% of what number?

e. Twelve is 100% of what number?

f. Fifteen is 15% of what number?

g. Seventy-five percent of what number is 6?

h. Twenty-seven is 90% of what number?

i. Write and solve a word problem for the equation below.

$$15\% \times n = 12$$

Written Practice

Distributed and Integrated

* 1. Three fifths of the townspeople voted. If 120 of the townspeople voted, how many
(110) people live in the town?

* 2. **Analyze** If 130 children are separated as equally as possible into four groups,
(104) how many will be in each group? (Write four numbers, one for each of the four groups.)

3. If the parking lot charges $1.25 per half hour, what is the cost of parking a car
(7, 89) from 11:15 a.m. to 2:45 p.m.?

4. **Analyze** If the area of the square is 400 m², then what is the area of
(84) the circle? (Use 3.14 for π.)

5. Only 4 of the 50 states have names that begin with the letter A.
(92) What percent of the states have names that begin with letters other
than A?

6. **Connect** The coordinates of the vertices of a triangle are (3, 6),
(RF27, 77) (5, 0), and (0, 0). What is the area of the triangle?

7. Write one hundred five thousandths as a decimal number.
RF19)

8. **Estimate** Round the quotient of $7.00 ÷ 9 to the nearest cent.
(46)

9. Arrange in order from least to greatest:
(96)

$$81\%, \frac{4}{5}, 0.815$$

Solve and check:

10. $6x - 12 = 60$
(85)

11. $\frac{9}{15} = \frac{m}{25}$
(83)

12. Six is $\frac{2}{5}$ of what number?
(110)

13. $\left(5 - 1\frac{2}{3}\right) - 1\frac{1}{2}$
(58)

14. $2\frac{2}{5} \div 1\frac{1}{2}$
(64)

15. 0.625×2.4
(33)

16. $-5 + -5 + -5$
(97)

17. The prime factorization of 24 is 2 · 2 · 2 · 3, which we can write as
(71) 2³ · 3. Write the prime factorization of 36 using exponents.

18. What is the total price of a $12.50 item plus 6% sales tax?
(35)

* 19. What is the area of this pentagon?
(101)

20. 60 is 30% of what number?
(111)

* 21. **Analyze** Calculate mentally:
(103)

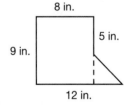

a. $\frac{-20}{-4}$

b. $\frac{-36}{6}$

c. $(-3)(8)$

d. $(-4)(-9)$

22. If each small cube has a volume of one cubic inch, then what is the
(RF29) volume of this rectangular solid?

23. **Represent** Draw a triangle in which each angle measures less than 90°.
(90) What type of triangle did you draw?

24. **Estimate** The mean, median, and mode of student scores on a test
(Inv. 4) were 89, 87, and 92 respectively. About half of the students scored what
score or higher?

25. Evaluate: $ab - a - b$ if $a = -3$ and $b = -1$
(108)

* 26. **Analyze** If the spinner is spun twice, what is the
(Inv. 6) probability that the arrow will stop on a number greater
than 1 on both spins?

27. a. $\dfrac{(-9)(+6)(-5)}{(-4) - (-1)}$
(107)

b. $-3(4) + 2(3) -1$

* 28. **Model** Draw a ratio box for this problem. Then solve the problem using
(100) a proportion.

*The ratio of cattle to horses on the ranch was 15 to 2. The combined number
of cattle and horses was 1020. How many horses were on the ranch?*

* 29. 63 is 90% of what number?
(111)

30. How far did the plane travel if it flew for 5 hours at 320 miles per hour?
(89, 93)

Early Finishers

Real-World Connection

You and your friends want to rent go-karts this Saturday. Go-Kart Track A
rents go-karts for a flat fee of $10 plus $7 per hour. Go-Kart Track B
rents go-karts for a flat fee of $7 plus $8 per hour.

a. If you and your friends plan to stay for 2 hours, which track has the better deal?
Show all your work.

b. Which track is the better value if you stay for four hours?

• **Percent Word Problems**

California Mathematics Content Standards

NS **1.0**, **1.3** Use proportions to solve problems (e.g., determine the value of *N* if $\frac{4}{7} = \frac{N}{21}$, find the length of a side of a polygon similar to a known polygon). Use cross-multiplication as a method for solving such problems, understanding it as the multiplication of both sides of an equation by a multiplicative inverse.

NS **1.0**, **1.4** Calculate given percentages of quantities and solve problems involving discounts at sales, interest earned, and tips.

MR 2.0, 2.2 Apply strategies and results from simpler problems to more complex problems.

facts

Power Up O

mental math

a. Fractions: $\frac{3}{4}$ of 24

b. Number Sense: 6×48

c. Percent: Chantoya found a coat that was on sale for 25% off the regular price of $44. How much is 25% of $44?

d. Money: $4.98 + $2.49

e. Decimals: $0.5 \div 10$

f. Geometry: A cube has sides that are 3 in. long. What is the volume of the cube?

g. Algebra: If $z^2 = 100$, what is z?

h. Calculation: $11 \times 4, + 1, \div 5, \sqrt{}, \times 4, - 2, \times 5, - 1, \sqrt{}$

problem solving

Choose an appropriate problem-solving strategy to solve this problem. When all the cards from a 52-card deck are dealt to three players, each player receives 17 cards, and there is one extra card. Dean invented a new deck of cards so that any number of players up to 6 can play and there will be no extra cards. How many cards are in Dean's deck if the number is less than 100?

New Concept

Percent word problems

Word problems that involve percents can be solved by writing the percent as a fraction or as a decimal.

For example, we can write an equation to help us answer this question in two different ways.

What is 20% of $32?	
Write 20% as a decimal:	Write 20% as a fraction:
$w = 0.20 \times \$32$	$w = \frac{1}{5} \times \$32$

Discuss Solve each equation for w. Which method do you prefer?

Whether we choose to use a decimal or a fraction depends on convenience. Sometimes the arithmetic is easier using a decimal. Other times the fraction has a simple form and is easier to use.

Example 1

The total of the dinner bill was $40. John wanted to leave a 15% tip. What amount should he leave?

Decide whether to use a decimal or fraction to solve the problem, and explain your choice.

John wonders,

"What is 15% of $40?"

We can write 15% as a decimal or as a fraction.

15% as a decimal: 15% as a fraction:

$$t = 0.15 \times \$40 \qquad t = \frac{3}{20} \times \$40$$

In this case, we will use the fraction, since we can cancel before multiplying.

$$t = \frac{3}{\cancel{20}} \times \frac{\cancel{\$40}^2}{1}$$

$$t = \$6$$

John should leave **$6.**

Some word problems can be solved mentally by thinking about the meaning of the percents before computation.

For example, to compute a 15% tip, we can add 10% and 5% of the amount.

15% of $40 = 10% of $40 + 5% of $40

Ten percent of $40 is easily computed mentally to be $4. Five percent is one-half of 10%.

15% of $40 = $4 + $2

15% of $40 = $6

Explain How would you mentally compute a 15% tip on $25?

When an amount is increased or decreased by a percent, we can add or subtract that percent from 100% to make the computation easier.

For example, if the price of an item is reduced by 40%, the new price is 100% − 40%, or 60%, of the original price.

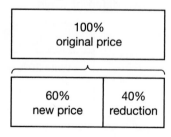

Example 2

A $110 watch was discounted by 40%. What is the sale price? Decide whether to use a decimal or fraction to solve the problem and explain your choice.

The sale price is 60% of $110.

$$s = 60\% \times \$110$$

We will write 60% as a decimal since 0.6 is easily multiplied by $110.

$$s = 0.60 \times \$110$$
$$s = \$66$$

The sale price is **$66.**

Example 3

Diana deposited $200 in a bank account that earns 5% simple interest per year. To compute the value of her account after one year, Diana multiplied 1.05 by $200. Explain why her method is reasonable, and compute the value.

After one year, Diana will have 100% of what she deposited plus 5% of that amount in interest. That is, she will have 105% of what she deposited in her account.

$$V = 105\% \times \$200$$
$$V = 1.05 \times \$200$$

Diana's method is reasonable because the decimal equivalent of 105% is 1.05.

$$V = \$210$$

Diana will have **$210** in her account.

Example 4

After an 8% sales tax, Marcello paid $21.33 for a game. What was the price of the game? Explain the method you use.

Marcello paid 100% of the price plus 8% more in tax. That is, he paid 108% of the price that was marked.

The question can be stated,

"108% of what is $21.33?"

$$108\% \times p = \$21.33$$

We write 108% as a decimal.

$$1.08p = \$21.33$$

To find p, we divide $21.33 by 1.08.

$$\begin{array}{r} 19.75 \\ 108\overline{)2133.00} \\ \underline{108} \\ 1053 \\ \underline{972} \\ 810 \\ \underline{756} \\ 540 \\ \underline{540} \\ 0 \end{array}$$

The price was **$19.75.**

Lesson Practice

a. How much money is a 20% tip on a $25 meal?

b. What is the sale price of a $32 shirt after a 20% discount?

c. What is the total price of a $46.00 item when 6% sales tax is added?

d. How much money is a 30% income tax on $100,000?

e. What is the sale price of a $25 pair of pants after a 10% discount?

f. What is the total price of a $300 item when 8% sales tax is added?

Written Practice *Distributed and Integrated*

1.
(104)
Tabari is giving out baseball cards. Seven students sit in a circle. If he goes around the circle giving out 52 of his baseball cards, how many students will get 8 cards?

* 2.
(76)
Multiple Choice (**Estimate**) About how long is a new pencil?

A 1.8 cm   **B** 18 cm   **C** 180 cm

3.
(110)
(**Estimate**) Texas is the second most populous state in the United States. About 6 million people under the age of 18 lived in Texas in the year 2000. This number was about $\frac{3}{10}$ of the total population of the state at that time. About how many people lived in Texas in 2000?

4.
(74)
Multiple Choice (**Verify**) The symbol $\neq$ means "is not equal to." Which statement is true?

A $\frac{3}{4} \neq \frac{9}{12}$ **B** $\frac{3}{4} \neq \frac{9}{16}$ **C** $\frac{3}{4} \neq 0.75$

5.
(35)
What is the total price, including 7% tax, of a $14.49 item?

6.
(16)
As Elsa peered out her window she saw 48 trucks, 84 cars, and 12 motorcycles go by her home. What was the ratio of trucks to cars that Elsa saw?

7. What is the mean of 17, 24, 27, and 28?
(Inv. 4)

8. Arrange in order from least to greatest:
(32, 74)
$$6.1, \sqrt{36}, 6\frac{1}{4}$$

* 9. **Analyze** Nine cookies were left in the package. That was $\frac{3}{10}$ of the original
(110) number of cookies. How many were in the package originally?

10. **Explain** Jamel measured the circumference of the trunk of the old oak tree.
(42) How can Jamel calculate the approximate diameter of the tree?

* 11. Twelve is $\frac{3}{4}$ of what number?
(110)

12. $2\frac{2}{3} + \left(5\frac{1}{3} - 2\frac{1}{2}\right)$
(55, 58)

13. $6\frac{2}{3} \div 4\frac{1}{6}$
(64)

14. $4\frac{1}{4} + 3.2$ (decimal answer)
(69)

15. The proposition was favored by 70% of the voters. If 1500 voters opposed the
(112) proposition, what is the total number of voters?

16. Fifty is 10% of what number?
(111)

* 17. On Earth a kilogram is about 2.2 pounds. Use a unit multiplier to
(105) convert 2.2 pounds to ounces. (Round to the nearest ounce.)

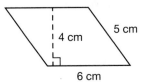

18. The quadrilateral at right is a parallelogram.
(66)
  a. What is the area of the parallelogram?

  b. If each obtuse angle measures 127°, then what is the measure of each acute angle?

19. What is the perimeter of this hexagon?
(101)

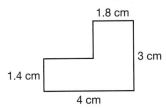

* 20. These triangles are similar. Find x.
(Inv. 10)

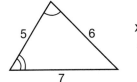

* 21. **Estimate** Each edge of a cube measures 4.11 feet. What is a good
(RF29, estimate of the cube's volume?
21)

22. Complete the proportion: $\frac{f}{12} = \frac{12}{16}$
(83)

23. If $x = -5$ and $y = 3x - 1$, then y equals what number?
(108)

24. **a.** $\frac{(-3)-(-4)(+5)}{(-2)}$
(108)
  b. $-3(+4)-5(+6)-7$

* 25. $-5 + +2 - +3 - -4 + -1$
(97)

26. Use the formula $C = \frac{5(F - 32)}{9}$ to find C when $F = 95°$.
(93)

* 27. Convert $7\frac{1}{2}\%$ to a fraction.
(106)

28. At noon the temperature was $-3°F$. By sunset the temperature had dropped
(9) another five degrees. What was the temperature at sunset?

29. There were three red marbles, three white marbles, and three blue marbles in
(54) a bag. Luis drew a white marble out of the bag and held it. If he draws another
marble out of the bag, what is the probability that the second marble also will
be white?

30. The probability that you will win a contest is $\frac{1}{1000}$. If you try the contest 300 times,
(Inv. 8) predict how many times you will win.

🖌 *California Mathematics Content Standards*

MG 1.0, 1.1 Understand the concept of a constant such as π; know the formulas for the circumference and area of a circle.

MG 1.0, 1.2 Know common estimates of π (3.14; $\frac{22}{7}$) and use these values to estimate and calculate the circumference and the area of circles; compare with actual measurements.

MG 1.0, 1.3 Know and use the formulas for the volume of triangular prisms and cylinders (area of base x height); compare these formulas and explain the similarity between them and the formula for the volume of a rectangular solid.

• Volume of a Cylinder

Power Up

facts	Power Up P
mental math	**a. Fractions:** $\frac{3}{10}$ of 40
	b. Fractions: $\frac{1}{3}$ of 15
	c. Fractions: $\frac{2}{3}$ of 15
	d. Percent: 10% of $110
	e. Measurement: Gina walked $\frac{1}{4}$ of a kilometer to the end of the block. How many meters is $\frac{1}{4}$ of a kilometer?
	f. Geometry: Pilar turned to her left one quarter of a full rotation. How many degrees did she turn?
	g. Geometry: A circle has a diameter of 10 mm. What is the circumference of the circle?
	h. Calculation: 6×8, $+ 2$, $\times 2$, $- 1$, $\div 3$, $- 1$, $\div 4$, $+ 2$, $\div 10$, $- 1$

problem solving

Choose an appropriate problem-solving strategy to solve this problem. A hexagon can be divided into four triangles by three diagonals drawn from a single vertex. How many triangles can a dodecagon be divided into using diagonals drawn from one vertex?

New Concept

Imagine pressing a quarter down into a block of soft clay.

Thinking Skills

What is the formula for the area of a circle?

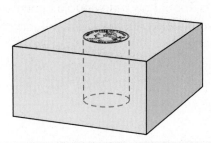

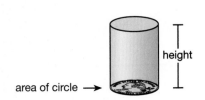

area of circle →

height

As the quarter is pressed into the block, it creates a hole in the clay. The quarter sweeps out a **cylinder,** a three-dimensional solid with two circular bases, as it moves through the clay. We can calculate the **volume,** the amount of space the shape occupies, of the cylinder by multiplying the area of the circular face of the quarter by the distance it moved through the clay. The distance the quarter moved is the **height** of the cylinder.

Example

Thinking Skills

Explain

Why is the volume of this cylinder called an approximation?

The diameter of this cylinder is 20 cm. Its height is 10 cm. What is its volume?

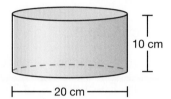

To calculate the volume of a cylinder, we find the area of the base end of the cylinder and multiply that area by the height of the cylinder—the distance between the circular ends.

Since the diameter of the cylinder is 20 cm, the radius is 10 cm. A square with a side the length of the radius has an area of 100 cm². So the area of the circle is about 3.14 times 100 cm², which is 314 cm².

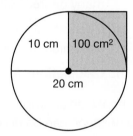

Now we multiply the area of the circular end of the cylinder by the height of the cylinder.

$$314 \text{ cm}^2 \times 10 \text{ cm} = 3140 \text{ cm}^3$$

We find that the volume of the cylinder is approximately **3140 cm³.**

Lesson Practice

a. A large can of soup has a diameter of about 8 cm and a height of about 12 cm. The volume of the can is about how many cubic centimeters? Round your answer to the nearest hundred cubic centimeters.

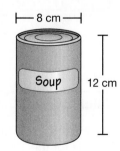

b. Find the volume of a cylinder with a radius of 10 mm and a height of 50 mm.

c. Find the volume of a cylinder with a radius of 1 m and a height of 1 m.

Written Practice *Distributed and Integrated*

1. Divide 555 by 12 and write the quotient
(18)
   a. with a remainder.

   b. as a mixed number.

2. The six gymnasts scored 9.75, 9.8, 9.9, 9.4, 9.9, and 9.95. The lowest score was
(41) not counted. What was the sum of the five highest scores?

* 3. (Analyze) Cantara said that the six trumpet players made up 10% of the band.
(102) The band had how many members?

* 4. Eight is $\frac{2}{3}$ of what number?
(110)

5. Find the volume of this cylinder. Dimensions are in cm. Use 3.14 for
(113) π.

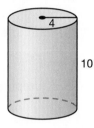
4

10

* 6. On Rhashan's scale drawing, each inch represents 8 feet. One of the
(Inv. 11) rooms in his drawing is $2\frac{1}{2}$ inches long. How long is the actual room?

* 7. Two angles of a triangle each measure 45°.
(95)
   a. (Generalize) What is the measure of the third angle?   °

   b. (Represent) Make a rough sketch of the triangle.

* 8. Convert $8\frac{1}{3}$% to a fraction.
(106)

9. Nine dollars is what percent of $12?
(73)

* 10. (Analyze) Twenty percent of what number is 12?
(111)

* 11. Three tenths of what number is 9?
(110)

12. $(-5) - (+6) + (-7)$ **13.** $(-15)(-6)$
(97)   (103)

14. Reduce: $\frac{60}{84}$ **15.** $2\frac{1}{2} - 1\frac{2}{3}$
(25)   (58)

16. (Analyze) Stephen competes in a two-event race made up of biking and running.
(100) The ratio of the length of the distance run to the length of the bike ride is 2 to 5.
If the distance run was 10 kilometers, then what was the total length of the
two-event race?

17. The area of the shaded triangle is 2.8 cm². What is the area of the parallelogram?
(66, 77)

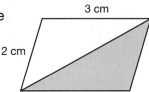

18. These triangles are similar. Find *x*.
(Inv. 10)

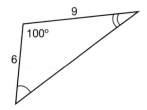 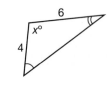

19. An account pays 9% simple interest per year. If $3000 is put in the account, how much interest will be earned in 1 year?
(112)

20. A regular solid has 20 faces numbered 1 through 20. What is the probability that a 1 is rolled. Leave the answer as a fraction.
(Inv. 8)

21. How many times would you predict that a 1 would result if the solid is rolled 180 times?
(Inv. 8)

22. Evaluate: $x + xy - xy$ if $x = 3$ and $y = -2$.
(108)

23. $\dfrac{-12-(6)(-3)}{(-12)-(-6)+(3)}$
(107)

* 24. **Multiple Choice** Which of these polygons is not a quadrilateral?
(59)

    A parallelogram   **B** pentagon   **C** trapezoid

25. Compare: area of the square ◯ area of the circle
(84)

26. Last year, 36 eggs were hidden and 33 eggs were found. If 4 dozen eggs are hidden this year, predict how many eggs will be found.
(Inv. 8)

27. Robert flipped a coin. It landed heads up. He flipped the coin a second time. It landed heads up. If he flips the coin a third time, what is the probability that it will land heads up?
(54)

28. James is going to flip a coin three times. What is the probability that the coin will land heads up all three times?
(Inv. 6)

29. The diameter of the circle is 10 cm.
(32, 84)

    a. What is the area of the square?

    b. What is the area of the circle?

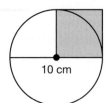

30. What is the mode and the range of this set of numbers?
(Inv. 4)
          4, 7, 6, 4, 5, 3, 2, 6, 7, 9, 7, 4, 10, 7, 9

California Mathematics Content Standards

MG 1.0, 1.1 Understand the concept of a constant such as π; know the formulas for the circumference and area of a circle.

MG 1.0, 1.2 Know common estimates of π (3.14; $\frac{22}{7}$) and use these values to estimate and calculate the circumference and the area of circles; compare with actual measurements.

MG 1.0, 1.3 Know and use the formulas for the volume of triangular prisms and cylinders (area of base × height); compare these formulas and explain the similarity between them and the formula for the volume of a rectangular solid.

• Volume of a Right Solid

 **Power Up**

facts	Power Up Q
mental math	**a. Fractions:** $\frac{7}{10}$ of 40
	b. Fractions: $\frac{1}{3}$ of 27
	c. Money: $4.99 + 65¢
	d. Decimals: 0.125 × 1000
	e. Geometry: What is the perimeter of a rectangle that is 3.5 cm by 2.5 cm?
	f. Statistics: The lengths of the fish were 9.6 cm, 9.9 cm, 10.3 cm, 18.0 cm, and 18.0 cm. What was the median length?
	g. Algebra: If $g = 7.5$, what is $2g$?
	h. Calculation: 5×7, $+ 1$, $\div 4$, $\sqrt{}$, $\times 7$, $- 1$, $\times 3$, $- 10$, $\times 2$, $\sqrt{}$

problem solving

Choose an appropriate problem-solving strategy to solve this problem. Four identical blocks marked *x*, a 250-gram mass, and a 500-gram mass were balanced on a scale as shown. Write an equation to represent this balanced scale, and find the mass of each block marked *x*.

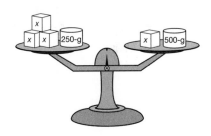

New Concept

A **right solid** is a geometric solid whose sides are perpendicular to the base. **Geometric solids** have length, width, and height; in other words, they occupy space. **The volume of a right solid equals the area of the base times the height.** This rectangular solid is a right solid. It is 5 m long and 2 m wide, so the area of the base is 10 m².

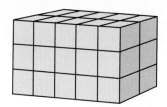

One **cube,** a three-dimensional solid with six square faces, will fit on each square meter of the base, and the cubes are stacked 3 m high, so

$$\text{Volume} = \text{area of the base} \times \text{height}$$
$$= 10 \text{ m}^2 \times 3 \text{ m}$$
$$= 30 \text{ m}^3$$

If the base of the solid is a polygon, the solid is called a **prism,** or **polyhedron.** If the base of a right solid is a circle, the solid is called a **right circular cylinder.**

Right square   Right triangular   Right circular
prism   prism   cylinder

(**Identify**) What is another name for a right square prism? What is the shape of the base of each figure shown?

Example 1

Find the volume of the right triangular prism below. Dimensions are in centimeters. We show two views of the prism.

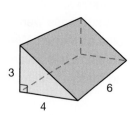
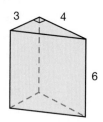

The area of the base is the area of the triangle.

$$\text{Area of base} = \frac{(4 \text{ cm})(3 \text{ cm})}{2} = 6 \text{ cm}^2$$

The volume equals the area of the base times the height.

$$\text{Volume} = (6 \text{ cm}^2)(6 \text{ cm}) = \textbf{36 cm}^3$$

Notice that the height of a prism is not necessarily measured vertically, as seen in the prism on the left. The height is the length of the edge stretching between the bases.

Example 2

The diameter of this right circular cylinder is 20 cm. Its height is 25 cm. What is its volume? Leave π as π.

First we find the area of the base. The diameter of the circular base is 20 cm, so the radius is 10 cm.

Area of base $= \pi r^2 = \pi(10 \text{ cm})^2 = 100\pi \text{ cm}^2$

The volume equals the area of the base times the height.

Volume $= (100\pi \text{ cm}^2)(25 \text{ cm}) = \mathbf{2500\pi \text{ cm}^3}$

The formula for the volume of a rectangular solid.

Volume = Area of base × height
$$V = Bh$$

We can apply this formula to cylinders and prisms as well.

(**Formulate**) The base of a cylinder is a circle. Create a specific formula for the volume of a cylinder by replacing B in the formula with the formula for the area of a circle.

(**Formulate**) Create a specific formula for the volume of a triangular prism. Many objects in the real world are composed of combinations of basic shapes like the solids in this lesson. We can calculate the volumes of such objects by dividing them into parts, finding the volume of each part, and then adding the volumes.

Example 3

Eric stores gardening tools in a shed with the dimensions shown. Find the approximate volume of the shed.

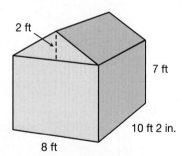

We separate the structure into a triangular prism and a rectangular prism. We round the dimensions. For example, the notation 10′2″ means 10 feet, 2 inches, which we round to 10 ft.

The volume of the triangular prism is about

$$V = Bh$$

$$V = \frac{1}{2}(2 \text{ ft})(8 \text{ ft})(10 \text{ ft}) = 80 \text{ ft}^3$$

The volume of the rectangular prism is about

$$V = lwh$$

$$V = (10 \text{ ft})(8 \text{ ft})(7 \text{ ft}) = 560 \text{ ft}^3$$

The volume of the shed is about

$$80 \text{ ft}^3 + 560 \text{ ft}^3 = \mathbf{640 \ ft^3}$$

Discuss Is there a different way of finding the volume of the shed from Example 3?

Lesson Practice

Analyze Find the volume of each right solid shown. Dimensions are in centimeters.

a.

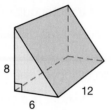

8

12

6

b.

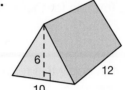

6

12

10

c.

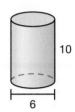

10

6

Leave π as π

d.

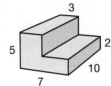

3

5

2

10

7

e.

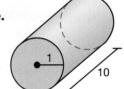

1

10

Leave π as π.

f.

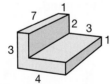

7

1

2 3

1

3

4

g.

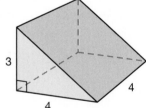

3

4

4

h.

The area of the base is 37 cm²

10

i. A farmer added an attached shed with the dimensions shown to the side of his barn. What is the approximate volume of the addition? List the steps needed to find the volume.

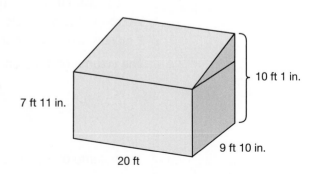

7 ft 11 in.

10 ft 1 in.

9 ft 10 in.

20 ft

Written Practice

1. Write the prime factorization of 750 using exponents.
(71)

2. **Multiple Choice** (**Estimate**) About how long is your little finger?
(76)
   A 0.5 mm   **B** 5 mm   **C** 50 mm   **D** 500 mm

3. (**Analyze**) If 3 parts is 24 grams, how much
(86) is 8 parts?

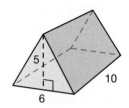

4. Complete the proportion: $\frac{3}{24} = \frac{8}{w}$
(83)

5. Find the volume of this right solid. Dimensions are in inches.
(114)

6. These triangles are similar. Find x.
(Inv. 10)

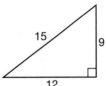

7. The mean of four numbers is 25. Three of the numbers are 17, 23, and 25.
(Inv. 4)
  a. What is the fourth number?

  b. What is the range of the four numbers?

8. Calculate mentally:
(98,
103) **a.** $-6 - -4$ **b.** $-10 + -15$ **c.** $(-10)(-10)$

* 9. Write $16\frac{2}{3}\%$ as a reduced fraction.
(106)

* 10. (**Analyze**) Twenty-four guests came to the party. This was $\frac{4}{5}$ of those who were
(110) invited. How many guests were invited?

11. $1\frac{1}{3} + 3\frac{3}{4} + 1\frac{1}{6}$ **12.** $\frac{5}{6} \times 3 \times 2\frac{2}{3}$
(56)   (67)

13. $5.62 + 0.8 + 4$ **14.** $0.08 \div (1 \div 0.4)$
(41)   (44)

15. $(-2) + (-2) + (-2)$
(97)

16. James always leaves a 15% tip. If he leaves a tip of $3.75, how much was
(112) his bill?

* 17. At $1.12 per pound, what is the price per ounce (1 pound = 16 ounces)?
(105)

18. The children held hands and stood in a circle. The diameter of the circle was
(42) 10 m. What was the circumference of the circle? (Use 3.14 for π.)

* 19. (**Analyze**) If the area of a square is 36 cm², what is the perimeter of
(32) the square?

20. The probability that you will win a contest is $\frac{1}{500}$.
(Inv. 8)

    a. Predict how many times you will win the contest in 500 tries.

    b. The contest pays $200 to the winner, and costs $1 to enter. Is it worth the money to enter the contest 500 times? Explain.

21. Sixty percent of the votes were cast for Janeka. If Janeka received 18 votes, how many votes were cast in all?
(102)

* 22. **Evaluate** Kareem has a spinner marked *A, B, C, D*. Each letter fills one fourth of the face of his spinner. If he spins the spinner three times, what is the probability he will spin *A* three times in a row?
(Inv. 6)

23. If the spinner from problem 22 is spun twenty times, how many times is the spinner likely to land on *C*?
(Inv. 7)

24.   **a.** $(-8) - (+7)$ **b.** $(-8) - (-7)$
(98)

25. $+3 + -5 - -7 - +9 + +11 + -7$
(97)

* 26. **Connect** The three vertices of a triangle have the coordinates $(0, 0)$, $(-8, 0)$, and $(-8, -8)$. What is the area of the triangle?
(RF27, 77)

27. Kaya tossed a coin and it landed heads up. What is the probability that her next two tosses of the coin will also land heads up?
(Inv. 6)

* 28. The inside diameter of a mug is 8 cm. The height of the mug is 7 cm. What is the capacity of the mug in cubic centimeters? (Think of the capacity of the mug as the volume of a cylinder with the given dimensions.)
(113)

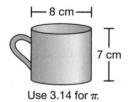

Use 3.14 for π.

29. **Estimate** A cubic centimeter of liquid is a milliliter of liquid. The mug in problem 28 will hold how many milliliters of hot chocolate? Round to the nearest ten milliliters.
(76)

* 30. **Model** Draw a ratio box for this problem. Then solve the problem using a proportion.
(102)

Ricardo correctly answered 90% of the trivia questions. If he incorrectly answered four questions, how many questions did he answer correctly?

Early Finishers
Real-World Connection

Lisa plans to use 20 tiles for a border around a square picture frame. She wants a two-color symmetrical design that has a 2 to 3 ratio of white tiles to gray tiles. What could her design look like?

A

acute angle
(20)

An angle whose measure is more than 0° and less than 90°.

acute angle   not **acute angles**

right angle   obtuse angle

*An **acute angle** is smaller than both a right angle and an obtuse angle.*

ángulo agudo

Ángulo que mide más de 0° y menos de 90°.

*Un **ángulo agudo** es menor que un ángulo recto y que un ángulo obtuso.*

acute triangle
(90)

A triangle whose largest angle measures less than 90°.

right triangle   obtuse triangle

acute triangle   not **acute triangles**

triángulo acutángulo

Triángulo cuyo ángulo mayor es menor que 90°.

addend
(5)

Any one of the numbers added in an addition problem.

$7 + 3 = 10$ *The **addends** in this problem are 7 and 3.*

sumando

Cualquiera de los números que se suman en una operación de suma.

*Los **sumandos** en este problema son el 7 y el 3.*

additive identity
(3)

The number 0. *See also* **identity property of addition.** $7 + 0 = 7$

$$7 + 0 = 7$$

additive identity

*We call zero the **additive identity** because adding zero to any number does not change the number.*

identidad de la suma

El número cero. Ver también **propiedad de identidad de la suma.**

*Le llamamos cero a la **identidad de la suma** porque sumar cero a cualquier número no cambia al número.*

adjacent angles
(20)

Two angles that have a common side and a common vertex. The angles lie on opposite sides of their common side.

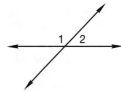

*∠1 and ∠2 are **adjacent angles**. They share a common side and a common vertex.*

ángulos adyacentes

Dos ángulos que tienen un lado y un vértice en común. Los ángulos se encuentran en regiones opuestas de su lado común.

*∠1 y ∠2 son **ángulos adyacentes**. Comparten un lado y un vértice en común.*

algebraic addition *(98)*	The combining of positive and negative numbers to form a sum. *We use **algebraic addition** to find the sum of −3, +2, and −11:* $$(-3) + (+2) + (-11) = -12$$
suma algebraica	La combinación de números positivos y negativos para formar una suma. *Usamos la **suma algebraica** para calcular la suma de −3, +2 y −11.*

alternate exterior angles *(94)*	A special pair of angles formed when a transversal intersects two lines. Alternate exterior angles lie on opposite sides of the transversal and are outside the two intersected lines. 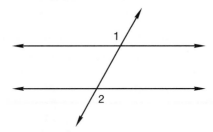 *∠1 and ∠2 are **alternate exterior angles.** When a transversal intersects parallel lines, as in this figure, **alternate exterior angles** have the same measure.*
ángulos alternos externos	Par especial de ángulos que se forman cuando una transversal interseca dos líneas rectas. Los ángulos alternos externos se ubican en lados opuestos de la transversal y fuera de las dos rectas intersecadas. *∠1 y ∠2 son **ángulos alternos externos.** Cuando una transversal interseca rectas paralelas, como en esta figura, los **ángulos alternos externos** tienen la misma medida.*

alternate interior angles *(94)*	A special pair of angles formed when a transversal intersects two lines. Alternate interior angles lie on opposite sides of the transversal and are inside the two intersected lines. 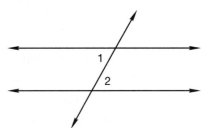 *∠1 and ∠2 are **alternate interior angles.** When a transversal intersects parallel lines, as in this figure, **alternate interior angles** have the same measure.*
ángulos alternos internos	Par especial de ángulos que se forman cuando una transversal interseca dos líneas rectas. Los ángulos alternos internos se ubican en lados opuestos de la transversal y entre las dos rectas intersecadas. *∠1 y ∠2 son **ángulos alternos internos.** Cuando una transversal interseca rectas paralelas, como en esta figura, los **ángulos alternos internos** tienen la misma medida.*

a.m. *(maintained)*	The period of time from midnight to just before noon. *I get up at 7 **a.m.**, which is 7 o'clock in the morning.*
a.m.	Período de tiempo desde la medianoche hasta justo antes del mediodía. *Me levanto a las 7 **a.m.** lo cual es las 7 de la mañana.*

angle *(20)*	The opening that is formed when two lines, rays, or segments intersect. 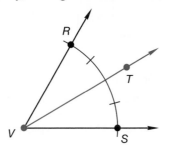 *These rays form an **angle.***
ángulo	Abertura que se forma cuando se intersecan dos rectas, rayos o segmentos de recta. *Estos rayos forman un **ángulo.***

angle bisector *(maintained)*	A line, ray, or segment that divides an angle into two congruent parts.  $\overrightarrow{VT}$ *is an **angle bisector.** It divides ∠RVS in half.*
bisectriz	Recta, rayo o segmento de recta que divide un ángulo en dos partes congruentes. *VT es una **bisectriz.** Divide el ∠RVS por la mitad.*

area *(23)*	The number of square units needed to cover a surface. 5 in. 2 in. *The **area** of this rectangle is 10 square inches.*
área	El número de unidades cuadradas que se necesita para cubrir una superficie. *El **área** de este rectángulo mide 10 pulgadas cuadradas.*

Associative Property of Addition *(2)*	The grouping of addends does not affect their sum. In symbolic form, $a + (b + c) = (a + b) + c$. Unlike addition, subtraction is not associative. <table><tr><td>$(8 + 4) + 2 = 8 + (4 + 2)$</td><td>$(8 - 4) - 2 \neq 8 - (4 - 2)$</td></tr><tr><td>*Addition is **associative.***</td><td>*Subtraction is not **associative.***</td></tr></table>
propiedad asociativa de la suma	La agrupación de los sumandos no altera la suma. En forma simbólica, $a + (b + c) = (a + b) + c$. A diferencia de la suma, la resta no es asociativa. <table><tr><td>$(8 + 4) + 2 = 8 + (4 + 2)$</td><td>$(8 - 4) - 2 \neq 8 - (4 - 2)$</td></tr><tr><td>*La suma es **asociativa.***</td><td>*La resta no es **asociativa.***</td></tr></table>

Associative Property of Multiplication *(2)*	The grouping of factors does not affect their product. In symbolic form, $a \times (b \times c) = (a \times b) \times c$. Unlike multiplication, division is not associative. <table><tr><td>$(8 \times 4) \times 2 = 8 \times (4 \times 2)$</td><td>$(8 \div 4) \div 2 \neq 8 \div (4 \div 2)$</td></tr><tr><td>*Multiplication is **associative.***</td><td>*Division is not **associative.***</td></tr></table>
propiedad asociativa de la multiplicación	La agrupación de los factores no altera el producto. En forma simbólica, $a \times (b \times c) = (a \times b) \times c$. A diferencia de la multiplicación, la división no es asociativa. <table><tr><td>$(8 \times 4) \times 2 = 8 \times (4 \times 2)$</td><td>$(8 \div 4) \div 2 \neq 8 \div (4 \div 2)$</td></tr><tr><td>*La multiplicación es **asociativa.***</td><td>*La división no es **asociativa.***</td></tr></table>

average (10)	*See* **mean.**
media	*Ver* **promedio.**

B

bar graph(s) (Inv. 3)	Displays numerical information with shaded rectangles or bars.

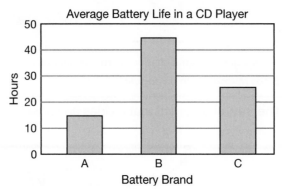

*This **bar graph** shows data for three different brands of batteries.*

gráfica(s) de barras	Muestra información numérica con rectángulos o barras sombreados. *Esta **gráfica de barras** muestra datos para tres marcas diferentes de baterías.*

base (32, 66)	**1.** A designated side or face of a geometric figure.

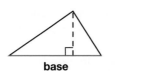 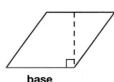

2. The lower number in an exponential expression.

base $\longrightarrow$ 5^3 $\longleftarrow$ exponent

5^3 *means* $5 \times 5 \times 5$, *and its value is 125.*

base	**1.** Lado (o cara) determinado de una figura geométrica. **2.** El número inferior en una expresión exponencial.

base $\longrightarrow$ 5^3 $\longleftarrow$ exponente

5^3 *significa* $5 \times 5 \times 5$ *, y su valor es 125.*

bias (Inv. 2)	A slant toward a particular point of view caused by a sampling procedure. *To avoid any **bias** in a survey about the candidates for mayor, the researchers chose a representative sample and used neutral wording for the survey questions.*
sesgo	Preferencia hacia un punto de vista en particular causado por un procedimiento de recolección de datos. *Para evitar **sesgo** en una encuesta acerca de los candidatos para alcalde, los investigadores escogieron una muestra representativa y usaron lenguaje neutral para las preguntas de la encuesta.*

bimodal *(Inv. 4)*	Having two modes.
	The numbers 5 and 7 are the *modes of the data at right.* 5, 1, 44, 5, 7, 13, 9, 7 *This set of data is **bimodal.***
bimodal	Que tiene dos modas.
	Los números 5 y 7 son las modas de los datos de arriba. Este conjunto de datos *es **bimodal.***

bisect *(maintained)*	To divide a segment or angle into two equal halves.

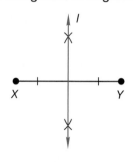 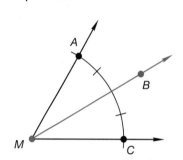

*Line l **bisects** $\overline{XY}$.*   *Ray MB **bisects** $\angle AMC$.*

bisecar	Dividir un segmento o un ángulo en dos mitades iguales.
	*La recta l **bisecta** $\overline{XY}$. El rayo MB **bisecta** $\angle AMC$*

C

cancel *(65)*	The process of reducing a fraction by matching equivalent factors from both the numerator and denominator.

$$\frac{14}{28} = \frac{\cancel{7} \cdot \cancel{2}}{\cancel{7} \cdot \cancel{2} \cdot 2} = \frac{1}{2}$$

cancelar	Proceso de reducción de una fracción mediante la identificación de factores equivalentes tanto en el numerador como el denominador.

capacity *(81)*	The amount of liquid a container can hold.
	*Cups, gallons, and liters are units of **capacity.***
capacidad	Cantidad de líquido que puede contener un recipiente.
	*Tazas, galones y litros son medidas de **capacidad.***

Celsius scale *(76)*	A scale used on some thermometers to measure temperature.
	*On the **Celsius scale,** water freezes at 0°C and boils at 100°C.*
escala Celsius	Escala que se usa en algunos termómetros para medir la temperatura.
	*En la **escala Celsius,** el agua se congela a 0°C y hierve a 100°C.*

central angle *(Inv. 4)*	An angle whose vertex is the center of a circle.
	*$\angle AOC$ is a **central angle.***

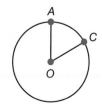

ángulo central	Ángulo cuyo vértice es el centro de un círculo. ∠*AOC es un **ángulo central**.*
chance (54)	A way of expressing the likelihood of an event; the probability of an event expressed as a percentage. *The **chance** of snow is 10%. It is not likely to snow.* *There is an 80% **chance** of rain. It is likely to rain.*
posibilidad	Modo de expresar la probabilidad de ocurrencia de un suceso; la probabilidad de un suceso expresada como porcentaje. *La **posibilidad** de nieve es del 10%. Es poco probable que caiga nieve.* *Hay un 80% de **posibilidad** de lluvia. Es muy probable que llueva.*
circle (42)	A closed, curved shape in which all points on the shape are the same distance from its center. 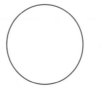 circle
círculo	Curva cerrada en la cual todos los puntos de la figura están a la misma distancia de su centro.
circle graph (30)	A method of displaying data, often used to show information about percentages or parts of a whole. A circle graph is made of a circle divided into sectors. **Class Test Grades** 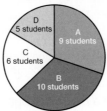 *This **circle graph** shows data for a class's test grades.*
gráfica circular	Método para representar datos. Se usa a menudo para mostrar información como porcentajes o partes de un todo. Una gráfica circular está formada por un círculo dividido en sectores. *Esta **gráfica circular** representa los datos de las calificaciones del examen de una clase.*
circumference (42)	The perimeter of a circle. 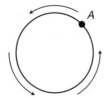 *If the distance from point A around to point A is 3 inches, then the **circumference** of the circle is 3 inches.*
circunferencia	Perímetro de un círculo. *Si la distancia desde el punto A alrededor del círculo hasta el punto A es 3 pulgadas, entonces la **circunferencia** del círculo mide 3 pulgadas.*

closed-option survey *(Inv. 2)*	A survey in which the possible responses are limited.

What is your favorite pet?
☐ dog
☐ cat
☐ bird
☐ fish

closed-option survey

encuesta de opción cerrada	Encuesta en la cual las respuestas posibles son limitadas.

coefficient *(Inv. 9)*	In common use, the number that multiplies the variable(s) in an algebraic term. If no number is specified, the coefficient is 1.

In the term $-3x$, *the* **coefficient** *is* -3.

In the term y^2, *the* **coefficient** *is 1.* |
| **coeficiente** | En uso común, el número que multiplica la(s) variable(s) en un término algebraico. Si no se especifica un número, el coeficiente es 1.

En el término $-3x$, *el* **coeficiente** *es* -3.

En el término y^2, *el* **coeficiente** *es 1.* |

common denominator *(52)*	A number that is the denominator of two or more fractions.

The fractions $\frac{2}{5}$ *and* $\frac{3}{5}$ *have* **common denominators.** |
| **denominador común** | Número que es el denominador de dos o más fracciones.

Las fracciones $\frac{2}{5}$ *y* $\frac{3}{5}$ *tienen* **denominadores comunes.** |

Commutative Property of Addition *(3)*	Changing the order of addends does not affect their sum. In symbolic form, $a + b = b + a$. Unlike addition, subtraction is not commutative.

$8 + 2 = 2 + 8$ $8 - 2 \neq 2 - 8$

Addition is **commutative.**   *Subtraction is not* **commutative.** |
| **propiedad conmutativa de la suma** | El orden de los sumandos no altera la suma. En forma simbólica, $a + b = b + a$. A diferencia de la suma, la resta no es conmutativa.

$8 + 2 = 2 + 8$ $8 - 2 \neq 2 - 8$

La suma es **conmutativa.**   *La resta no es* **conmutativa.** |

Commutative Property of Multiplication *(3)*	Changing the order of factors does not affect their product. In symbolic form, $a \times b = b \times a$. Unlike multiplication, division is not commutative.

$8 \times 2 = 2 \times 8$ $8 \div 2 \neq 2 \div 8$

Multiplication is **commutative.**   *Division is not* **commutative.** |
| **propiedad conmutativa de la multiplicación** | El orden de los factores no altera el producto. En forma simbólica, $a \times b = b \times a$. A diferencia de la multiplicación, la división no es conmutativa.

$8 \times 2 = 2 \times 8$ $8 \div 2 \neq 2 \div 8$

La multiplicación es **conmutativa.**   *La división no es* **conmutativa.** |

compass *(maintained)*	A tool used to draw circles and arcs.

radius gauge

pivot point

marking point

compás	Instrumento para dibujar círculos y arcos.

---

complementary angles
(60)

Two angles whose sum is 90°.

∠A and ∠B are
complementary angles.

ángulos complementarios

Dos ángulos cuya suma es 90°.

∠A y ∠B son ***ángulos complementarios.***

---

complement of an event
(54)

The opposite of an event. The complement of event B is "not B." The probability of an event and the probability of its complement add up to 1.

complemento de un suceso

El opuesto de un suceso. El complemento del suceso B es "no B". La probabilidad de un suceso y la probabilidad de su complemento suman 1.

---

composite number
(61)

A counting number greater than 1 that is divisible by a number other than itself and 1. Every composite number has three or more factors and can be expressed as a product of two or more prime factors.

*9 is divisible by 1, 3, and 9. It is **composite.***

*11 is divisible by 1 and 11. It is not **composite.***

número compuesto

Número natural mayor que 1, divisible entre algún otro número distinto de sí mismo y de 1. Cada número compuesto tiene tres o más factores y puede ser expresado como producto de dos o más factores primos.

*9 es divisible entre 1, 3 y 9. Es **compuesto.***

*11 es divisible entre 1 y 11. No es **compuesto.***

---

compound events
(80)

In probability, the result of combining two or more simple events.

*An outcome of one coin flip is a simple event. An outcome of more than one flip is a **compound event.***

sucesos compuestos

En probabilidad, el resultado de combinar dos o más sucesos simples.

*El resultado de lanzar una moneda es un suceso simple. El resultado de más de un lanzamiento es un **suceso compuesto.***

---

compound experiments
(Inv. 6)

Experiments that contain more than one part performed in order.

experimentos compuestos

Experimentos que contienen más de una parte que se realiza en orden.

---

compound interest _(maintained)_	Interest that pays on previously earned interest.

Compound Interest	**Simple Interest**
$100.00 principal	$100.00 principal
+ $6.00 first-year interest (6% of $100)	$6.00 first-year interest (6% of $100)
$106.00 total after one year	+ $6.00 second-year interest (6% of $100)
+ $6.36 second-year interest (6% of $106)	$112.00 total after two years
$112.36 total after two years	

interés compuesto	Interés que se paga sobre el interés obtenido previamente.

compound outcomes _(Inv. 6)_ **resultados compuestos**	The outcomes to a compound experiment.
	Los resultados de un experimento compuesto.

concentric circles _(maintained)_	Two or more circles with a common center.

common
center
of four
**concentric
circles**

círculos concéntricos	Dos o más círculos con un centro en común.

cone _(maintained)_	A three-dimensional solid with a circular base and a single vertex.

cone

cono	Sólido tridimensional de base circular y con un sólo vértice.

congruent _(77)_	Having the same size and shape.

These polygons are **congruent.** They have the same size and shape.

congruentes	Que tienen el mismo tamaño y forma.

Estos polígonos son **congruentes.** Tienen igual tamaño y forma.

coordinate(s) _(Inv. 5)_

1. A number used to locate a point on a number line.

A

-3   -2   -1   0   1   2   3

The **coordinate** of point A is −2.

2. An ordered pair of numbers used to locate a point in a coordinate plane.

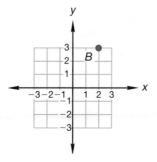

The **coordinates** of point B are (2, 3). The x-coordinate is listed first, the y-coordinate second.

coordenada(s)	**1.** Número que se utiliza para ubicar un punto sobre una recta numérica.

*La **coordenada** del punto A es −2.*

2. Par ordenado de números que se utiliza para ubicar un punto sobre un plano coordenado.

*Las **coordenadas** del punto B son (2, 3). La coordenada x se escribe primero, seguida de la coordenada y.*

---

coordinate plane
(Inv. 5)

A grid on which any point can be identified by an ordered pair of numbers.

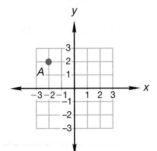

Point A is located at (−2, 2) on this **coordinate plane.**

plano coordenado

Cuadrícula en que cualquier punto se puede identificar con un par ordenado de números.

*El punto A está ubicado en la posición (−2, 2) sobre este **plano coordenado.***

---

corresponding angles
(94)

A special pair of angles formed when a transversal intersects two lines. Corresponding angles lie on the same side of the transversal and are in the same position relative to the two intersected lines.

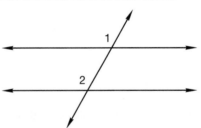

∠1 and ∠2 are **corresponding angles.** *When a transversal intersects parallel lines, as in this figure, **corresponding angles** have the same measure.*

ángulos correspondientes

Par de ángulos especiales que se forma cuando una transversal corta dos rectas. Los ángulos correspondientes se encuentran al mismo lado de la transversal y en la misma posición con respecto a las dos rectas que corta.

*∠1 y ∠2 son **ángulos correspondientes.** Cuando una transversal corta dos rectas paralelas, como en esta ilustración, los **ángulos correspondientes** tienen la misma medida.*

---

corresponding parts
(Inv. 10)

Sides or angles that occupy the same relative positions in similar polygons.

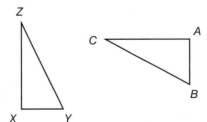

$\overline{BC}$ **corresponds** to $\overline{YZ}$.
∠A **corresponds** to ∠X.

partes correspondientes	Lados o ángulos que ocupan la misma posición relativa en polígonos semejantes. $\overline{BC}$ es **correspondiente** con $\overline{YZ}$. $\angle A$ es **correspondiente** con $\angle X$.

---

counting numbers
(1)

The numbers used to count; the members of the set {1, 2, 3, 4, 5, ...}. Also called *natural numbers*.

*1, 24, and 108 are **counting numbers.***

*−2, 3.14, 0, and $2\frac{7}{9}$ are not **counting numbers.***

números de conteo

Números que se utilizan para contar; los elementos del conjunto {1, 2, 3, 4, 5, ...}. También llamados *números naturales*.

*1, 24 y 108 son **números de conteo.***

*−2, 3.14, 0, y $2\frac{7}{9}$ no son **números de conteo.***

---

cross product
(52)

The product of the numerator of one fraction and the denominator of another.

$5 \times 16 = 80 \qquad\qquad 20 \times 4 = 80$

$$\frac{16}{20} = \frac{4}{5}$$

*The **cross products** of these two fractions are equal.*

productos cruzados

Producto del numerador de una fracción por el denominador de otra.

*Los **productos cruzados** de estas dos fracciones son iguales.*

---

cube
(114)

A three-dimensional solid with six square faces. Adjacent faces are perpendicular and opposite faces are parallel.

cube

cubo

Sólido tridimensional con seis caras cuadradas. Las caras adyacentes son perpendiculares y las caras opuestas son paralelas.

---

cylinder
(113)

A three-dimensional solid with two circular bases that are opposite and parallel to each other.

cylinder

cilindro

Sólido tridimensional con dos bases circulares que son opuestas y paralelas entre sí.

D

---

data
(10)

Information that is gathered and organized in a way that conclusions can be drawn from it.

datos

Información que se reune y organiza de manera que se pueda llegar a conclusiones a partir de ella.

---

data points
(Inv. 4)

Individual measurements or numbers in a set of data.

puntos de datos

Medidas individuales o números en un conjunto de datos.

decimal number (28)	A numeral that contains a decimal point. *23.94 is a **decimal number** because it contains a decimal point.*
número decimal	Número que contiene un punto decimal. *23.94 es un **número decimal**, porque tiene punto decimal.*
decimal places (28)	Places to the right of a decimal point. *5.47 has two **decimal places.*** *6.3 has one **decimal place.*** *8 has no **decimal places.***
cifras decimales	Números ubicados a la derecha del punto decimal. *5.47 tiene dos **cifras decimales.*** *6.3 tiene una **cifra decimal.*** *8 no tiene **cifras decimales.***
decimal point (28)	The symbol in a decimal number used as a reference point for place value. *34.15* ↑ **decimal point**
punto decimal	Símbolo en un número decimal, que se usa como punto de referencia para el valor posicional.

degree (°)
(60, 76)

1. A unit for measuring angles.

*There are 90 **degrees** (90°) in a right angle.*

*There are 360 **degrees** (360°) in a circle.*

2. A unit for measuring temperature.

— 100   Water boils.

°C

— 0   Water freezes.

*There are 100 **degrees** (100°) between the freezing and boiling points of water on the Celsius scale.*

grado (°)

1. Unidad para medir ángulos.

*Un ángulo recto mide 90 **grados** (90°).*   *Un círculo mide 360 **grados** (360°).*

2. Unidad para medir la temperatura.

*Hay 100 **grados** (100°) de diferencia entre los puntos de ebullición y congelación del agua en la escala Celsius.*

denominator *(Inv. 1)*	The bottom term of a fraction. $$\frac{5}{9} \begin{matrix} \leftarrow \textit{numerator} \\ \leftarrow \textbf{\textit{denominator}} \end{matrix}$$
denominador	El número inferior de una fracción.
dependent events *(80)*	In probability, events that are not independent because the outcome of one event affects the probability of the other event. *If a bag contains 4 red marbles and 2 blue marbles and a marble is drawn from the bag twice without replacing the first draw, then the probabilities for the second draw is **dependent** upon the outcome of the first draw.*
sucesos dependientes	En probabilidad, sucesos que no son independientes porque la ocurrencia de un suceso afecta la probabilidad del otro suceso. *Si una bolsa contiene 4 canicas rojas y 2 canicas azules y se saca una canica de la bolsa dos veces sin que la primera sea reemplazada, entonces la probabilidad del segundo suceso es **dependiente** del resultado del primer suceso.*
diameter *(42)*	The distance across a circle through its center. 3 in. *The **diameter** of this circle is 3 inches.*
diámetro	Distancia entre dos puntos opuestos de un círculo a través de su centro. *El **diámetro** de este círculo mide 3 pulgadas.*
difference *(4)*	The result of subtraction. $12 - 8 = 4$ *The **difference** in this problem is 4.*
diferencia	Resultado de una resta. $12 - 8 = 4$ *La **diferencia** en este problema es 4.*
digit *(1)*	Any of the symbols used to write numbers: 0, 1, 2, 3, 4, 5, 6, 7, 8, 9. *The last **digit** in the number 7862 is 2.*
dígito	Cualquiera de los símbolos que se utilizan para escribir números: 0, 1, 2, 3, 4, 5, 6, 7, 8, 9. *El último **dígito** del número 7862 es 2.*
disjoint events *(70)*	Two or more events in a probability experiment that cannot happen at the same time. *If a bag of marbles contains 6 red, 8 blue and 10 green marbles and if one marble will be drawn, the events "draw a red" and "draw a blue" are **disjoint events** since both colors cannot be selected in one draw.*
sucesos excluyentes	Dos o más sucesos en un experimento de probabilidad que no pueden ocurrir al mismo tiempo. *Si una bolsa de canicas contiene 6 rojas, 8 azules y 10 canicas verdes y si se saca una canica, el suceso de "sacar una roja y "sacar una azul" son **sucesos excluyentes** ya que ambos colores no pueden ser seleccionados al mismo tiempo.*

Distributive Property
(88)

A number times the sum of two addends is equal to the sum of that same number times each individual addend:

$$a \times (b + c) = (a \times b) + (a \times c).$$

$$8 \times (2 + 3) = (8 \times 2) + (8 \times 3)$$

*Multiplication is **distributive** over addition.*

propiedad distributiva

Un número multiplicado por la suma de dos sumandos es igual a la suma de los productos de ese número por cada uno de los sumandos:

$$a \times (b + c) = (a \times b) + (a \times c).$$

$$8 \times (2 + 3) = (8 \times 2) + (8 \times 3)$$

*La multiplicación es **distributiva** con respecto a la suma.*

---

dividend
(5)

A number that is divided.

$$12 \div 3 = 4 \qquad 3\overline{)12} \;\; {}^{4} \qquad \frac{12}{3} = 4$$

The **dividend** is 12 in each of these problems.

dividendo

Número que se divide.

$$12 \div 3 = 4 \qquad 3\overline{)12} \;\; {}^{4} \qquad \frac{12}{3} = 4$$

El **dividendo** es 12 en cada una de estas operaciones.

---

divisible
(5)

Able to be divided by a whole number without a remainder.

$$4\overline{)20} \;\; {}^{5}$$

The number 20 is **divisible** by 4, since 20 ÷ 4 has no remainder.

$$3\overline{)20} \;\; {}^{6\,R\,2}$$

The number 20 is not **divisible** by 3, since 20 ÷ 3 has a remainder.

divisible

Número que se puede dividir entre un número entero sin dejar residuo.

$$4\overline{)20} \;\; {}^{5}$$

El número 20 es **divisible** entre 4, ya que 20 ÷ 4 no tiene residuo.

$$3\overline{)20} \;\; {}^{6\,R\,2}$$

El número 20 no es **divisible** entre 3, ya que 20 ÷ 3 tiene residuo.

---

divisor
(5)

1. A number by which another number is divided.

$$12 \div 3 = 4 \qquad 3\overline{)12} \;\; {}^{4} \qquad \frac{12}{3} = 4$$

The **divisor** is 3 in each of these problems.

2. A factor of a number.

*2 and 5 are **divisors** of 10.*

divisor

1. Número que divide a otro en una división.

$$12 \div 3 = 4 \qquad 3\overline{)12} \;\; {}^{4} \qquad \frac{12}{3} = 4$$

El **divisor** es 3 en cada una de estas operaciones.

2. Factor de un número.

*2 y 5 son **divisores** de 10.*

double bar graph *(Inv. 4)*	Displays numerical information with shaded rectangles or bars like a regular bar graph, but gives two sets of information for each item.

*This **double bar graph** shows the data for new boy and girl club members in the different grade levels.*

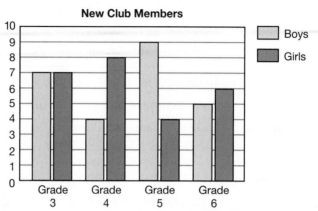

gráfica de barra doble	Muestra información numérica con rectángulos sombreados o barras como una gráfica de barras común, pero da dos conjuntos de información para cada objeto.

*Esta **gráfica de barra doble** muestra los datos de los distintos grados escolares de los nuevos miembros de un club de niñas y niños.*

E

edge *(maintained)*	A line segment formed where two faces of a polyhedron intersect.

*One **edge** of this cube is colored blue. A cube has 12 **edges.***

arista	Segmento de recta formado por la intersección de dos caras de un poliedro.

*Una **arista** de este cubo está coloreada en azul. Un cubo tiene 12 **aristas.***

endpoint *(maintained)*	A point at which a segment ends.

*Points A and B are the **endpoints** of segment AB.*

extremo	Punto donde termina un segmento de recta.

*Los puntos A y B son los **extremos** del segmento AB.*

equation *(5)*	A statement that uses the symbol "=" to show that two quantities are equal.

$$x = 3 \qquad 3 + 7 = 10 \qquad\qquad 4 + 1 \qquad x < 7$$

equations   not **equations**

ecuación	Enunciado que usa el símbolo "=" para indicar que dos cantidades son iguales.

equilateral triangle *(90)*	A triangle in which all sides are the same length.

*This is an **equilateral triangle.**
All of its sides are the same length.*

triángulo equilátero	Triángulo que tiene todos sus lados de la misma longitud.

*Éste es un **triángulo equilátero.** Todos sus lados tienen la misma longitud.*

equivalent fractions *(37)*	Different fractions that name the same amount.

$$\frac{1}{2}\ \boxed{\ \blacksquare\ \ \square\ } = \boxed{\ \blacksquare\ \blacksquare\ \square\ \square\ }\ \frac{2}{4}$$

$\frac{1}{2}$ and $\frac{2}{4}$ are **equivalent fractions.**

fracciones equivalentes

Fracciones diferentes que representan la misma cantidad.

$\frac{1}{2}$ y $\frac{2}{4}$ son **fracciones equivalentes.**

---

estimate *(21)*

To determine an approximate value.

*We **estimate** that the sum of 199 and 205 is about 400.*

estimar

Encontrar un valor aproximado.

Estimamos *que la suma de 199 y 205 es aproximadamente 400.*

---

evaluate *(6)*

To find the value of an expression

*To **evaluate** a + b for a = 7 and b = 13, we replace a with 7 and b with 13:*

$$7 + 13 = 20$$

evaluar

Calcular el valor de una expresión.

*Para **evaluar** a + b, con a = 7 y b = 13, se reemplaza a con 7 y b con 13:*

$$7 + 13 = 20$$

---

even numbers *(1)*

Numbers that can be divided by 2 without a remainder; the members of the set $\{..., -4, -2, 0, 2, 4, ...\}$.

***Even numbers** have 0, 2, 4, 6, or 8 in the ones place.*

números pares

Números que se pueden dividir entre 2 sin dar un residuo; los elementos del conjunto $\{..., -4, -2, 0, 2, 4, ...\}$.

*Los **números pares** tienen 0, 2, 4, 6 u 8 en el lugar de las unidades.*

---

event *(54)*

Outcome(s) resulting from an experiment or situation.

- *Events that are certain to occur have a probability of 1.*
- *Events that are certain not to occur have a probability of zero.*
- *Events that are uncertain have probabilities that fall anywhere between zero and one.*

suceso

Resultado(s) de un experimento o situación.

- *Los sucesos que seguro ocurrirán tienen una probabilidad de 1.*
- *Los sucesos que seguro no ocurrirán tienen una probabilidad de cero.*
- *Los sucesos que son inciertos tienen una probabilidad que varía del 0 al 1.*

---

expanded form *(maintained)*

*See **expanded notation.***

forma desarrollada

*Ver **notación desarrollada.***

expanded notation *(maintained)*	A way of writing a number that shows the value of each digit. Also called *expanded form.* *Two ways to write 234 in expanded form are* <center>200 + 30 + 4 and</center> <center>(2 × 100) + (3 × 10) + (4 × 1)</center>
notación desarrollada	Una forma de escribir un número que muestra el valor de cada dígito. También llamada *forma desarrollada.* *Dos formas de escribir 234 en forma desarrollada son* <center>200 + 30 + 4 y</center> <center>(2 × 100) + (3 × 10) + (4 × 1)</center>
experimental probability *(Inv. 7)*	The probability of an event occurring as determined by experimentation. *If we roll a number cube 100 times and get 22 threes, the* **experimental probability** *of getting three is* $\frac{22}{100}$*, or* $\frac{11}{50}$*.*
probabilidad experimental	Probabilidad de ocurrencia de un suceso en base a la experimentación. *Si se lanza un dado 100 veces y salen 22 veces tres, la* **probabilidad experimental** *de obtener tres es* $\frac{22}{100}$*,* $\frac{11}{50}$*.*
exponent *(32)*	The upper number in an exponential expression; it shows how many times the base is to be used as a factor. *base* ⟶ 5^3 ⟵ *exponent* *5^3 means 5 × 5 × 5, and its value is 125.*
exponente	El índice de una expresión exponencial; indica cuántas veces debe usarse la base como factor. *5^3 significa 5 × 5 × 5 y su valor es 125.*
exponential expression *(71)*	An expression containing a base and an exponent such as 4^3 or s^2. *The* **exponential expression** *4^3 is evaluated by using 4 as a factor 3 times. Its value is 64.* <center>$4^3 = 4 × 4 × 4 = 64$</center>
expresión exponencial	Expresión que contiene una base y un exponente como 4^3 ó s^2. *La* **expresión exponencial** *4^3 se calcula usando 3 veces el 4 como factor. Su valor es 64.*
expression *(6)*	A combination of numbers and/or variables by operations, but not including an equal or inequality sign. <center>equation inequality</center><center>$3x + 2y (x − 1)^2$ $y = 3x − 1$ $x < 4$</center><center>**expressions** not **expressions**</center>
expresión	Una combinación de números o valores en operaciones, pero que no incluyen un signo de igual o de desigualdad.
exterior angle *(94)*	In a polygon, the supplementary angle of an interior angle. 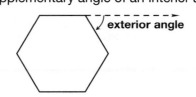 <center>**exterior angle**</center>
ángulo exterior	En un polígono, el ángulo suplementario de un ángulo interior.

F

face
(Inv. 5)

A flat surface of a geometric solid.

*One **face** of the cube is shaded.*
*A cube has six **faces.***

cara

Superficie plana de un sólido geométrico.

*Una **cara** del cubo está sombreada.*

*Un cubo tiene seis **caras.***

---

fact family
(3)

A group of three numbers related by addition and subtraction or by multiplication and division.

*The numbers 3, 4, and 7 are a **fact family.** They make these four facts:*

$$3 + 4 = 7 \quad 4 + 3 = 7 \quad 7 - 3 = 4 \quad 7 - 4 = 3$$

familia de operaciones

Grupo de tres números relacionados por sumas y restas o por multiplicaciones y divisiones.

*Los números 3, 4 y 7 forman una **familia de operaciones.** Con ellos se pueden formar estas cuatro operaciones:*

$$3 + 4 = 7 \quad 4 + 3 = 7 \quad 7 - 3 = 4 \quad 7 - 4 = 3$$

---

factor
(5)

1. Noun: One of two or more numbers that are multiplied.

$$3 \times 5 = 15 \qquad \textit{The **factors** in this problem are 3 and 5.}$$

2. Noun: A whole number that divides another whole number without a remainder.

*The numbers 3 and 5 are **factors** of 15.*

3. Verb: To write as a product of factors.

*We can **factor** the number 15 by writing it as 3×5.*

factor
(factorizar)

1. Sustantivo: Uno de dos o más números en una operación de multiplicación.

$$3 \times 5 = 15 \qquad \textit{Los **factores** en esta operación son el 3 y el 5.}$$

2. Sustantivo: Número entero que divide a otro número entero sin residuo.

*Los números 3 y 5 son **factores** de 15.*

3. Verbo: Escribir como un producto de factores.

*Se puede **factorizar** el número 15 escribiéndolo como 3×5.*

---

factor tree
(61)

A method of finding all the prime factors of a number.

*The numbers on each branch of this **factor tree** are factors of the number 210. Each number at the end of a branch is a prime factor of 210.*

The prime factors of 210 are 2, 3, 5 and 7.

árbol de factores

Un método para encontrar todos los factores primos de un número.
Los factores primos de 210 son 2, 3, 5 y 7.

Fahrenheit scale (81)	A scale used on some thermometers to measure temperature.

Fahrenheit scale *(81)*

A scale used on some thermometers to measure temperature.

*On the **Fahrenheit scale,** water freezes at 32°F and boils at 212°F.*

escala Fahrenheit

Escala que se usa en algunos termómetros para medir la temperatura.

*En la **escala Fahrenheit,** el agua se congela a 32°F y hierve a 212°F.*

---

formula *(7)*

An equation used to calculate a desired result.

*The **formula** for the area of a circle is $A = \pi r^2$, where A is the area and r is the length of the radius.*

fórmula

Una ecuación que se utiliza para calcular un resultado dado.

*La **fórmula** para el área de un círculo es $A = \pi r^2$, donde A es el área y r es la longitud del radio.*

---

fraction *(Inv. 1)*

1. A number that names part of a whole.

$\frac{1}{4}$ of the circle is shaded.

$\frac{1}{4}$ is a **fraction.**

2. A number in the form $\frac{numerator}{denominator}$.

fracción

1. Número que representa una parte de un entero.

$\frac{1}{4}$ del círculo está sombreado.

$\frac{1}{4}$ es una **fracción.**

2. Un número en la forma de $\frac{numerador}{denominador}$.

---

frequency table *(Inv. 2)*

Displays the number of times a value occurs in data.

Daily Temperature Highs in October

Temperature (°F)	Tally	Frequency
81–85	||||| ||||| |||||		16
76–80	||||| ||||	9
71–75	|||||	5
65–70			1

*This **frequency table** shows data for temperatures in October.*

tabla de frecuencias

Muestra el número de veces que se repite un dato.

*Esta **tabla de frecuencias** muestra datos de las temperaturas que hubo en octubre.*

---

function *(Inv. 5)*

A rule for using one number (an input) to calculate another number (an output). Each input produces only one output.

$y = 3x$

x	y
3	9
5	15
7	21
10	30

*There is exactly one resulting number for every number we multiply by 3. Thus, $y = 3x$ is a **function.***

función

Regla para usar un número (una entrada) para calcular otro número (una salida). Cada valor de entrada origina sólo un valor de salida.

*Hay exactamente un número que le corresponde a cada número que se multiplica por 3. Por lo tanto, $y = 3x$ es una **función.***

G

geometric solid
(114)

A three-dimensional geometric figure.

geometric solids

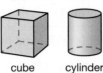

cube   cylinder

not **geometric solids**

 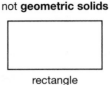

circle   rectangle   hexagon

sólido geométrico

Una figura geométrica tridimensional.

---

graph
(Inv. 3, 5)

1. Noun: A diagram, such as a bar graph, a circle graph (pie chart), or a line graph, that displays quantitative information.

Rainy Days

Days

8
6
4
2

Jan.   Feb.   Mar.   Apr.

bar **graph**

Hair Colors of Students

Brown 4   Red 2   Black 6   Blond 4

circle **graph**

2. Noun: A point, line, or curve on a coordinate plane.

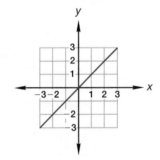

The **graph** of the equation $y = x$

3. Verb: To draw a point, line, or curve on a coordinate plane.

gráfica
(graficar)

1. Sustantivo: Diagrama, tal como una gráfica de barras, una gráfica circular o una gráfica lineal, que representa información cuantitativa.

2. Sustantivo: Punto, recta o curva sobre un plano coordenado.

Gráfica de la ecuación $y = x$

3. *Verbo:* Dibujar un punto, recta o curva sobre un plano coordenado.

---

greatest common factor (GCF)
(12)

The largest whole number that is a factor of two or more given numbers.

The factors of 12 are 1, 2, 3, 4, 6, and 12.

The factors of 18 are 1, 2, 3, 6, 9, and 18.

*The **greatest common factor** of 12 and 18 is 6.*

máximo común divisor (MCD)

Es el número entero más grande que es factor de dos o más números.

Los factores de 12 son 1, 2, 3, 4, 6 y 12.

Los factores de 18 son 1, 2, 3, 6, 9 y 18.

*El **máximo común divisor** de 12 y 18 es 6.*

H

height
(66)

The perpendicular distance from the base to the opposite side of a parallelogram or trapezoid; from the base to the opposite face of a prism or cylinder; or from the base to the opposite vertex of a triangle, pyramid, or cone.

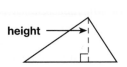 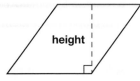 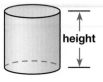

altura

Distancia perpendicular desde la base al lado opuesto de un paralelogramo o trapecio; desde la base a la cara opuesta de un prisma o cilindro; o desde la base al vértice opuesto de un triángulo, pirámide o cono.

---

histogram
(Inv. 2)

A method of displaying a range of data. A histogram is a special type of bar graph that displays data in intervals of equal size with no space between bars.

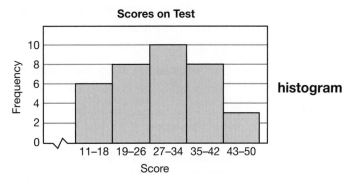

histograma

Método para representar un conjunto de datos. Un histograma es un tipo especial de gráfica de barras que muestra los datos a intervalos de igual tamaño y sin espacio entre las barras.

---

horizontal
(Inv. 5)

Parallel to the horizon; perpendicular to vertical.

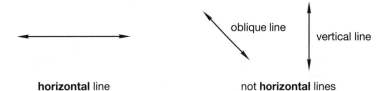

horizontal

Paralelo al horizonte; perpendicular a una vertical.

I

Identity Property of Addition
(3)

The sum of any number and 0 is equal to the initial number. In symbolic form, $a + 0 = a$. The number 0 is referred to as the *additive identity*.

The **Identity Property of Addition** is shown by this statement:

$$13 + 0 = 13$$

propiedad de identidad de la suma

La suma de cualquier número más 0 es igual al número mismo. En forma simbólica, $a + 0 = a$. El 0 se conoce como identidad aditiva.

La **propiedad de identidad de la suma** se muestra en el siguiente enunciado:

$$13 + 0 = 13$$

Identity Property of Multiplication *(3)*	The product of any number and 1 is equal to the initial number. In symbolic form, $a \times 1 = a$. The number 1 is referred to as the *multiplicative identity*. The **Identity Property of Multiplication** is shown by this statement: $$94 \times 1 = 94$$
propiedad de identidad de la multiplicación	El producto de cualquier número por 1 es igual al número mismo. En forma simbólica, $a \times 1 = a$. El número 1 se conoce como *identidad multiplicativa*. La **propiedad de identidad de la multiplicación** se muestra en el siguiente enunciado: $$94 \times 1 = 94$$
improper fraction *(Inv. 1)*	A fraction with a numerator equal to or greater than the denominator. $\frac{12}{12}$, $\frac{57}{3}$, and $2\frac{15}{2}$ are **improper fractions.** All **improper fractions** are greater than or equal to 1.
fracción impropia	Fracción con el numerador igual o mayor que el denominador. $\frac{12}{12}$, $\frac{57}{3}$ y $2\frac{15}{2}$ son **fracciones impropias.** Todas las **fracciones impropias** son iguales o mayores que 1.
independent events *(80)*	Two events are *independent* if the outcome of one event does not affect the probability that the other event will occur. *If a number cube is rolled twice, the outcome (1, 2, 3, 4, 5, or 6) of the first roll does not affect the probability of rolling 1, 2, 3, 4, 5, or 6 on the second roll. The first and second rolls are **independent events.***
sucesos independientes	Dos sucesos son independientes si el resultado de uno no afecta la probabilidad de que el otro ocurra. *Si se lanza un cubo de números dos veces, el resultado (1, 2, 3, 4, 5, 6) del primer lanzamiento no afecta la probabilidad de obtener 1, 2, 3, 4, 5, 6 en el segundo lanzamiento. El primer y segundo lanzamiento son **sucesos independientes.***
integers *(9)*	The set of counting numbers, their opposites, and zero; the members of the set $\{..., -2, -1, 0, 1, 2, ...\}$. -57 and 4 are **integers.** $\frac{15}{8}$ and -0.98 are not **integers.**
enteros	Conjunto de números de conteo, sus opuestos y el cero; los elementos del conjunto $\{..., -2, -1, 0, 1, 2, ...\}$. -57 y 4 son **enteros.** $\frac{15}{8}$ y -0.98 no son **enteros.**
interest *(maintained)*	An amount added to a loan, account, or fund, usually based on a percentage of the principal. *If we borrow $500.00 from the bank and repay the bank $575.00 for the loan, the **interest** on the loan is $575.00 − $500.00 = $75.00.*
interés	Cantidad adicional agregada a un préstamo, cuenta o fondo de inversión. Generalmente se calcula como un porcentaje del capital. *Si pedimos al banco un préstamo de $500.00 y pagamos al banco $575.00 para saldar la deuda, el **interés** del préstamo es $575.00 – $500.00 = $75.00.*

interior angle (94)	An angle that opens to the inside of a polygon.

This hexagon has six **interior angles.**

ángulo interior	Ángulo que se abre hacia el interior de un polígono.

*Este hexágono tiene seis **ángulos interiores.***

International System (76)	*See **metric system.***
Sistema internacional	*Ver **sistema métrico.***

intersect (20)	To share a point or points.

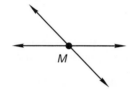

*These two lines **intersect.**
They share the point M.*

intersecarse	Tener uno o más puntos en común.

*Estas dos rectas se **intersecan.** Tienen el punto común M.*

inverse operations (3)	Operations that "undo" one another.

$a + b - b = a$	*Addition and subtraction are*
$a - b + b = a$	***inverse operations.***
$a \times b \div b = a \quad (b \neq 0)$	*Multiplication and division are*
$a \div b \times b = a \quad (b \neq 0)$	***inverse operations.***
$\sqrt{a^2} = a \quad (a \geq 0)$	*Squaring and finding square*
$(\sqrt{a})^2 = a \quad (a \geq 0)$	*roots are **inverse operations.***

operaciones inversas	Operaciones que se "cancelan" mutuamente.

$a + b - b = a$	*La suma y la resta son*
$a - b + b = a$	***operaciones inversas.***
$a \times b \div b = a \quad (b \neq 0)$	*La multiplicación y la división son*
$a \div b \times b = a \quad (b \neq 0)$	***operaciones inversas.***
$\sqrt{a^2} = a \quad (a \geq 0)$	*Elevar a una potencia y calcular la raíz cuadrada*
$(\sqrt{a})^2 = a \quad (a \geq 0)$	*son **operaciones inversas.***

irrational numbers (maintained)	Numbers that cannot be expressed as a ratio of two integers. Their decimal expansions are nonending and nonrepeating.

π and $\sqrt{3}$ are **irrational numbers.**

números irracionales	Números que no pueden expresarse como la razón de dos enteros. Sus expansiones decimales no terminan y no se repiten.

π y $\sqrt{3}$ son **números irracionales.**

isosceles triangle *(90)*	A triangle with at least two sides of equal length. 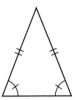 *Two of the sides of this **isosceles triangle** have equal lengths.*
triángulo isósceles	Triángulo que tiene por lo menos dos lados de igual longitud. *Dos de los lados de este **triángulo isósceles** tienen igual longitud.*

K

Kelvin Scale *(76)*	Method of temperature measurement that places 0 K at absolute zero, or $-273°C$.
escala Kelvin	Escala de temperatura que coloca el 0 K en el cero absoluto ó $-273°$ C.

L

least common denominator (LCD) *(53)*	The least common multiple of all the denominators. $\frac{1}{2} + \frac{1}{4} + \frac{1}{8}$ *The **least common denominator** of 2, 4, and 8 is 8.*
mínimo denominador común	El mínimo común múltiplo de todos los denominadores. $\frac{1}{2} + \frac{1}{4} + \frac{1}{8}$ *El mínimo denominador común de 2, 4 y 8 es 8.*

least common multiple (LCM) *(26)*	The smallest whole number that is a multiple of two or more given numbers. *Multiples of 6 are 6, 12, 18, 24, 30, 36, ...* *Multiples of 8 are 8, 16, 24, 32, 40, 48, ...* *The **least common multiple** of 6 and 8 is 24.*
mínimo común múltiplo (mcm)	El menor número entero que es múltiplo común de dos o más números. *Los múltiplos de 6 son 6, 12, 18, 24, 30, 36, ...* *Los múltiplos de 8 son 8, 16, 24, 32, 40, 48, ...* *El **mínimo común múltiplo** de 6 y 8 es 24.*

legend *(Inv. 11)*	A notation on a map, graph, or diagram that describes the meaning of the symbols and/or the scale used. 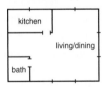 $\frac{1}{4}$ inch = 5 feet *The **legend** of this scale drawing shows that $\frac{1}{4}$ inch represents 5 feet.*
rótulo	Nota en un mapa, gráfica o diagrama, que describe el significado de los símbolos y/o la escala usados. *El **rótulo** de este dibujo a escala muestra que $\frac{1}{4}$ pulg representa 5 pies.*

line *(20)*	A straight collection of points extending in opposite directions without end. A B ←———●————————————●———→ **line** *AB* or **line** *BA*
línea recta	Sucesión recta de puntos que se extiende indefinidamente en ambas direcciones.

line graph *(Inv. 4)*	A method of displaying numerical information as points connected by line segments. 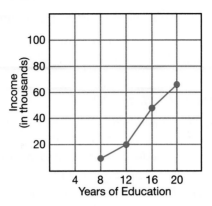 *This **line graph** has a horizontal axis that shows the number of completed years of education and a vertical axis that shows the average yearly income.*
gráfica lineal	Un método para representar información numérica como puntos conectados por segmentos de recta. *Esta **gráfica lineal** tiene un eje horizontal que muestra el número de años de educación terminados y un eje vertical que muestra el salario anual promedio.*
line of symmetry *(maintained)*	A line that divides a figure into two halves that are mirror images of each other. 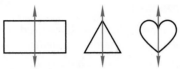 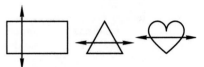 **lines of symmetry** not **lines of symmetry**
eje de simetría	Recta que divide una figura en dos mitades que son imágenes especulares la una de la otra.
line plot *(10)*	A method of plotting a set of numbers by placing a mark above a number on a number line each time it occurs in the set. 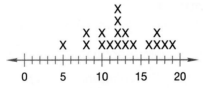 *This is a **line plot** of the numbers 5, 8, 8, 10, 10, 11, 12, 12, 12, 12, 13, 13, 14, 16, 17, 17, 18, and 19.*
diagrama de puntos	Método para representar un conjunto de números, que consiste en colocar una marca sobre un número de una recta numérica cada vez que dicho número ocurre en el conjunto. *Éste es un **diagrama de puntos** de los números 5, 8, 8, 10, 10, 11, 12, 12, 12, 12, 13, 13, 14, 16, 17, 17, 18 y 19.*

M

mass *(maintained)*	The amount of matter in an object. *Grams and kilograms are units of **mass.***
masa	Cantidad de materia en un objeto. *Gramos y kilogramos son unidades de **masa.***

mean *(10)*	The number found when the sum of two or more numbers is divided by the number of addends in the sum; also called *average*. *To find the **mean** of the numbers 5, 6, and 10, first add.* $$5 + 6 + 10 = 21$$ *Then, since there were three addends, divide the sum by 3.* $$21 \div 3 = 7$$ *The **mean** of 5, 6, and 10 is 7.*
media	Número que se obtiene al dividir la suma de dos o más números entre la cantidad de sumandos en la suma; también se le llama *media*. *Para calcular la **media** de los números 5, 6 y 10, primero se suman.* $$5 + 6 + 10 = 21$$ *Luego, como hay tres sumandos, se divide la suma entre 3.* $$21 \div 3 = 7$$ *La **media** de 5, 6 y 10 es 7.*

median *(Inv. 4)*	The middle number (or the mean of the two central numbers) of a list of data when the numbers are arranged in order from the least to the greatest. *In the data at right,* *1, 1, 2, 5, 6, 7, 9, 15, 24, 36, 44* *7 is the **median.***
mediana	Número que está en medio (o la media de los dos números centrales) dentro de una lista de datos, cuando los números se ordenan de menor a mayor. *En los datos a la derecha, 7 es la **mediana.***

metric system *(76)*	An international system of measurement based on multiples of ten. Also called *International System*. *Centimeters and kilograms are units in the **metric system.***
sistema métrico	Sistema internacional de medidas basado en múltiplos de diez. También se le llama *Sistema internacional*. *Centímetros y kilogramos son unidades del **sistema métrico.***

minuend *(5)*	A number from which another number is subtracted. $12 - 8 = 4$ *The **minuend** in this problem is 12.*
minuendo	Número del cual se resta otro número. $12 - 8 = 4$ *El **minuendo** en este problema es el 12.*

mixed number *(Inv. 1)*	A whole number and a fraction together. *The **mixed number** $2\frac{1}{3}$ means "two and one third."*
número mixto	Número formado por un número entero y una fracción. *El **número mixto** $2\frac{1}{3}$ significa "dos y un tercio".*

mode *(Inv. 4)*	The number or numbers that appear most often in a list of data. *In the data at right,* *5, 12, 32, 5, 16, 5, 7, 12* *5 is the **mode.***
moda	El número o números que aparecen con más frecuencia en una lista de datos. *En esta lista de datos,* *5, 12, 32, 5, 16, 5, 7, 12* *el número 5 es la **moda.***

multiple
(19)

A product of a counting number and another number.

*The **multiples** of 3 include 3, 6, 9, and 12.*

múltiplo

Producto de un número de conteo por otro número.

*Los **múltiplos** de 3 incluyen 3, 6, 9 y 12.*

multiplicative identity
(3)

The number 1. *See also* **Identity Property of Multiplication.**

$-2 \times 1 = -2$

*The number **1** is called the **multiplicative identity** because multiplying any number by 1 does not change the number*

identidad en la multiplicación

El número 1. Ver también **propiedad de identidad de la multiplicación.**

$-2 \times 1 = -2$

*El número **1** es llamado la **identidad de la multiplicación** porque multiplicar un número por 1 no cambia el número.*

N

negative numbers
(9)

Numbers less than zero.

*-15 and -2.86 are **negative numbers.***

*19 and 0.74 are not **negative numbers.***

números negativos

Los números menores que cero.

*-15 y -2.86 son **números negativos.***

*19 y 0.74 no son **números negativos.***

net
(maintained)

A two-dimensional representation of a three-dimensional figure.

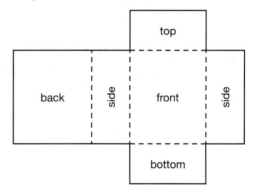

red

Representación bidimensional de una figura tridimensional.

nonexample
(maintained)

A nonexample is the opposite of an example. Nonexamples can be used to prove that a fact or a statement in mathematics is incorrect.

*The integer 7 is a **nonexample** of an even number, and a circle is a non-example of a polygon.*

contraejemplo

Un contraejemplo es lo opuesto de un ejemplo. Los contraejemplos se pueden usar para demostrar en matemáticas que una operación o enunciado es incorrecto,

*El entero 7 es un **contraejemplo** de un número par, y un círculo es un contraejemplo de un polígono.*

number line *(9)*	A line for representing and graphing numbers. Each point on the line corresponds to a number.

recta numérica	Recta para representar y graficar números. Cada punto de la recta corresponde a un número.

numerator *(Inv. 1)*	The top term of a fraction.

$$\frac{9}{10} \xleftarrow{\text{\ }} \textbf{numerator}$$
$$\xleftarrow{\text{\ }} \text{denominator}$$

numerador	El término superior de una fracción.

O

oblique line(s) *(20)*	**1.** A line that is neither horizontal nor vertical.

oblique line   not **oblique lines**

2. Lines in the same plane that are neither parallel nor perpendicular.

oblique lines   not **oblique lines**

recta(s) oblicua(s)	**1.** Recta que no es ni horizontal ni vertical. **2.** Rectas ubicadas en un mismo plano, que no son ni paralelas ni perpendiculares entre sí.

obtuse angle *(20)*	An angle whose measure is more than 90° and less than 180°.

obtuse angle   not **obtuse angles**

*An **obtuse angle** is larger than both a right angle and an acute angle.*

ángulo obtuso	Ángulo que mide más de 90° y menos de 180°. *Un **ángulo obtuso** es más grande que, ambos, un ángulo recto y que un ángulo agudo.*

obtuse triangle *(90)*	A triangle whose largest angle measures more than 90° and less than 180°.

obtuse triangle   not **obtuse triangles**

triángulo obtusángulo	Triángulo cuyo ángulo mayor mide más que 90° y menos que 180°.

odd numbers *(1)*	Numbers that have a remainder of 1 when divided by 2; the members of the set {..., −3, −1, 1, 3, ...}. ***Odd numbers** have 1, 3, 5, 7, or 9 in the ones place.*
números impares	Números que cuando se dividen entre 2 tienen residuo de 1; los elementos del conjunto {..., −3, −1, 1, 3, ...}. *Los **números impares** tienen 1, 3, 5, 7 ó 9 en el lugar de las unidades.*
open-option survey *(Inv. 2)*	A survey that does not limit the possible responses. What is your favorite sport? _____ **open-option survey**
encuesta de opción abierta	Encuesta que no limita las respuestas posibles.
operations of arithmetic *(4)*	The four basic mathematical operations: addition, subtraction, multiplication, and division. $1 + 9 \qquad 21 − 8 \qquad 6 \times 22 \qquad 3 \div 1$ the **operations of arithmetic**
operaciones aritméticas	Las cuatro operaciones matemáticas básicas: suma, resta, multiplicación y división.
opposites *(9)*	Two numbers whose sum is 0. $(−3) + (+3) = 0$ *The numbers +3 and −3 are **opposites.***
opuestos	Dos números cuya suma es 0. $(−3) + (+3) = 0$ *Los números +3 y −3 son **opuestos.***
order of operations *(2)*	The order in which the four fundamental operations occur. **1.** Simplify powers and roots. **2.** Multiply or divide in order from left to right. **3.** Add and subtract in order from left to right. With parentheses, we simplify within the parentheses, from innermost to outermost, before simplifying outside the parentheses.
orden de las operaciones	El orden en el cual ocurren las cuatro operaciones fundamentales. **1.** Simplificar potencias y raíces. **2.** Multiplicar o dividir en orden de izquierda a derecha. **3.** Sumar o restar en orden de izquierda a derecha. Cuando hay paréntesis, simplificamos dentro del paréntesis, de dentro hacia afuera, antes de simplificar afuera del paréntesis.
ordered pair *(Inv. 5)*	A pair of numbers, written in a specific order, that are used to designate the position of a point on a coordinate plane. *See also* **coordinate(s).** $(0, 1) \qquad (2, 3) \qquad (3.4, 5.7) \qquad \left(\frac{1}{2}, -\frac{1}{2}\right)$ **ordered pairs**
par ordenado	Un par de números, escritos en un orden específico, que se usan para determinar la posición de un punto en un plano coordenado. *Ver también* **coordenada(s).**

origin
(Inv. 5)

1. The location of the number 0 on a number line.

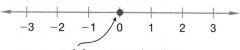

origin on a number line

2. The point (0, 0) on a coordinate plane.

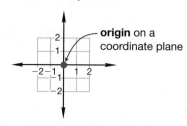

origin on a coordinate plane

origen

1. Posición del número 0 en una recta numérica.

2. El punto (0, 0) en un plano coordenado.

---

outlier
(10)

A number in a list of data that is distant from the other numbers in the list.

*1, 5, 4, 3, 6, 28, 7, 2   In this list, the number 28 is an **outlier** because it is distant from the other numbers in the list.*

valor lejano

Número en una lista de datos que es distante de los otros números en la lista.

*1, 5, 4, 3, 6, 28, 7, 2   En esta lista el número 28 es un **valor lejano** porque es distante de los otros números en la lista.*

P

---

parallel lines
(20)

Lines in the same plane that do not intersect.

parallel lines

rectas paralelas

Rectas ubicadas en un mismo plano y que nunca se intersecan.

---

parallelogram
(59)

A quadrilateral that has two pairs of parallel sides.

parallelograms   not a parallelogram

paralelogramo

Cuadrilátero que tiene dos pares de lados paralelos.

---

percent
(27)

A fraction whose denominator of 100 is expressed as a percent sign (%).

$$\frac{99}{100} = 99\% = 99 \textbf{ percent}$$

porcentaje

Fracción cuyo denominador 100 se expresa con un signo (%), que se lee *por ciento*.

---

perfect square
(32)

The product when a whole number is multiplied by itself.

*The number 9 is a **perfect square** because 3 × 3 = 9.*

cuadrado perfecto

El producto de un número entero cuando se multiplica por sí mismo.

*El número 9 es un **cuadrado perfecto** porque 3 × 3 = 9.*

perimeter
(23)

The distance around a closed, flat shape.

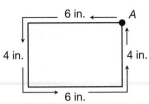

*The **perimeter** of this rectangle (from point A around to point A) is 20 inches.*

perímetro

Distancia alrededor de una figura cerrada y plana.

*El **perímetro** de este rectángulo (desde el punto A alrededor del rectángulo hasta el punto A) es 20 pulgadas.*

perpendicular bisector
(maintained)

A line, ray, or segment that intersects a segment at its midpoint at a right angle, thereby dividing the segment into two congruent parts.

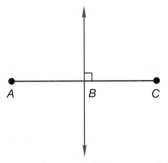

*This vertical line is a **perpendicular bisector** of $\overline{AC}$.*

mediatriz

Recta, rayo o segmento de recta, que interseca un segmento de recta en su punto medio, en un ángulo recto. Divide el segmento en dos partes congruentes.

*Esta recta vertical es una **mediatriz** de $\overline{AC}$.*

perpendicular lines
(20)

Two lines that intersect at right angles.

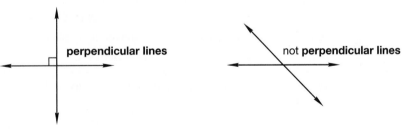

perpendicular lines   not **perpendicular lines**

rectas perpendiculares

Dos rectas que se intersecan en ángulos rectos.

pi (π)
(42)

The number of diameters equal to the circumference of a circle.

*Approximate values of **pi** are 3.14 and $\frac{22}{7}$.*

pi (π)

Número de diámetros que equivalen a la circunferencia de un círculo.

*Dos valores aproximados de **pi** son 3.14 y $\frac{22}{7}$.*

pictograph
(Inv. 4)

A method of displaying data that involves using pictures to represent the data being counted.

*This is a **pictograph**. It shows how many stars each person saw.*

pictograma	Método de presentar información por medio de dibujos que representan datos que se quieren contar.
	*Éste es un **pictograma**. Muestra el número de estrellas que vió cada persona.*

---

pie graph *(30)*	*See **circle graph**.*
gráfica circular	*Ver **gráfica circular**.*

---

place value *(21)*	The value of a digit based on its position within a number.

$$\begin{array}{r} 341 \\ 23 \\ + \ \ 7 \\ \hline 371 \end{array}$$

***Place value** tells us that 4 in 341 is worth four tens.*

*In addition and subtraction problems we align digits with the same **place value**.*

valor posicional	Valor de un dígito de acuerdo al lugar que ocupa en el número.

---

plane *(20)*	A flat surface that has no boundaries.
	*The flat surface of a desk is part of a **plane**.*
plano	Superficie plana ilimitada.
	*La superficie plana de un escritorio es parte de un **plano**.*

---

p.m. *(maintained)*	The period of time from noon to just before midnight.
	*I go to bed at 9 **p.m.** I go to bed at 9 o'clock at night.*
p.m.	Período de tiempo desde el mediodía hasta la medianoche.
	*Me voy a dormir a las 9 **p.m.** Me voy a dormir a las 9 en punto de la noche.*

---

point *(14)*	An exact position on a line, on a plane, or in space.
	•$_A$ *This dot represents **point** A.*
punto	Una posición exacta en una recta, un plano o en el espacio.
	*Esta marca representa el **punto** A.*

---

polygon *(59)*	A closed, flat shape with straight sides.

polygons   not **polygons**

polígono	Figura cerrada y plana que tiene lados rectos.

---

polyhedron *(114)*	A geometric solid whose faces are polygons.

polyhedrons   not **polyhedrons**

cube   triangular   pyramid   sphere   cylinder   cone
        prism

poliedro	Solido geométrico cuyas caras son polígonos.

population *(Inv. 3)*	A certain group of people that a survey is about.
población	Un cierto grupo de personas de los que trata una encuesta.
positive numbers *(9)*	Numbers greater than zero. *0.25 and 157 are* **positive numbers.** *−40 and 0 are not* **positive numbers.**
números positivos	Números mayores que cero. *0.25 y 157 son* **números positivos.** *−40 y 0 no son* **números positivos.**
power(s) *(71)*	**1.** The value of an exponential expression. *16 is the fourth* **power** *of 2 because* $2^4 = 16$. **2.** An exponent. *The expression* 2^4 *is read "two to the fourth* **power.**"
potencia(s)	**1.** El valor de una expresión exponencial. *16 es la cuarta* **potencia** *de 2, porque* $2^4 = 16$. **2.** Un exponente. *La expresión* 2^4 *se lee "dos a la cuarta* **potencia**".
prime factorization *(61)*	The expression of a composite number as a product of its prime factors. **The prime factorization** *of 60 is* $2 \times 2 \times 3 \times 5$.
factorización prima	La expresión de un número compuesto como el producto de sus factores primos. *La* **factorización prima** *de 60 es* $2 \times 2 \times 3 \times 5$.
prime number *(11)*	A counting number greater than 1 whose only two factors are the number 1 and itself. *7 is a* **prime number.** *Its only factors are 1 and 7.* *10 is not a* **prime number.** *Its factors are 1, 2, 5, and 10.*
número primo	Número de conteo mayor que 1, cuyos dos únicos factores son el 1 y el número mismo. *7 es un* **número primo.** *Sus únicos factores son 1 y 7.* *10 no es un* **número primo.** *Sus factores son 1, 2, 5 y 10.*
principal *(maintained)*	The amount of money borrowed in a loan, deposited in an account that earns interest, or invested in a fund. *If we borrow $750.00, the* **principal** *is $750.00.*
capital	Cantidad de dinero obtenida en un préstamo, que se deposita en una cuenta que gana intereses, o que se invierte en un fondo. *Si pedimos un préstamo por $750.00, nuestro* **capital** *es $750.00.*

prism *(Inv. 5)*	A polyhedron with two congruent parallel bases.

rectangular **prism**   triangular **prism**

prisma	Poliedro con dos bases paralelas congruentes.

---

probability
(54)

A way of describing the likelihood of an event; the ratio of favorable outcomes to all possible outcomes.

The **probability** of rolling a 3 with a standard number cube is $\frac{1}{6}$.

probabilidad

Manera de describir la ocurrencia de un suceso; la razón de resultados favorables a todos los resultados posibles.

La **probabilidad** de obtener 3 al lanzar un cubo de números estándar es $\frac{1}{6}$.

---

product
(4)

The result of multiplication.

$5 \times 4 = 20$ The **product** of 5 and 4 is 20.

producto

Resultado de una multiplicación.

El **producto** de 5 por 4 es 20.

---

proportion
(82)

A statement that shows two ratios are equal.

$$\frac{6}{10} = \frac{9}{15}$$

These two ratios are equal, so this is a **proportion.**

proporción

Enunciado que expresa que dos razones son iguales.

Estas dos razones son iguales, por lo tanto es una **proporción.**

---

protractor
(maintained)

A tool used to measure and draw angles.

protractor

transportador

Instrumento que sirve para medir y trazar ángulos.

---

pyramid
(maintained)

A three-dimensional solid with a polygon as its base and triangular faces that meet at a vertex.

pyramid

pirámide

Sólido tridimensional con un polígono como su base y caras triangulares que se encuentran en un vértice.

quadrilateral
(59)

Any four-sided polygon.

*Each of these polygons has 4 sides. They are all **quadrilaterals.***

cuadrilátero

Polígono de cuatro lados.

*Cada uno de estos polígonos tiene 4 lados. Todos son **cuadriláteros.***

qualitative
(Inv. 3)

Expressed in or relating to categories rather than quantities or numbers.

***Qualitative data** are categorical: Examples include the month in which someone is born and a person's favorite flavor of ice cream.*

cualitativo

Que está expresado en función de categorías en vez de cantidades o números.

*Los **datos cualitativos** expresan categorías: el mes en que alguien nació, o el sabor favorito de helados de una persona.*

quantitative
(Inv. 3)

Expressed in or relating to quantities or numbers.

***Quantitative data** are numerical: Examples include the population of a city, the number of pairs of shoes someone owns, and the number of hours per week someone watches television.*

cuantitativo

Que está expresado en función de cantidades o números.

*Los **datos cuantitativos** se expresan con números: la población de una ciudad, el número de pares de zapatos que tiene una persona, o el número de horas por semana que alguien ve la televisión.*

quotient
(4)

The result of division.

$$12 \div 3 = 4 \qquad 3\overline{)12}^{\,4} \qquad \frac{12}{3} = 4$$

*The **quotient** is 4 in each of these problems.*

cociente

Resultado de una división.

$$12 \div 3 = 4 \qquad 3\overline{)12}^{\,4} \qquad \frac{12}{3} = 4$$

*El **cociente** es 4 en cada una de estas operaciones.*

radius
(42)

(Plural: *radii*) The distance from the center of a circle to a point on the circle.

*The **radius** of circle A is 2 inches.*

radio

Distancia desde el centro de un círculo hasta un punto del círculo.

*El **radio** del círculo A es 2 pulgadas.*

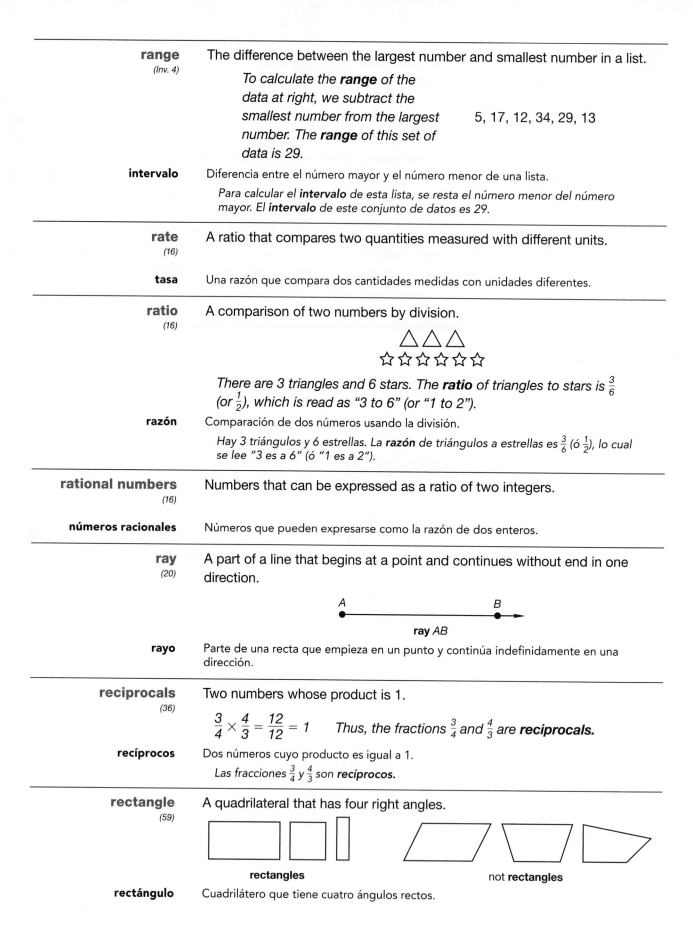

range *(Inv. 4)*	The difference between the largest number and smallest number in a list. *To calculate the **range** of the data at right, we subtract the smallest number from the largest number. The **range** of this set of data is 29.* 5, 17, 12, 34, 29, 13
intervalo	Diferencia entre el número mayor y el número menor de una lista. *Para calcular el **intervalo** de esta lista, se resta el número menor del número mayor. El **intervalo** de este conjunto de datos es 29.*
rate *(16)*	A ratio that compares two quantities measured with different units.
tasa	Una razón que compara dos cantidades medidas con unidades diferentes.
ratio *(16)*	A comparison of two numbers by division. *There are 3 triangles and 6 stars. The **ratio** of triangles to stars is $\frac{3}{6}$ (or $\frac{1}{2}$), which is read as "3 to 6" (or "1 to 2").*
razón	Comparación de dos números usando la división. *Hay 3 triángulos y 6 estrellas. La **razón** de triángulos a estrellas es $\frac{3}{6}$ (ó $\frac{1}{2}$), lo cual se lee "3 es a 6" (ó "1 es a 2").*
rational numbers *(16)*	Numbers that can be expressed as a ratio of two integers.
números racionales	Números que pueden expresarse como la razón de dos enteros.
ray *(20)*	A part of a line that begins at a point and continues without end in one direction. **ray** *AB*
rayo	Parte de una recta que empieza en un punto y continúa indefinidamente en una dirección.
reciprocals *(36)*	Two numbers whose product is 1. $\frac{3}{4} \times \frac{4}{3} = \frac{12}{12} = 1$ *Thus, the fractions $\frac{3}{4}$ and $\frac{4}{3}$ are **reciprocals.***
recíprocos	Dos números cuyo producto es igual a 1. *Las fracciones $\frac{3}{4}$ y $\frac{4}{3}$ son **recíprocos.***
rectangle *(59)*	A quadrilateral that has four right angles. **rectangles** **not** rectangles
rectángulo	Cuadrilátero que tiene cuatro ángulos rectos.

rectanglar prism *(Inv. 5)*	*See* **prism.**
prisma rectangular	*Ver* **prisma.**
reduce *(22)*	To rewrite a fraction in lowest terms. *If we **reduce** the fraction $\frac{9}{12}$, we get $\frac{3}{4}$.*
reducir	Escribir una fracción en su mínima expresión. *Al **reducir** la fracción $\frac{9}{12}$, se obtiene $\frac{3}{4}$.*

reflection
(maintained)

Flipping a figure to produce a mirror image.

reflection

reflexión

Inversión de una figura para producir una imagen especular.

regular polygon
(59)

A polygon in which all sides have equal lengths and all angles have equal measures.

regular polygons   not **regular polygons**

polígono regular

Polígono en el cual todos los lados tienen la misma longitud y todos los ángulos tienen la misma medida.

relative frequency
(Inv. 7)

Used to estimate a probability by dividing the frequency by the total amount.

Sport	Frequency	Relative Frequency
Basketball	20	$\frac{20}{50} = 0.40$
Football	5	$\frac{5}{50} = 0.10$
Baseball	25	$\frac{25}{50} = 0.50$

*To estimate the probability of what sport a student would prefer to play, the PE teacher uses **relative frequency.***

frecuencia relativa

Se utiliza para estimar una probabilidad al dividir la frecuencia entre la cantidad total.

*Para estimar la probabilidad del deporte que un estudiante preferiría, el maestro de educación física utiliza la **frecuencia relativa.***

rhombus
(59)

A parallelogram with all four sides of equal length.

rhombuses   not **rhombuses**

rombo

Paralelogramo con sus cuatro lados de igual longitud.

right angle (20)	An angle that forms a square corner and measures 90°. It is often marked with a small square.

right angle   not **right angles**

ángulo recto	Ángulo que forma una esquina cuadrada y mide 90°. Se indica con frecuencia con un pequeño cuadrado.

right solid (114)	A solid with bases that are parallel and sides that are perpendicular to the bases.

right solids   not a
 right solid

right solids   *not a* **right solid**

sólido recto	Un sólido con bases que son paralelas y lados que son perpendiculares a la bases.

right triangle (90)	A triangle whose largest angle measures 90°.

right triangle   not **right triangles**

triángulo rectángulo	Triángulo cuyo ángulo mayor mide 90°.

rotation (maintained)	To rotate, or turn a figure about a specified point is called the *center of rotation.*

rotation

rotación	Girar o voltear una figura alrededor de un punto específico llamado *centro de rotación.*

rotational symmetry (maintained)	A figure has rotational symmetry when it does not require a full rotation for the figure to look as if it re-appears in the same position as when it began the rotation, for example, a square or a triangle.

original position   45° turn   **90° turn**   150° turn   **180° turn**   210° turn   **270° turn**

simetría rotacional	Una figura tiene simetría rotacional cuando no requiere de una rotación completa para que la figura se vea como si reapareciera en la misma posición que tenía antes de la rotación, por ejemplo, un cuadrado o triángulo.

round (21)	A way of estimating a number by increasing or decreasing it to a certain place value. Example: 517 **rounds** to 520

redondear	Una manera de estimar un número al aumentarlo o disminuirlo hasta un cierto valor posicional. Ejemplo: 517 se **redondea** a 520.

sales tax
(35)

The tax charged on the sale of an item and based upon the item's purchase price.

*If the **sales-tax** rate is 7%, the **sales tax** on a $5.00 item will be $5.00 × 7% = $0.35.*

impuesto sobre la venta

Impuesto que se carga al vender un objeto y que se calcula como un porcentaje del precio del objeto.

*Si la tasa de impuesto es 7%, el **impuesto sobre la venta** de un objeto que cuesta $5.00 es: $5.00 x 7% = $0.35.*

---

sample
(Inv. 2)

A smaller group of a population that a survey focuses on.

muestra

Un grupo menor de una población en el cual se enfoca una encuesta.

---

sample space
(54)

Set of all possible outcomes of a particular event.

*The **sample space** of a 1–6 number cube is {1, 2, 3, 4, 5, 6}.*

espacio muestral

Conjunto de todos los resultados posibles de un suceso en particular.

*El **espacio muestral** de un cubo de números del 1–6 es {1,2,3,4,5,6}.*

---

scale
(Inv. 11)

A ratio that shows the relationship between a scale drawing or model and the actual object.

*If a drawing of the floor plan of a house has the legend 1 inch = 2 feet, the **scale** of the drawing is $\frac{1 \text{ in.}}{2 \text{ ft}} = \frac{1}{24}$.*

escala

Razón que muestra la relación entre un dibujo o modelo a escala y el objeto real.

*Si el dibujo del plano de una casa tiene la clave 1 pulgada = 2 pies, la **escala** del dibujo es $\frac{1 \text{ pulg}}{2 \text{ pies}} = \frac{1}{24}$.*

---

scale drawing
(Inv. 11)

A two-dimensional representation of a larger or smaller object.

*Blueprints and maps are examples of **scale drawings**.*

dibujo a escala

Representación bidimensional de un objeto más grande o más pequeño.

*Los planos y los mapas son ejemplos de **dibujos a escala**.*

---

scale factor
(82)

The number that relates corresponding sides of similar geometric figures.

25 mm

10 mm

10 mm

4 mm

*The **scale factor** from the smaller rectangle to the larger rectangle is 2.5.*

factor de escala

Número que relaciona los lados correspondientes de dos figuras geométricas semejantes.

*El **factor de escala** del rectángulo más pequeño al rectángulo más grande es 2.5.*

---

scale model
(Inv. 11)

A three-dimensional rendering of a larger or smaller object.

*Globes and model airplanes are examples of **scale models**.*

modelo a escala

Representación tridimensional de un objeto más grande o más pequeño.

*Los globos terráqueos y los modelos de aviones son ejemplos de **modelos a escala**.*

scalene triangle *(90)*	A triangle with three sides of different lengths.

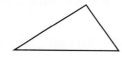

All three sides of this **scalene triangle** *have different lengths.*

triángulo escaleno	Triángulo con todos sus lados de diferente longitud.

Los tres lados de este **triángulo escaleno** *tienen diferente longitud.*

---

sector *(Inv. 4)*	A region bordered by part of a circle and two radii.

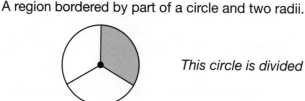

This circle is divided into 3 **sectors.**

sector	Región de un círculo limitada por parte de un círculo y dos radios.

Este círculo esta dividido en 3 **sectores.**

---

segment *(maintained)*	A part of a line with two distinct endpoints.

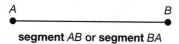

segment *AB* or **segment** *BA*

segmento	Parte de una línea recta con dos extremos definidos.

---

sequence *(1)*	A list of numbers arranged according to a certain rule.

The numbers 2, 4, 6, 8, ... form a **sequence.** *The rule is "count up by twos."*

secuencia	Lista de números ordenados de acuerdo a una regla.

Los números 2, 4, 6, 8, ... forman una **secuencia.** *La regla es "contar hacia adelante de dos en dos".*

---

similar *(Inv. 10)*	Having the same shape but not necessarily the same size. Corresponding angles of similar figures are congruent. Corresponding sides of similar figures are proportional.

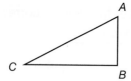

$\triangle ABC$ *and* $\triangle DEF$ *are* **similar.** *They have the same shape but not the same size.*

semejante	Que tiene la misma forma, pero no necesariamente el mismo tamaño. Los ángulos correspondientes de dos figuras semejantes son congruentes. Los lados correspondientes de dos figuras semejantes son proporcionales.

$\triangle ABC$ *y* $\triangle DEF$ *son* **semejantes.** *Tienen la misma forma, pero diferente tamaño.*

simple interest *(maintained)*	Interest calculated as a percentage of the principal only.

	Simple Interest	**Compound Interest**
	$100.00 principal	$100.00 principal
	$6.00 first-year interest (6% of $100)	+ $6.00 first-year interest (6% of $100)
	+ $6.00 second-year interest (6% of $100)	$106.00 total after one year
	$112.00 total after two years	+ $6.36 second-year interest (6% of $106)
		$112.36 total after two years

interés simple	Interés que se calcula como porcentaje solamente del capital.

solid *(114)*	*See* **geometric solid.**
sólido	*Ver* **sólido geométrico.**

sphere *(maintained)*	A round geometric solid in which every point on the surface is at an equal distance from its center.

sphere

esfera	Superficie geométrica cuyos puntos están todos a la misma distancia de su centro.

square *(32, 59)*	**1.** The product of a number and itself.

*The **square** of 4 is 16.*

2. A rectangle with all four sides of equal length.

2 in.

2 in.   2 in.   *All four sides of this **square** are 2 inches long.*

2 in.

cuadrado	**1.** El producto de un número por sí mismo.

*El **cuadrado** de 4 es 16.*

2. Un retángulo que tiene todos sus lados de igual longitud.

*Los cuatro lados de este **cuadrado** miden 2 pulgadas de longitud.*

square root *(32)*	One of two equal factors of a number. The symbol for the principal, or positive, square root of a number is $\sqrt{\ }$.

*A **square root** of 49 is 7 because $7 \times 7 = 49$.*

raíz cuadrada	Uno de dos factores iguales de un número. El símbolo de la raíz cuadrada de un número es $\sqrt{\ }$.

*La **raíz cuadrada** de 49 es 7, porque $7 \times 7 = 49$.*

square units *(23)*	A unit of measurement for area. Examples of square units include: square centimeter, square inches, square foot, square yards, and square meters.

1 ft

This is 1 square foot.

unidades cuadradas	Una unidad de medida para área. Ejemplos de unidades cuadradas incluyen: centímetros cuadrados, pies cuadrados, yardas cuadradas y metros cuadrados.

Este es un pie cuadrado.

statistics *(Inv. 2)*	The science of gathering and organizing data in such a way that conclusions can be made; the study of data.
estadística	La ciencia de reunir y organizar datos de tal manera que se puedan llegar a conclusiones; el estudio de datos.

straight angle
(20)

An angle that forms a straight line.

straight ∠ABC

straight angle

ángulo llano Un ángulo que forma una línea recta.

subtrahend
(5)

A number that is subtracted.

$12 - 8 = 4$ The **subtrahend** in this problem is 8.

sustraendo Número que se resta de otro.

$12 - 8 = 4$ El **sustraendo** en este problema es 8.

sum
(4)

The result of addition.

$7 + 6 = 13$ The **sum** of 7 and 6 is 13.

suma total Resultado de una suma.

$7 + 6 = 13$ La **suma total** de 7 más 6 es 13.

supplementary angles
(60)

Two angles whose sum is 180°.

∠AMB and ∠CMB are **supplementary.**

ángulos suplementarios Dos ángulos cuya suma es 180°.

surface area
(Inv. 5)

The total area of the surface of a geometric solid.

Area of top	= 5 cm × 6 cm =	30 cm²
Area of bottom	= 5 cm × 6 cm =	30 cm²
Area of front	= 3 cm × 6 cm =	18 cm²
Area of back	= 3 cm × 6 cm =	18 cm²
Area of side	= 3 cm × 5 cm =	15 cm²
+ Area of side	= 3 cm × 5 cm =	15 cm²
Total **surface area**		= 126 cm²

área superficial Área total de la superficie de un sólido geométrico.

survey
(Inv. 2)

A method of collecting data about a particular population.

*Mia conducted a **survey** by asking each of her classmates the name of his or her favorite television show.*

encuesta Método de reunir información acerca de un grupo de personas.

*Mia hizo una **encuesta** entre sus compañeros para averiguar cuál era su programa favorito de televisión.*

symbols of inclusion *(72)*	Symbols that are used to set apart portions of an expression so that they may be evaluated first: ( ), [ ], { }, and the division bar in a fraction. *In the statement (8 − 4) ÷ 2, the **symbols of inclusion** indicate that 8 − 4 should be calculated before dividing by 2.*
símbolos de inclusión	Símbolos que se utilizan para separar partes de una expresión para que sean evaluadas antes que otras: (), [], {}, y la barra de división en una fracción.

T

term(s) *(1)*	**1.** A number that serves as a numerator or denominator of a fraction. $\dfrac{5}{6}$ terms **2.** A number in a sequence. *1, 3, 5, 7, 9, 11, ...* *Each number in this sequence is a **term.***
término(s)	**1.** Número que se usa como numerador o denominador de una fracción. **2.** Un número en una secuencia. *Cada número de esta secuencia es un **término.***
theoretical probability *(Inv. 7)*	The probability that an event will occur, as determined by analysis rather than by experimentation. *The **theoretical probability** of rolling a three with a standard number cube is $\frac{1}{6}$.*
probabilidad teórica	Probabilidad de ocurrencia de un suceso, determinada por análisis en vez de experimentación. *La **probabilidad teórica** de obtener un tres al lanzar un cubo de números estándar es $\frac{1}{6}$.*
transformation *(maintained)*	The changing of a figure's position through rotation, reflection, or translation. **Transformations** | Movement	Name
Flip	Reflection
Slide	Translation
Turn	Rotation	
transformación	Cambio en la posición de una figura por medio de una rotación, reflexión o traslación.
translation *(maintained)*	Sliding a figure from one position to another without turning or flipping the figure. 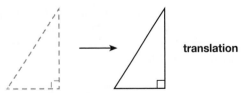 translation
traslación	Deslizamiento de una figura de una posición a otra, sin rotar ni voltear la figura.

transversal *(94)*	A line that intersects one or more other lines in a plane. ⟵ transversal
transversal	Línea recta que interseca una o más líneas rectas en un plano.

trapezium *(59)*	A quadrilateral with no parallel sides. **trapezium** not **trapeziums**
trapezoide	Cuadrilátero que no tiene lados paralelos.

trapezoid *(59)*	A quadrilateral with exactly one pair of parallel sides. **trapezoids** not **trapezoids**
trapecio	Cuadrilátero que tiene exactamente un par de lados paralelos.

tree diagram *(Inv. 6)*	A visual representation of a compound experiment. Spinner Coin *This **tree diagram** shows all 6 possible outcomes for a spinner with 3 sectors being spun and a coin being flipped.*
diagrama de árbol	Una representación visual de un experimento compuesto. *Este **diagrama de árbol** muestra todos los 6 resultados posibles de una rueda giratoria con tres sectores girando y los lanzamientos de una moneda.*

triangular prism *(maintained)*	*See* **prism.**
prisma triangular	*Ver* **prisma.**

U

unit multiplier (105)	A ratio equal to 1 that is composed of two equivalent measures. $$\frac{12\ inches}{1\ foot} = 1$$ *We can use this **unit multiplier** to convert feet to inches.*
factor de conversión	Razón igual a 1, compuesta de dos medidas equivalentes. *Este **factor de conversión** puede usarse para convertir pies a pulgadas:* $\frac{12\ pulg}{1\ pie} = 1$
unknown (5)	A value that is not given. A letter is frequently used to stand for an unknown number.
incógnita	Un valor no dado. Frecuentemente se usa una letra para representar una incógnita o valor desconocido.
U.S. Customary System (81)	A system of measurement used almost exclusively in the United States. *Pounds, quarts, and feet are units in the **U.S. Customary System.***
Sistema usual de EE.UU.	Unidades de medida que se usan casi exclusivamente en EE.UU. *Libras, cuartos y pies son unidades del **Sistema usual de EE.UU.***

V

variable (3)	A quantity that can change or assume different values. Also, a letter used to represent an unknown in an expression or equation. *In the statement $x + 7 = y$, the letters x and y are **variables.***
variable	Una cantidad que cambia o toma diferentes valores. También, una letra que se utiliza para representar una incógnita en una expresión o ecuación. *En el enunciado $x + 7 = y$, la letras x e y son **variables.***
vertex (20)	(Plural: *vertices*) A point of an angle, polygon, or polyhedron where two or more lines, rays, or segments meet. *A dot is placed at one **vertex** of this cube. A cube has eight **vertices.***
vértice	Punto de un ángulo, polígono o poliedro, donde se unen dos o más rectas, rayos o segmentos de recta. *Un punto está en un **vértice** de este cubo. Un cubo tiene ocho **vértices.***
vertical (Inv. 5)	Upright; perpendicular to horizontal. horizontal line oblique line **vertical** line not **vertical** lines
vertical	Perpendicular a la horizontal.

vertical angles (20)	A pair of nonadjacent angles formed by a pair of intersecting lines. Vertical angles share the same vertex and are congruent.

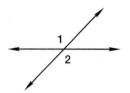

*Angles 1 and 2 are **vertical angles.***

ángulos opuestos por el vértice	Un par de ángulos no adyacentes formados por un par de líneas que se intersecan. Los ángulos opuestos por el vértice comparten el mismo vértice y son congruentes.

vertices (maintained)	*See* **vertex.**

vértices	*Ver* **vértice.**

volume (113)	The amount of space a solid shape occupies. Volume is measured in cubic units.

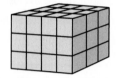

*This rectangular prism is 4 units wide, 3 units high, and 3 units deep. Its **volume** is $4 \cdot 3 \cdot 3 = 36$ cubic units.*

volumen	Cantidad de espacio ocupado por una figura sólida. El volumen se mide en unidades cúbicas.

W

weight (maintained)	The measure of how heavy an object is.

*The **weight** of the car was about 1 ton.*

peso	Medida que indica cuanto pesa un objeto.

*El **peso** del carro era de 1 tonelada.*

whole numbers (1)	The members of the set {0, 1, 2, 3, 4, …}.

*0, 25, and 134 are **whole numbers.***

-3, 0.56, and $100\frac{3}{4}$ are not **whole numbers.**

números enteros	Los elementos del conjunto {0, 1, 2, 3, 4, …}.

*0, 25 y 134 son **números enteros.***

-3, 0.56 y $100\frac{3}{4}$ no son **números enteros.**

X

x-axis (Inv. 5)	The horizontal number line of a coordinate plane.

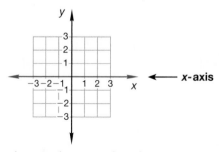

eje de las x	Recta numérica horizontal en un plano coordenado.

Y

y-axis
(Inv. 5)

The vertical number line of a coordinate plane.

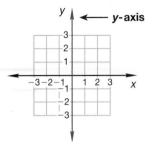

eje de las y

La recta numérica vertical en un plano coordenado.

Z

Zero Property of Multiplication
(3)
propiedad del cero en la multiplicación

Zero times any number is zero. In symbolic form, $0 \times a = 0$.
*The **Zero Property of Multiplication** tells us that $89 \times 0 = 0$.*

El cero mutiplicado por cualquier número es cero. En forma simbólica, $0 \times a = 0$.
*La **propiedad del cero en la multiplicación** nos dice que $89 \times 0 = 0$.*

A

Abbreviations, *See also* **Symbols and signs**
Celsius (C), 471-472, 580
centimeter (cm), 208, 470
Fahrenheit (F), 472
foot (ft), 505
Greatest Common Factor (GCF), 78, 172
inch (in), 505
kilometer (km), 470
kilowatt hours (kwh), 551
length (l), *See* Length
meter (m), 470
mile (mi.), 505
millimeter (mm), 470
ounce (oz), 504
per, 490
square centimeters (cm²), 208, 489
square (sq.), 149
yard (yd), 505

Act It Out or Make a Model, 5

Activities
area of parallelogram, 414
area of triangle, 476
circumference, 270
fraction manipulatives, 65
perimeter, 544-545
prime numbers, 74-75
probability experiment, 442
protractors, using, 255
rulers, inch, 90
Sign Game, 600
triangles, area of, 476

Acute angles, 122

Acute triangles, 553–555

Addends, 31, 43

Addition
algebraic, 599–604, 613
Associative Property, 12, 17, 611
checking answers by, 30–31
with common denominators, 105–110, 141–146,
335–341
Commutative Property, 16, 611
of fractions, 105–110, 141–146, 335–341, 351–356,
357–360, 419–424
Identity Property of, 16
of integers, 605–611
of mixed measures, 488–489
mixed numbers, 351–354
of negative numbers, 599–604, 669–672
order of operations, 11–13, 15–18, 450–452,
663–668
of positive numbers, 599–604, 669–672
rule, 564
SOS memory aid, 419–421
subtraction as inverse of, 565
sums, *See* **Sums**
three or more fractions, 357–360
of units of measure, 488–489
unknown numbers in, 29–34
word problems, 41–47

Algebra
adding integers, 605–608, 611
comparing integers, 605–608
dividing integers, 642–644
multiplying integers, 642–644, 647–648
ordering integers, 605–608
solving equations for unknown numbers,
See **Equations, solving for unknown numbers**
subtracting integers, 612–614
writing algebraic equations, 35–40

Algebraic addition of integers, 599–602, 608,
612–616

Alternate angles, exterior and interior, 584–586,
589–592

Angles
acute, 120-126
adjacent, in parallelograms, 377, 412–418, 584–586
alternate interior, 584–586, 589–592
central, 255
classifying, 120–126
complementary, 376–379, 584–586
exterior, 584–586, 589–592
interior, 584–586, 589–592, 633–634
naming, 120–126
obtuse, 122
opposite, in parallelograms, 413, 416
in parallelograms, 413, 416
in quadrilaterals, sum of measures, 589–592
right, 121
supplementary, 376–379, 584–586
of transversals, 583–588
sum of measures of angles in triangles, 589–594
vertex, 121

Approximately equal to (≈), 520

Approximation
pi and, 270-271, 523–527
volume of cylinder, 702

Area
of circles, 523–527, 633
of complex shapes, 631–633
of parallelograms, 414, 537–542
of rectangles, 537–542, 633
of right triangles, 475–478, 480
of squares, 207–211, 537–539
of triangles, 475–482, 537–542, 633
units of measure, 488–491

Arithmetic mean, *See* **Mean**

Arithmetic operations, *See also* **Addition; Division; Multiplication; Subtraction**
    alignment in, 217–224
    order of operations, *See* **Order of Operations**
    rules for decimals, 175–180
    SOS memory aid, 419–421
    terms for answers of, 24–27
    with units of measure, 488–491
    words that indicate, 23–28

Associative Property, 12, 15–22, 214, 611, 647

Axis (Axes), 258, 321-324

B

Bar graphs
    double-bar graphs, 255–256
    misleading data, 196, 199

Base(s) and exponents, 445–449

Bases of geometric figures, *See also* specific figure
    parallelograms, 371–375, 414, 538

Bias, in surveys, 130, 267

C

Calculators, order of operations, 452

Canceling
    in reducing fractions, 406–408, 419–421, 549–552
    in the Sign Game, 599–602
    unit multipliers and, 654–656

Celsius (°C), 471-472, 580

Centimeter (cm), 470, 472

Centimeter squared (cm²), 208

Central angle, 255

Chance, *See* **Probability**

Checking answers, *See also* **Inverse operations**
    in addition problems, 30–31
    in division problems, 31–32
    in multiplication problems, 31–32
    in subtraction problems, 31
    in unknown-number problems, 29–34

Circle graphs
    qualitative data display, 254–255
    representing a whole, 186–191
    sampling errors, 130, 132

Circles
    area of, 523–527
    circumference of, 269–274
    diameter of, 269–274
    fractional parts of, 65-67
    measures of, 523–526
    radius of, 269–274, 524–526

Circumference, 269–274, 270, *See also* **pi**

Classification
    parallelograms, 372
    of quadrilaterals, 371–375
    of triangles, 553–558

Coefficients, 562

Combining, in word problems, 41–47

Common denominators, *See also* **Denominators**
    in addition and subtraction of fractions, 201–206, 329–334, 335–341
    in addition and subtraction of mixed numbers, 141–146
    choosing, 342
    least common, 335–341
    in multiplication and division of fractions, 155–160, 285–288, 329–334
    renaming, 329–334, 335–341

Common factors, *See* **Factors; Greatest Common Factor (GCF)**

Common multiples, 116–119

Commutative Property
    of addition, 16, 611
    of multiplication, 16, 647
    subtraction and the, 16

Comparing
    decimals, 175–180, 181–185, 461
    fractions, 66, 461
    integers, 605–608
    number lines for, 52–56, 87–94
    percentages, usefulness for, 229–230
    word problems about, 41–47

Complementary angles, 376–379

Complex shapes, 629-636

Composite numbers, 385–388

Compound events, 495

Compound probability experiments, 380–384, *See also* **Probability**

Convenience sampling, 128

Conversions, *See also* **Equivalent forms; Mixed measures**
    centimeters to millimeters, 470
    of decimals by multiplication,
    decimals to fraction equivalents, 181–185, 425–426, 429–431, 573–577, 595–598
    decimals to percent equivalents, 223-228, 455-457, 595-598
    feet to yards, 505
    fractions to decimal equivalents, 181–185, 425–426, 429–431, 573–577, 595–598
    fractions to percent equivalents, 171-174, 223-228, 455-457, 573-577, 595-596, 659-660
    of fractions by division, 285–288, 307–312, 325–328
    improper fractions to mixed numbers, 325–328, 361–366, 392
    meters to centimeters, 470
    metric system, 469–474
    mixed numbers to improper fractions, 325–328, 361–366, 392
    of mixed numbers by multiplication, 391-393
    ratios to decimals, 429–431
    unit multipliers for, 654–658
    units of measure, 488–491
    U.S. Customary System, 504–510

Coordinate plane, graphing, 320–321

Counting numbers
    integers as, 605–628
    multiples and, 116–119

Cross products and proportions, 516–522

Cylinders, volume of, 701–703, 706-708

D

Data, *See also* **Graphs**; **Statistics**
    collecting, 132–133
    collection errors, 132–133
    data points, 256
    defined, 127
    display errors, 132–133, 196
    double-bar graphs, 254–256
    histograms, 131
    line graphs, 258
    line plots, 193–195, 256
    mean, 57–64, 256–257
    measures of central tendency, 256, 260
    median, 256, 260
    misleading display, 132–133, 196, 199–200
    mode, 256, 260
    pictographs, 254
    qualitative, 194, 254–255
    quantitative, 194, 256
    sampling, 193–198
    statistics and sampling, 254–259
    supporting claims, 199–200
    surveys, 127–129, 132–133, 193-198, 267, 441

Decimal division, 261–266, 279–284

Decimal numbers, *See* Decimals

Decimal place values, 176

Decimal points
    aligning by, 218
    in decimal division, 261–266, 279–284
    money and, 176
    place value, 176
    purpose of, 176
    shifting by division or multiplication, 212–216

Decimals
    comparing, 175–180, 461
    converting by multiplication, 212–216
    converting ratios to decimals, 429-431
    dividing, 261–266, 279–284
    dividing by whole numbers, 261–266
    fraction equivalents, 181–185, 425–426, 429–431,
        573–575, 595–596
    multiplication, 217–224
    negative, 295–300
    on number lines, 177, 295–300
    percent equivalents, 223–228, 455–457, 573-575,
        595–597
    positive, 295–300
    reading, 181–185
    rounding, 290–294
    rules for dividing by, 261–266, 279–284
    unknown numbers in, 245–248
    writing, alignment of decimal points, 181–185, 218

    writing as ratios, 429-431

Decimals Chart, 261–266

Degrees, measuring turns by, 590

Denominators, *See also* **Common denominators**
    defined, 65
    fractions, 141–146, 330–332, 335–339
    least common, 335–341, 342
    mixed numbers to improper fractions, 141–146,
        361–366
    multiplying, 329–334
    representation of, 69

Dependent events, 494–498

Diameter, 270

Difference, *See also* **Subtraction**
    defined, 31
    in arithmetic operations, 23-27

Digits, place value of, 135

Disjoint events, 435–437

Distance, *See* **Length**

Distributive Property, 543–547, 647

Divisibility, 82–86

Division
    answers as mixed numbers, 401–404
    applications, 649–653
    checking answers, 31–32
    common denominators, *See* **Common**
        denominators
    converting fractions to decimals using, 325–328,
        425–426, 429–431, 573–577
    decimals, rules for, 261–266, 279–284
    divisors, 261–264
    equations, 32
    equivalent, 239–244
    factors and, 82–86
    by fractions, 69–70, 285–288, 314–318
    of fractions, 307-312
    of integers, 642–644
    of mixed numbers, 401–405
    multiplication as inverse of, 289, 313, 565–572
    order of operations, 11–14, 15–22, 450–451,
        663–666
    by primes, 71–76, 385–388
    quotients, *See* **Quotients**
    remainders, 111–115, 649
    rules for, 261–266, 279–284
    by ten and by one hundred, 261–266
    of units of measure, 488–491
    unknown numbers in, 29–34
    by whole numbers, 261-266, 279–284
    of whole numbers, 285–288
    word problems, 48–51
    words that indicate, 100, 111-115

Divisors as decimals, 279-286

Double-bar graphs, 254, 255–256

Draw a Picture or Diagram, 5

Drawing
    to compare fractions, 66

Drawing, *continued*
protractors for, 255
scale, 684–688

E

Early Finishers/Real-World Connection, 14, 27, 56, 76, 81, 86, 94, 99, 110, 115, 140, 153, 160, 180, 192, 206, 222, 253, 274, 278, 284, 306, 312, 334, 341, 360, 370, 396, 410, 424, 434, 440, 459, 468, 482, 510, 521, 527, 532, 542, 548, 558, 582, 604, 610, 616, 678, 683, 694, 710

Elapsed-time, 44

Electrical-charge model in Sign Game, 599–604

Endpoints, 121

Enrichment
Early Finishers/Real-World Connection, *See* **Early Finishers/Real-World Connection**
Investigations, *See* **Investigations**

Equal groups, 58–59

Equalities, *See* **Equivalent forms**

Equality, properties of, 565–572

Equations, *See also* **Representation**
addition, *See* **Addition**
balanced, 559–564
divisions, *See* **Division**
evaluating and translating, 35–40, 673–678
exponential, *See* **Exponents**
formulas, *See* **Formulas**
formulating, *See* **Representation**
multiplication, *See* **Multiplication**
order of operations, *See* **Order of Operations**
with percents, *See* **Percents**
proportion, *See* **Proportions**
ratios, *See* **Ratios**
solving for unknown numbers, *See* **Equations, solving for unknown numbers**
subtraction, *See* **Subtraction**
using inverse operations, 565–572
writing for problem-solving, 35–40

Equations, solving for unknown numbers
in addition, 29-34
in division, 29-34
in equal groups, 95-99
in fractions and decimals, 181-185, 245-248, 530
in mixed numbers, 529-530
in multiplication, 29-34
in proportions, 511-513
in rate problems, 100-104
SOS method, 419-421
in subtraction, 29-34
in word problems, 29-34, 41-51

Equilateral triangles, 553–555

Equivalent
cross products for determining, 516–521
division problems, 239–244
equal fractions as, 164, 303–306
fraction-decimal-percent, 224–228, 455–457,

573–576, 595–598
ratios, 429–431

Estimation
diameters in a circumference, 269–274
in probability, 442–443, 501
and rounding, 134–140
words that indicate, 134

Evaluations
expressions and equations, 35–40, 673–678
negative and positive numbers, 669–672

Even number sequences, 8–10

Expanded notation with exponents, 445–447

Experimental probability, *See* **Probability**

Explain, *See* **Writing**

Exponential expressions, 446

Exponents, *See also* **Powers of ten; Squared numbers**
bases and, 445–447
correct form for, 446
expanded notation with, 446–447
finding values of, 446–447
function of, 446–447
order of operations with, 450–452
powers of ten and, 446
reading and writing, 445–446, 489

Expressions, evaluating and translating, 35–40, 673–678

Exterior angles, 583–588, 589–592

F

Fact families, 16

Factors, *See also* **Prime factorization**
in addition, 29–34
calculating, 529–532
constant, 475–482
decimal numbers, 245–248
defined, 72, 78
divisibility tests for, 82–86
division, 29–34, 82–86
equal to one, 303–306
in estimation, 134–140
greatest common, *See* **Greatest Common Factor (GCF)**
method of solving for unknown factors, 529–532
unknown factors in mixed numbers, 529–532
unknown factors in multiplication, 29–34
of positive integers, 642–644
and prime numbers, 71–76, 385–390
reducing fractions and, 163–166, 303–306, 406–408, 549–552
in subtraction, 29–34
unknown, 245–248, 529–532

Fahrenheit (°F), 472, 506–507, 580

Figures, *See* specific geometric figures

Find/Extend a Pattern, 5

Fluid ounces, 505–507

Foot (ft), 505-508

Formulas
    area, *See* **Formulas for area**
    definition, 578
    for diameter, 269–274
    distance, rate, and time, 579
    Distributive Property, 543–548
    length, *See* **Length**
    perimeter, *See* **Formulas for perimeter**
    and substitutions, 578-580
    temperature conversion, 580
    volume, *See* **Formulas for volume**

Formulas for area
    circles, 523–527
    cylinders, 701–704
    parallelogram, 414, 537-542
    rectangles, 537-42
    squares, 537–542
    triangles, 475–482, 537–542

Formulas for perimeter
    parallelograms, 416, 537–542
    rectangles, 537–542
    squares, 537–542
    triangles, 475–482, 537–542, 589–594

Formulas for volume
    cylinders, 701-703
    prisms, 705-708

Four-step problem-solving process, 1–6

Fraction-decimal-percents, 223–228, 573–576, 595–598

Fractional parts of the whole, 314–318

Fractions, *See also* **Denominators; Mixed numbers; Numerators**
    adding with common denominators, 105–110, 141–146, 335–341
    adding with different denominators, 329–334
    adding with mixed numbers, 351-356
    adding three or more fractions, 357–360
    canceling terms, 406–408, 420-421
    comparing, 66, 461
    cross products, 516–518, 521–522
    decimal equivalents, 181–185, 425–428, 429–431, 573–576
    dividing by fractions, 69–70, 285–288, 289–290, 313, 314–316
    dividing by whole numbers, 261–266, 279–284
    drawing to compare, 66
    equal to one, 303–306
    equivalent, 224–228, 455–457, 516–521, 595–598
    finding the whole using a fraction, 679–683
    fraction-decimal-percent, 224–228, 455–459, 595–598
    by fractions, 285–288, 307–312, 314–318
    improper fractions, 66, 325–326, 361–364, 391–394
    lowest terms, defined, 326
    mixed numbers, 201-206, 351–356, 361–366, 391–394
    multiplying, common denominators for, 155–160, 329–334
    multiplying by one, 236–237
    multiplying three or more fractions, 419-424
    multiplying, reducing before, 163–166, 407–408, 549–551, *See also* **Reducing Fractions**
    on number lines, 67–68, 87–94, 161–162
    percent equivalents, 171–174, 223–228, 455–457, 573–576, 595–596, 659–660
    purpose of, 353
    reading, 112
    reciprocals, *See* **Reciprocals**
    renaming, 235–238, 329–334, 354
    simplifying, 325–328
    SOS memory aid, 419–424
    subtracting, 105–110, 141–146, 201–206, 329–334, 335–341, 351–356
    unknown numbers, 245–248
    whole numbers, 201–206, 285–288
    without common denominators, 236–238, 329–334
    writing decimal equivalents 181-185, 425–426, 429-431
    writing fractions as percents, 171-174, 223-228, 455-457, 659-660

Frequency tables, 127, 130, 131

Functions, 319, 507

G

GCF (Greatest Common Factor), *See* **Greatest Common Factor (GCF)**

Geometric figures
    bases of, *See* **Bases of geometric figures**
    circles, *See* **Circles**
    cylinders, *See* **Cylinders**
    perimeter of, *See* **Perimeter**
    polygons, *See* **Polygons**
    prisms, *See* **Prisms**
    quadrilaterals, *See* **Quadrilaterals**
    rectangles, *See* **Rectangles**
    similar figures, 623–627
    solids, *See* **Solids**
    squares, *See* **Squares**
    triangles, *See* **Triangles**

Graphs
    bar, 193
    circle, 186–191, 254–255
    coordinate plane, 319-324
    double-bar, 254–256
    functions, 319–322
    histograms, 127
    line graphs, 258
    line plots, 193–195, 256
    misleading display, 129–133, 196, 199–200
    on number lines, *See* **Number lines**
    pictographs, 254

Greatest Common Factor (GCF), 77–81, 164, 172, 398, 411

Grouping factors, 303–306

Grouping property, *See* **Associative Property;**
Parentheses

Guess and Check, 5

H

Height
    of parallelograms, 414-416, 538-539
    ratios, 431
    of triangles, 475–478, 480, 538–539

Higher order thinking skills, *See* **Thinking skills**

Histograms, 127

Hundredths, 176

I

Identity Property of Addition, 16

Identity Property of Multiplication, 16, 236, 647

Improper fractions, converting mixed numbers to,
    325–328, 361–366, 392–394, 563

Inch (in.), 90, 505

Independent events, 383, 494–498

Indirect information, 464–466

Information, finding unstated, 464–466, *See also* **Data**

Integers
    adding, 605–610, 611
    algebraic addition, 599–602, 608, 613
    comparing, 605-610
    consecutive, 605–610
    counting numbers as, 605–608
    defined, 88, 605
    dividing, 642–644
    multiplying, 642–644, 647–648
    negative, 611, 647–648
    on number lines, 52–56, 605–608, 612–614
    ordering, 605–608
    subtracting, 612–616
    Interior angles
    complex shapes, 633-634
    corresponding, 584
    definition, 584
    forming, 583–588
    sum of measures of, 589–594
    of transversals, 583–586

Intersecting lines, angle pairs formed by, 583–588

Inverse operations
    addition and subtraction, 565–570
    division and multiplication, 289, 313, 565–570
    solving proportions, 522
    squaring and square roots, 207–211

Investigations
    angles, drawing and measuring with protractors,
      255
    balanced equations, 559–564
    compasses, 255
    compound probability experiments, 380–384
    data, 127–131, 193–198, 254–259
    drawing using protractors, 255

    experimental and theoretical probability, 441–444
    fraction manipulatives, 65–68
    frequency tables, 127–131
    graphing, 193–198, 254–259, 319–324
    histograms, 127–131
    models and scale drawings, 684–688
    probability and prediction, 501–503
    protractors, 255
    scale drawings and models, 684–688
    scale factor, 684–688
    samples, 127-131
    similar figures, 623–628
    statistics and sampling, 254–259
    surveys, 127–131, 193–198, 441–444

Isolation of the variable, 560

J

Jordan curve, 23–24

K

Kilometer (km), 469–474

L

Language, math, *See* **Reading math; Math language**

Last-digit divisibility test, 82–86

Least common denominator, 105–110, 335–341

Least common multiple (LCM)
    adding three or more fractions, 357–360
    choosing a common denominator, 342
    defined, 167–170, 336

Legend, 684

Length, *See also* **Circumference; Perimeter**
    abbreviation for, 505
    conversion of units of measure, 469-474, 489–491,
      505–507
    Distributive Property, 543–546
    ratio, 431
    sides of a square, 207–211, 538–539
    units of measure, 489–491

Letters, points designated with, 599-04

Line graphs, 258

Line plots, 193–195, 256

Line segments, *See* **Segments**

Lines, *See also* **Number lines; Segments**
    intersecting, 583–586
    oblique, 583
    parallel, 121, 583, 586
    perpendicular, 120-121, 586
    properties of, 586
    segments and rays, 121
    transversals, 583–588

Lowest terms of fractions, defined, 326

M

Make a Model, 5

Make an Organized List, 5

Make It Simpler, 5

Make or Use a Table, Chart, or Graph, 5

Math language, *See also* **Glossary** in this book; 8, 30-32, 53, 78, 85, 88, 106, 121, 135, 142, 149, 223, 225, 236, 239, 269, 326, 336, 346, 372, 377, 402, 413, 420, 446, 475, 495, 511–512, 517, 550-551, 583, 605-606, 612, 618, 638, 705

Mean, calculating the, 57–64, 256–257, 260

Measurement, *See also* **Units of measure**
    of area, *See* **Area**
    of circles, *See* **Circles**
    of height, *See* **Height**
    of length, *See* **Length**
    linear, *See* **Length**
    parallax, 57
    of perimeters, *See* **Perimeter**
    protractors for, 255
    of rectangles, *See* **Rectangles**
    of surface area, *See* **Area**
    of temperature, 506-507, 580
    of turns, 590
    of volume, *See* **Volume**

Measures of central tendency, *See* **Mean; Median; Mode**

Median, 256–257, 260

Memory aids
    decimal number chart, 261–266
    Please Excuse My Dear Aunt Sally (PEMDAS), 663–668
    SOS, 419–421

Meter (m), 470

Metric system, 469–474, *See also* **Units of measure**

Mile (mi), 505

Millimeter (mm), 470, 472

Minus sign, *See* **Negative numbers; Signed numbers**

Missing numbers, *See* **Unknown numbers**

Mixed measures, 489–491, *See also* **Units of measure**

Mixed numbers, *See also* **Improper fractions**
    with common denominators, 141–146
    converting to improper fractions, 325–328, 361–366, 392–394
    division, 401–405
    fractions, 351–356
    multiplication, 392–394
    on number lines, 87–94
    ratios as, 100–104
    regrouping, 275–278, 367–370
    subtracting, with common denominators, 141–146
    subtracting, with regrouping, 275–278, 367–370
    unknown factors in, 245–248, 529–532

Mode, 256–257, 260

Models
    of fractions, 65-68
    of parallelograms, 412–418
    scale factor, 684–688
    scale models, 684-688

Money, rounding with, 137

Multiple outcomes, 499–503

Multiples, 116–119, *See also* **Least Common Multiple (LCM)**

Multiplication, *See also* **Exponents**
    Associative Property, 12, 15–22, 214, 647
    checking answers, 31–32
    common denominators and, 155–160, 329–332
    Commutative Property, 15–22, 219, 647
    cross product of fractions, 516–522
    of decimals, 212–216, 217–224
    division as inverse of, 289, 313, 568-569
    equations, 31
    fractions with common denominators, 155–160, 314–318, 329–334, 419–424, 516–521
    Identity Property, 16, 236, 647
    of integers, 642–644
    of mixed numbers, 392–394
    negative integers, 647
    number lines, 161–162
    "of" as term for, 155
    order of operations, 11–14, 450–452, 664–666
    process of multiplying fractions, 155–160, 314–318
    properties, 647–648
    reducing rates before multiplying, 163–166, 406–408, 411, 549–552
    three or more fractions, 419–424
    of units of measure, 488–493
    unknown numbers in, 29–34
    words that indicate multiplication, 48–51

Multiplication sequences, 8–10

Multiplicative inverse, 522

Multiplier, unit *See* **Unit multiplier**

Multistep problems, 23–28

N

Naming, *See also* **Renaming**
    angles, 120–126
    complex shapes, 632
    rays, 121

Negative numbers, *See also* **Signed numbers**
    addition of, 669–670
    algebraic addition of, 599–604, 612–614
    evaluations, 669–670
    integers, 605–611, 643–644
    on number lines, 52–56, 295–300, 605–608
    order of operations, 663-666
    simplifying, 663-666
    symbol for, 52, 599–602, 670–671

Negative signs (–), 52, 599–604, 601, 669–672

Nonprime numbers, *See* **Composite numbers**

Number cubes and probability, 346

INDEX

Number lines, *See also* **Graphs**
comparing using, 52–56, 87–94
counting numbers on, 52–56
decimals on, 177, 181–185, 295–300
fractions on, 67–68, 87–94, 161–162
integers on, 52–56, 605–608, 612–614
mixed numbers on, 87–94
multiplication, 161–162
negative numbers on, 52–56
opposite numbers on, 53
ordering with, 52–56, 87–94
positive numbers on, 52–56, 295–300
rounding with, 134–140
tick marks on, 52–56
using, 301–302
whole numbers on, 52–56

Number sentences, *See* **Equations**

Numbers
counting, *See* **Counting numbers**
decimal, *See* **Decimals**
equal to one, 303–306
missing, *See* **Unknown numbers**
mixed, *See* **Mixed numbers**
negative, *See* **Negative numbers**
nonprime, *See* **Composite numbers**
percents of, *See* **Percents**
positive, *See* **Positive numbers**
prime, *See* **Prime numbers**
signed, *See* **Signed numbers**
whole, *See* **Whole numbers**

Numerators, *See also* **Fractions**
defined, 65
mixed numbers to improper fractions, 361–366, 392–394
representation of, 69

O

Oblique lines, 121

Obtuse angles
naming, 122
of transversals, 584–586

Obtuse triangles, 476–478, 554–555

Odd number sequences, 8–10

One, numbers equal to, 303–306

Operation Properties of Equality, 566

Operations
of arithmetic, *See* **Arithmetic operations**
inverse, *See* **Inverse operations**
order of, *See* **Order of Operations**

Opposite numbers on number lines, 606

Order of Operations
addition, 11–14, 15–22, 450–452, 663–666
calculators for, 452
division, 11–14, 15–22, 450–452, 663–666
with exponents, 450–452, 664–666
memory aid, *See* **Memory aids**
multiplication, 11–14, 15–22, 450–452, 664–666

parentheses in, 11–14, 450–452, 664–666
Please Excuse My Dear Aunt Sally, *See* **PEMDAS**
process, 11–14, 15–22, 450–452, 663–666
properties, 15–22, 450–452, 664–666
rules for, 11–14, 15–22, 450–452, 664–666
for simplification, 450–452, 664–666
subtraction, 11–14, 15–22, 450–452, 664–666
using calculators, 452

Ordered pairs, 321

Ordering decimals, 175–180

Ordering integers, 605–608

Ounce (oz.), 504–510

Outcomes, multiple, 499–500

P

Parallax, 57

Parallel lines, 121, 583

Parallelograms
angles of, 371–372, 413
area of, 414-416
base of, 413-416
characteristics of, 371–372, 413–416, 538
height of, 413–416, 538
model of, 413–415, 538
perimeter of, 414–416, 538–539
properties of, 413–416, 538–539
as quadrilateral, 371–375
rectangles as, 372, 538
rhombus as, 372
sides of, 413–416, 538

Parentheses
in Order of Operations, 11–14, 450–452, 665
with signed numbers, 606
symbol for multiplication, 643

PEMDAS, 664, *See also* **Order of Operations**

Percents
comparison, 299–300
converting to decimals, 455–457
decimal equivalents, 223–228, 455–457, 595–598
defined, 455
finding the whole using, 223–228, 690–692
fractional equivalents, 171–174, 223–228, 455–457, 573–576, 595–598, 659–660
multiplication, 455–459
parts out of, 100, 229
properties of, 171–174, 455–457
proportions, 223–228, 455–457, 637–640
symbols for, 171–174, 455
usefulness, 229–230
word problems, 224–228, 455–457, 637–640, 695–700
writing decimals as percents, 223–228, 455–457, 596

Perfect squares, 207–211

Perimeter
activity, 544
of circles, *See* **Circumference**
of complex shapes, 629–634

Distributive Property, 543–548
of parallelograms, 414–416, 538–539
of rectangles, 538–539
of squares, 207–211, 538–539
of triangles, 538–539

Perpendicular lines
angles formed by, 120–126, 584–586
in area of parallelograms, 371–375, 413–416, 584

pi, 270, 523–528

Pictographs, 254

Pie charts, *See* **Circle graphs**

Pie graphs, *See* **Circle graphs**

Place value
in decimals, 176
in whole numbers, 135

Placeholder, zero as, 217–224

Please Excuse My Dear Aunt Sally (PEMDAS), 209,
664

Plots in word problems, 42

Points, *See* **Decimal points; Number lines; Graphs**

Polygons, *See also* specific polygons
similar figures, 624
triangles as, 553–555

Population, 129–130, 132, 195

Positive numbers
algebraic addition of, 599–604, 612–614
evaluations, 669–670
integers, 642–644
on number lines, 52–56, 605–608
Sign Game, 599–602
symbol for, 599–602

Powers, 446, *See also* **Exponents**

Powers of ten, *See also* **Exponents**
multiplying, 445–447
place value and, 445–447
whole number place values, 445–447

Predicting, *See* **Probability**

Prime factorization, 385–388, 397–399, 446,
See also **Factors**

Prime numbers
activity with, 74
composite numbers compared, 385
defined, 74, 385
division by, 385-388
factors of, 71–76
greatest common factor (GCF), 77–81

Prisms
rectangular, *See* **Rectangular prisms**
volume of, 705–708

Probability
activity, 442
chance and, 343–350
compound experiments, 495
dependent and independent events, 383, 494–496
disjoint events, 435–437
of events, 343–350, 435–437, 494–496

events and their complement, 436–437, 494–496
experimental and theoretical, 441–444
multiple outcomes, 499–500
and predicting, 501–503
theoretical, 441–444

Problem solving
four-step process, 1-6
real world, *See* **Real-world connection**
strategies, *See* **Problem-solving strategies**

Problem-solving strategies
Act It Out or Make a Model, *See* **Act It Out or
Make a Model**
Draw a Picture or Diagram, *See* **Draw a Picture or
Diagram**
Find/Extend a Pattern, *See* **Find/Extend a Pattern**
Guess and Check, *See* **Guess and Check**
Make an Organized List, *See* **Make an Organized
List**
Make It Simpler, *See* **Make It Simpler**
Make or Use a Table, Chart, or Graph, *See* **Make or
Use a Table, Chart, or Graph**
Use Logical Reasoning, *See* **Use Logical
Reasoning**
Work Backwards, *See* **Work Backwards**
Write a Number Sentence or Equation, *See* **Write a
Number Sentence or Equation**

Products, *See also* **Multiplication**
defined, 31
multiples and, 116–119
operations of arithmetic, 23–27
of reciprocals, 231–234
unknown numbers, 29–34, 48–51

Proportions, *See also* **Rates; Ratios**
cross products, 516–522
defined, 511, 611, 618
multiplicative inverse, 522
percent problems with, 637–640
ratios, 100–104, 511–515, 517–518, 533–536,
617–622, 627
in scale drawings and models, 511–513, 684-688
solving using a constant factor, 484-485
solving using cross products, 516–521
unknown numbers in, 511–515, 685
using a constant factor, 484-485
writing, 511–513

Protractors, measuring and drawing with, 255

Q _____

Quadrilaterals
classifying, 371–375
defined, 372
parallelograms as, *See* **Parallelograms**
rectangles, *See* **Rectangles**
squares, *See* **Squares**
sum of angle measures in, 589–592

Qualitative data, 194, 254–255

Quantitative data, 194, 256

Quotients, *See also* **Division**

calculating, 111–115
in decimal division, 261–266, 279–284
defined, 111–115
in equivalent division problems, 239–244
missing, 48–51
writing, 111–115

R

Radius (radii) of circle, 269–274, 523-527

Random sampling, 128-129

Range, 256, 260

Rates, reducing before multiplying, 549–552

Ratio boxes, 533–536, 618–620, 685

Ratios
actual-count numbers, 483, 533
as comparisons, 100–104
constant factors, 484–485
constant factors word problems, 476-478
converting to decimals, 429–431
defined, 100
equivalents, writing, 429–431, 534–535
fractional form of, 100–104, 429–431
problems involving totals, 618–620
proportions and, 100–104, 511–513, 517–519, 627
proportions word problems, 100-104, 517-518, 533-535
ratio boxes word problems, 533–535, 618-620
ratio numbers, 483, 533
reducing, 550
won-loss, 101
writing decimal equivalents, 429–431

Rays, 121, 322

Reading
decimal points, 176
decimals, 181–185
exponents, 445–449

Reading math, *See also* **Math language**; 16, 52, 112, 122, 155, 172, 207, 208, 256, 392, 413, 446, 472, 489, 490, 520, 601, 643, 664

Real-World Connection, *See* **Early Finishers/ Real-World Connection**

Reciprocals, 231-234

Rectangles
area, 543-633
area formula, 537-542
characteristics of, 372, 538–539
as parallelograms, 371–375, 539
perimeter of, 537–542
similar, 625
as squares, 371–375, 537-542

Rectangular prisms, volume of, 705–708

Reducing fractions
by canceling, 163–166, 406–408, 420–421, 549–551
common factors in, 163-166, 406–408
by grouping factors equal to one, 303–306

before multiplying, 163–166, 406–408, 411, 549–551
prime factorization for, 397–400
rules for, 163–166, 406–408
and units of measure, 490

Regrouping, in subtraction of mixed numbers, 275–278, 367–370

Relationships
inverse operations, *See* **Inverse operations**
ratios and proportions, 100–104, 429–434, 511–513, 517–519, 533–535
sides to angles in triangles, 475–478, 480

Relative frequency, 441

Remainder, 111-115

Reminders, *See* **Memory aids**

Renaming, *See also* **Naming**
fractions, 236–238, 329–334, 353, 419–424
mixed measures, 489–491
multiplying by one, 236–238
SOS memory aid, 419–424
with common denominators, 201-206, 329–334, 335-341
without common denominators, 236–238, 329–334

Rhombus, 371–375

Right angles, naming, 120–126

Right circular prism, volume 706

Right solids, 705–710

Right triangles, 476–478, 553–555

Roots, *See* **Square roots**

Rotations (turns) of geometric figures, 590
See also **Transformations**

Rounding, *See also* **Estimation**
decimals, 290–294
estimating, 134–140
money, 137
with number lines, 134–140
whole numbers, 134–140

Rounding up, 134–140

Rules, *See also* **Order of Operations**
for decimal division, 261–266
of functions, 506–507
for reducing fractions, 163–166, 406–408, 549–551
of sequences, 7–10

S

samples and surveys, 127–131

sampling
and data, 193–198
errors, 132–133
and statistics, 254–259

Scale, temperature, 506–508

Scale factor, 512, 684–688

Scalene triangles, 553–555

Segments, *See also* **Lines**
in creating complex shapes, 632

Sequences, 7-10

Shapes, complex, *See* **Complex shapes**

Sides
    of parallelograms, 412–416, 537-39
    of quadrilaterals, 589–590, 592
    of squares, 207–211, 537–539
    of triangles, 475–478, 481, 537–539

Signed numbers, 599–602, 611

Signs, *See* **Symbols and signs**

Simplifying
    fractions, 325–328
    order of operations for, 450–452, 663–666
    SOS memory aid, 419–421

Solids, volume of, 705–708

Solving equations, *See* **Algebra**

SOS method, 419–421

Square (sq.), 207–211, 538

Square angles, 120–126

Square centimeters (cm²), 208

Square roots, 207–211

Square units, 207–211

Squared numbers, 207–211, *See also* **Exponents**

Squares, *See* **Rectangles;** *See also* **Area; Perimeter**

Statistics, *See also* **Data**
    data, 193–198
    defined, 127
    operations, 57–64, 256
    and sampling, 254–259

Substitution, formulas and, 578–580

Subtraction, *See also* **Difference**
    addition as inverse of, 565–568
    checking answers, 30–31
    comparing, 41–47
    elapsed-time, 44, 47
    equation, 31
    fractions with common denominators, 105–110, 141–146, 201–206, 329–334, 351–354, 419–424
    fractions with different denominators, 201–206, 335-341
    integers, 612–614
    mixed measures, 491
    mixed numbers, 141–146, 275–278, 351–356, 367–370
    negative numbers, 669–670
    order of operations, 11–14, 15–22, 450–452, 663–666
    with regrouping, 275–277, 367–368
    separating, 41–45
    SOS memory aid, 419–421
    subtrahends role, 29–34
    of units of measure, 489, 491
    unknown numbers in, 29–34
    from whole numbers, 201–206
    word problems, 41–47
    words that indicate subtraction, 41–47

Subtrahends, 31

Sum-of-digits divisibility tests, 82–86

Sums, 24–27, 30, 43

Supplementary angles, 376–379, 583–588
    definition, 584
    See also **Exterior angles**
    See also **Interior angles**
    See also **Transversals**

Surveys
    data and sampling, 193–195
    factors affecting, 267–268
    population, 129–130, 132, 195, 268
    probability, 441-444
    samples, 127–131
    sampling errors, 129–133
    statistics and sampling, 258–259

Symbols and signs, *See also* **Abbreviations**
    approximately equal to ($\approx$), 520
    comparison, 460–463, 520
    decimal points, 176
    degrees (°), 472
    dot (•), 489
    equal to (=), 461
    greater than (>), 461
    less than (<), 461
    multiplication dot (•), 489, 643
    multiplication ×, 643
    negative numbers (-), 599–604, 601, 669–672
    parentheses ( ), 643
    percent (%), 171–174, 229, 455–457
    pi (π), 523-528
    positive numbers (+), 599–602, 669–670
    ratios (:), 429–431
    square root ($\sqrt{}$), 208

Symmetric Property of Equality, 568

Systematic sampling, 129

T

Temperature, 471-472, 506–507, 580, 606, *See also* **Thermometers**

Tens
    in decimals, 181–185
    multiplying decimals by, mentally, 217–224

Terms of sequences, 8–10

Theoretical probability, 441-444

Thermometers, 506, 606

Thinking skills
    analyze, 11, 49, 72, 113, 121, 134, 372, 579, 606, 670
    classify, 632
    conclude, 16, 74, 225, 630
    connect, 91, 176, 270, 426, 507, 568
    discuss, 82, 172, 187, 203, 213, 275, 281, 353, 363, 379, 398, 407, 524, 599, 643, 679, 691
    explain, 338, 377, 446, 451, 456, 461, 478, 495, 690, 702
    formulate, 545

Thinking Skills, *continued*

generalize, 8, 250, 291, 553, 607, 613, 664

infer, 96

justify, 142, 403, 407, 472, 530, 534, 554, 574, 660

list, 386

model, 361, 618

predict, 358

represent, 236

summarize, 280

verify, 30, 141, 225, 262, 271, 276, 353, 391, 426, 530, 584, 590

Tick marks on number lines, 52–56, 87–94, 295–300

Totals, ratio problems involving, 618–620

Transformations of figures

rotations (turns), 590

Translating, expressions and equations, 35–40, 673–676

Transversals, 583–588

Trapezoids, 628

Triangles

activity, 476

acute, 554

angles, sum of measures, 589–594

angles used to classify, 553-558

area of, 475–482, 538, 633

base, 475-482

classifying, 553–558

equilateral, 322, 553–558

height, 475-482

isosceles, 554–555

obtuse, 475–482, 554

perimeter, 537–542

polygons, 553

relationship to angles, 475–482

right, area of, 475–482, 554

scalene, 554

sides used to classify, 553–558

similar, 623–626

supplementary angles, 584

Turns, measuring, 590

U

Unit multipliers, 654–658

Units of measure, *See also* **Measurement**

adding, 489

converting, 488–491

dividing, 490

length, 488–491, 504–510

mass and weight, 504–510

metric system, 469–472

mixed measures, 488–491

multiplying, 489–490

reducing, 488–491

subtracting, 491

uniform, 488–491

U.S. Customary System, 504–510

Unknown numbers

in addition, 29–34

angles, 591

checking answers, 30-31

defined, 30

in division and multiplication, 29–34

in fractions and decimals, 181–185, 245–248, 429–431, 529–532

missing products, 29–34

in mixed numbers, 529–532

in proportions, 511–515, 685

in rate problems, 100–104

in subtraction, 29–34

in word problems, 29–34, 41–47, 48–51, 529–532

U.S. Customary System, 504–510, 655-656, *See also* **Units of measure**

Use Logical Reasoning, 5

V

Vertex (vertices), 121

Vertical angles, 586

Vocabulary, *See* Glossary in this book

Volume

bases of solids and, 705–708

of cylinders, 701–703, 706

defined, 702

of prisms, 705–708

of right solids, 705–708

Volunteer sampling, 129

W

Whole numbers
    decimal points, 175–180
    dividing decimals, 261–266, 279–284
    division, 261–266, 279–284
    find whole number when fraction is known,
       679–681
    find whole number when percent is known,
       223-228, 690-692
    on number lines, 52–56, 605–608
    rounding, 134–140
    subtracting fractions from, 201–206
    writing with decimal points, 175–180

Win-loss ratio, 101

Work Backwards, 5

Write a Number Sentence or Equation, 5

Writing
    decimal points, 175–180
    decimals, 181-185, *See also* **Decimals**
    equations, 35–40
    exponents, 445–447
    improper fractions, 361–366
    mixed numbers, 361–366, 392–394
    percents, *See* **Percents**
    proportions, 511–513, 627
    quotients, 111–115
    ratios, 100–104, 429–434, 533–536, 618–620
    remainders, 111–115
    whole numbers, 175–180

X

x-axis, 321-322

Y

y-axis, 321-322

Z

Zero
    as placeholder, 217–224
    in rounding, 134–140
Zero Property, 647